CONSERVATION BIOLOGY

CONSERVATION BIOLOGY
Voices from the Tropics

Navjot S. Sodhi
National University of Singapore

Luke Gibson
National University of Singapore

Peter H. Raven
Missouri Botanical Garden

WILEY Blackwell

Library of Congress Cataloging-in-Publication Data
Conservation biology : voices from the tropics / [edited by] Navjot S. Sodhi, Luke Gibson, Peter H. Raven.
       pages cm
   Includes bibliographical references and index.
   ISBN 978-0-470-65863-5 (cloth)
1. Nature conservation–Tropics.   2. Conservation biology–Tropics.   I. Sodhi, Navjot S.
   QH77.T78C655 2014
   333.720913–dc23
                              2013009680

A catalogue record for this book is available from the British Library.

Cover image: Paraguay's Chaco forest, one of the last wilderness frontiers in South America, is home to jaguars, giant anteaters, and several groups of indigenous people. But this remote habitat is now being plundered by rapid agricultural development, and much of the forest already has fallen to make way for widespread cattle pastures. Feeding the world's growing population will present challenging problems for conservationists, as highlighted in the essay by Alberto Yanosky and other voices from the tropics. Photo by Alberto Yanosky.
Cover design by Design Deluxe

Set in 9/11 pt Photina MT by Toppan Best-set Premedia Limited

1   2013

CONTENTS

LIST OF CONTRIBUTORS

Sylvanus Abua
Calabar, Cross River State, Nigeria
Email: s_abua@yahoo.co.uk

Germán Ignacio Andrade-Pérez
School of Management, Universidad de Los Andes,
Bogotá, Colombia
Email: giandradep@gmail.com

Carter T. Atkinson
US Geological Survey, Pacific Island Ecosystems
Research Center, P.O. Box 44, Hawaii National Park,
Hawaii 96718
Email: catkinson@usgs.gov

Paul C. Banko
US Geological Survey, Pacific Island Ecosystems
Research Center, P.O. Box 44, Hawaii National Park,
Hawaii 96718
Email: pbanko@usga.gov

Hans Bauer
WildCRU, University of Oxford, Tubney House,
Tubney OX13 5QL, UK
Email: hans.bauer@zoo.ox.ac.uk

Kamaljit S. Bawa
Department of Biology, University of Massachusetts,
Boston, MA 02478, USA;
Sustainability Science Program, Harvard University,
Cambridge, MA 02138, USA;
Ashoka Trust for Research in Ecology and the
Environment, Bangalore 560 024, India
Email: kamal.bawa@gmail.com

Gilianne Brodie
School of Biological and Chemical Sciences,
University of the South Pacific, Private Bag, Suva, Fiji
Islands
Email: brodie_g@usp.ac.fj

Gerardo Ceballos
Instituto de Ecología, Universidad Nacional
Autónoma de México, 3er Circuito Exterior S/N
Ciudad Universitaria, 04510 México, D.F. México
Email: gceballo@ecologia.unam.mx

Vijak Chimchome
Department of Forest Biology, Faculty of Forestry,
Kasetsart University, Ngamwongwan Road, Bangkok
10900, Thailand
Email: fforvjc@ku.ac.th

Richard T. Corlett
Xishuangbanna Tropical Botanical Garden, Chinese
Academy of Sciences, Menglun, Mengla, Yunnan
666303, China
Email: corlett@xtbg.org.cn

Wondmagegne Daniel
Texas Tech University, Department of Natural
Resources Management, Goddard Building, Box
42125, Lubbock, TX 79409, USA
Email: wondud.whibesilassie@ttu.edu or
wondu97@yahoo.com

Charbel El-Hani
Programa de Pós-Graduação em Ecologia e
Biomonitoramento, Instituto de Biologia,
Universidade Federal da Bahia, R. Barão de Jeremoabo
S/N, CEP 40170-290, Salvador, BA, Brazil
Email: charbel@ufba.br

F.B. Vincent Florens
Department of Biosciences, University of Mauritius,
Réduit, Mauritius;
UMR 53 PVBMT, Université de la Réunion, St Denis,
La Réunion, France
Email: Vin.Florens@uom.ac.mu

Andrés García
Estación de Biología Chamela, Instituto de Biología, Universidad Nacional Autónoma de México, Apdo. Postal 21. San Patricio, Melaque, Jalisco, México 48980
Email: chanoc@ibiologia.unam.mx

Fikirte Gebresenbet
Oklahoma State University, Department of Zoology, 311 D Life Sciences West, Stillwater, OK 74078, USA
Email: fikirte.erda@okstate.edu or kememgst@gmail.com

Roy E. Gereau
Missouri Botanical Garden, P.O. Box 299, St. Louis, MO 63166-0299, USA
Email: Roy.Gereau@mobot.org

Luke Gibson
Department of Biological Sciences, National University of Singapore, 14 Science Drive 4, Singapore 117543, Singapore
Email: lggibson@nus.edu.sg

Mwangi Githiru
Department of Zoology, National Museums of Kenya, P.O. Box 40658-00100, Nairobi, Kenya
Wildlife Works, P.O. Box 310-80300, Voi, Kenya
Email: mwangi_githiru@yahoo.co.uk

Amleset Haile
Wageningen University and Research, The Netherlands; CASCAPE project, Addis Ababa, Ethiopia
Email: amleset.haile@cascape.org or amlihaile@gmail.com

Bila-Isia Inogwabini
Durrell Institute of Conservation and Ecology (DICE), University of Kent at Canterbury, Canterbury, Kent CT2 7NR, UK
Email: bi4@kentforlife.net

James D. Jacobi
US Geological Survey, Pacific Island Ecosystems Research Center, P.O. Box 44, Hawaii National Park, Hawaii 96718
Email: jjacobi@usgs.gov

Richard K.B. Jenkins
Durrell Institute of Conservation and Ecology, School of Anthropology, University of Kent, Canterbury, UK; School of Environment, Natural Resources and Geography, Bangor University, Bangor, Gwynedd, UK; Madagasikara Voakajy, B.P. 5181, Antananarivo 101, Madagascar
Email: rkbjenkins@gmail.com

Michelle Kalamandeen
Department of Biology, Faculty of Natural Sciences, University of Guyana, Turkeyen Campus, Georgetown, Guyana
Email: michellek@bbgy.com

Gustavo H. Kattan
Departamento de Ciencias Naturales y Matemáticas, Pontificia Universidad Javeriana-Cali, Cali, Colombia
Email: gustavokattan@gmail.com

Lian Pin Koh
Department of Environmental Sciences, ETH Zurich, CHN G 73.1, Universitatstrasse 16, 8092 Zurich, Switzerland
Email: lian.koh@env.ethz.ch

Inza Koné
Centre Suisse de Recherches Scientifiques en Côte d'Ivoire, 01 BP 1303 Abidjan 01; Laboratory of Zoology, Université Félix Houphouet-Boigny, Abidjan, Côte d'Ivoire, 22 BP 582 Abidjan 22
Email: inza.kone@csrs.ci

Virendra Kumar
Centre for Interdisciplinary Studies of Mountain and Hill Environment, University of Delhi, Delhi – 110007, India
Email: k.virendra7@gmail.com

William F. Laurance
Centre for Tropical Environmental and Sustainability Science (TESS) and School of Marine and Tropical Biology, James Cook University, Cairns, Queensland 4870, Australia
Email: bill.laurance@jcu.edu.au

Nigel Leader-Williams
Durrell Institute of Conservation and Ecology (DICE), University of Kent at Canterbury, Canterbury, Kent CT2 7NR, UK
Email: nl293@cam.ac.uk

Jianguo Liu
Center for Systems Integration and Sustainability,
Michigan State University, East Lansing, MI 48823-
5243, USA
Email: liuji@msu.edu

Narong Mahannop
Director of Wildlife Conservation Office, Department
of National Parks, Wildlife and Plant Conservation,
Phahonyothin Road, Bangkhen, Bangkok 10900,
Thailand
Email: nmahannop@yahoo.com

Dino J. Martins
Insect Committee of Nature Kenya, The East Africa
Natural History Society, P.O. Box 44486 GPO 00100,
Museum Hill, Nairobi, Kenya;
Museum of Comparative Zoology, Harvard University
Cambridge, MA, USA;
Turkana Basin Institute, Stony Brook University,
Kenya
Email: dino.martins@gmail.com

Sittichai Mudsri
Director of Forest Fire Control Division, Department
of National Parks, Wildlife and Plant Conservation,
Thailand
Email: sitichai.budo@hotmail.com

Carolina Murcia
Department of Biology, University of Florida,
Gainesville, Florida, USA
Email: carolinamurcia01@gmail.com

Maharaj K. Pandit
Department of Environmental Studies, University of
Delhi, Delhi – 110007, India;
Centre for Interdisciplinary Studies of Mountain &
Hill Environment, University of Delhi, Delhi –
110007, India
Email: rajkpandit@gmail.com

Flavia Pardini
Página 22, Rua Itararé, 123, CEP 01308-030, São
Paulo, SP, Brazil
Email: flapar@gmail.com

Renata Pardini
Departamento de Zoologia, Instituto de Biociências,
Universidade de São Paulo, Rua do Matão – travessa
14, 101, CEP 05508-090, São Paulo, SP, Brazil
Email: renatapardini@uol.com.br

Kelvin S.-H. Peh
Conservation Science Group, Department of Zoology,
University of Cambridge, Downing Street, Cambridge
CB2 3EJ, UK
Email: kelvin.peh@gmail.com

Carlos A. Peres
Centre for Biodiversity Research, School of
Environmental Sciences, University of East Anglia,
Norwich NR4 7TJ, UK
Email: c.peres@uea.ac.uk

Rohan Pethiyagoda
Ichthyology Section, Australian Museum, 6 College
Street, Sydney, NSW 2010, Australia
Email: rohanpet@gmail.com

Patrick Pikacha
Solomon Islands Community Conservation
Partnership, P.O. Box R82, Ranadi, Honiara, Solomon
Islands
Email: patrick.pikacha@gmail.com

Pilai Poonswad
Department of Microbiology, Faculty of Science,
Mahidol University, Rama 6 Road, Bangkok 10400,
Thailand
Email: pilai.poo@mahidol.ac.th

Thane K. Pratt
US Geological Survey, Pacific Island Ecosystems
Research Center, P.O. Box 44, Hawaii National Park,
Hawaii 96718
Email: tkpratt@usgs.gov

Dewi M. Prawiradilaga
Division of Zoology, Research Centre for Biology,
Indonesian Institute of Sciences, Cibinong-Bogor,
Indonesia
Email: dewi005@lipi.go.id or dewi.malia.
prawiradilaga@lipi.go.id

Hajanirina Rakotomanana
Department of Animal Biology, Faculty of Science,
University of Antananarivo, Antananarivo 101,
Madagascar
Email: rakotomh@refer.mg

Jonah Ratsimbazafy
Groupe d'Etude et de Recherche sur les Primates de
Madagascar 34, Cité des Professeurs, Fort Duschenne,
Antananarivo 101, Madagascar;
Department of Paleontology and Biological
Anthropology, Faculty of Science, University of
Antananarivo, Antananarivo 101, Madagascar;
Durrell Wildlife Conservation Trust – Madagascar
Programme, B.P. 8511, Antananarivo (101)
Madagascar
Email: jonah.ratsimbazafy@durrell.org

Peter H. Raven
President Emeritus
Missouri Botanical Garden
P.O. Box 299
St. Louis, MO 63166-0299, USA
Email: Peter.Raven@mobot.org

Pedro L.B. da Rocha
Programa de Pós-Graduação em Ecologia e
Biomonitoramento, Instituto de Biologia,
Universidade Federal da Bahia, R. Barão de Jeremoabo
S/N, CEP 40170-290, Salvador, BA, Brazil
Email: peurocha@ufba.br

Phil Shearman
University of Papua New Guinea, Papua New Guinea
Email: shearman@ozemail.com.au

Navjot S. Sodhi
Department of Biological Sciences, National
University of Singapore, 14 Science Drive 4,
Singapore 117543, Singapore

Herwasono Soedjito
Division of Botany, Research Centre for Biology,
Indonesian Institute of Sciences (LIPI), Cibinong
Science Centre, Jl. Raya Jakarta Bogor KM 46,
Cibinong-Bogor 16911, Indonesia
Email: herwasono.soedjito@lipi.go.id

Dimitrina Spencer
Department of Education, University of Oxford, UK
Email: spencer.dimitrina@gmail.com

Robert Spencer
Environmental Consultant, UK
Email: Robert.Spencer@Urs.com

Flora I. Tibazarwa
Department of Botany, University of Dar es Salaam,
P.O Box 35060, Dar es Salaam, Tanzania
Email: ismailf@udsm.ac.tz

Marika Tuiwawa
South Pacific Regional Herbarium, University of the
South Pacific, Private Bag, Suva, Fiji Islands
Email: tuiwawa_m@usp.ac.fj

Bethany L. Woodworth
Department of Environmental Studies,
University of New England,
11 Hills Beach Road, Biddeford, ME 04005, Maine,
USA
Email: bwoodworth@une.edu

Alberto Yanosky
Guyra Paraguay, Gaetano Martino 215, Asunción,
Paraguay
Email: yanosky@guyra.org.py

NOTES ON CONTRIBUTORS

Sylvanus Abua, with a background in geography, lives and works in Cross River State where he has observed first-hand the donor-supported conservation efforts presented as case studies in this paper. He has extensive social research experience with a focus on local communities' capacities for participatory forestry management. He holds a masters' degree in environment and development from the University of Reading, United Kingdom. His education also includes a masters' degree in geographical information systems from the University of Ibadan, Nigeria.

Germán Ignacio Andrade-Pérez is a Professor of Ecosystem Management and Global Environmental Change at the School of Management at Universidad de los Andes in Bogotá, Colombia. He is a biologist with a Master's in Environmental Studies from Yale University. He has been scientific director and executive director of the Colombian non-governmental organization (NGO) Fundación Natura, and coordinator of the Conservation Biology Program of the Colombian Alexander von Humboldt Institute. He has advised conservation programs for international organizations in Colombia, Ecuador, Peru, Bolivia, and Venezuela with the World Conservation Union, World Bank, International Development Bank, United Nations, and international NGOs.

Carter T. Atkinson is a Research Microbiologist with US Geological Survey, Pacific Island Ecosystems Research Center and has worked in Hawaii and the Pacific Basin for over 20 years on the ecology and impacts of introduced diseases and disease vectors on native and indigenous forest birds. He has either served on or advised on a number of US Fish and Wildlife Service Recovery Teams on disease issues that may affect recovery of threatened and endangered island birds.

Paul C. Banko is a Research Wildlife Biologist with US Geological Survey, Pacific Island Ecosystems Research Center, where his research is focused on the conservation biology of Hawaiian birds. Most of his research has focused on the ecology of endangered forest bird species, with special emphasis on historical population trends, feeding ecology and specialization, threats to food webs, and species restoration.

Hans Bauer has lived in Ethiopia since 2008, building research capacity at several universities. Previously he studied conservation of large carnivores, with a focus on West and Central Africa. Dr Hans Bauer specializes in biodiversity conservation and capacity building in Africa. With an interdisciplinary conservation science background, he has focused on higher education development, human–wildlife conflict, management of protected areas, sustainable development, and community-based management of natural resources. He has ample experience with sectoral support in conservation and in education, with knowledge management, and with international policy dialogue. He coordinates lion conservation and research in several African countries for WildCRU, University of Oxford.

Kamal S. Bawa is a Distinguished Professor of Biology at the University of Massachusetts Boston, and Founder-President of the Bangalore-based Ashoka Trust for Research in Ecology and the Environment (ATREE). He has published more than 190 scientific papers and 10 authored or edited books and monographs. Among the many awards he has received are: Giorgio Ruffolo Fellowship at Harvard University, Guggenheim Fellowship, Pew Scholar in Conservation and the Environment, the Distinguished Service Award from the Society for Conservation Biology, and the Gunnerus Award in Sustainability Science from the Royal Norwegian Society of Sciences and Letters. His latest book, *Himalaya: Mountains of Life* (www.Himalayabook.com), a sequel to *Sahyadris: India's Western Ghats* was published earlier this year. www.kbawa.com

Gilianne Brodie has lived and worked in the tropics and on islands for almost 30 years. She has predominantly studied invertebrate biodiversity in marine, freshwater, and terrestrial ecosystems. Gilianne is currently biodiversity and conservation research group leader for the Faculty of Science, Technology and Environment at the University of the South Pacific (USP) where since 2008 she has taught invertebrate biology and conservation biology. Gilianne received her PhD from James Cook University, Australia, where she also lectured for 16 years in the School of Marine Biology and Aquaculture. Gilianne has a long history with the Pacific Islands having also been a postgraduate student at USP in the late 1980s and more recently having worked with the Secretariat of Pacific Community Land Resources Division from 2005 to 2007.

Gerardo Ceballos received his Master's degree in Ecology from the University of Wales and his PhD in Ecology from the University of Arizona. He is a member of the Mexican Academy of Sciences and has spent a couple of sabbaticals at Stanford University. His research is on population and community ecology, macroecology, and conservation of vertebrates. He has published 350 scientific and outreach papers and 35 technical books. His passion is the conservation of nature.

Vijak Chimchome is an Assistant Professor in Wildlife Science and Head of the Department of Forest Biology Faculty of Forestry, at Kasetsart University, Thailand. Dr Chimchome teaches courses in wildlife ecology and management to both undergraduates and graduate students, such as Principles of Wildlife Ecology and Management, Ornithology, Population Ecology, and Quantitative Ecology. He and some graduate students continue carrying on research on hornbill biology, ecology, and population aspects.

Richard T. Corlett is Director of the Center for Integrative Conservation at the Chinese Academy of Science's Xishuangbanna Tropical Botanical Garden in Yunnan. Until recently, he was a Professor at the National University of Singapore. He received his PhD from the Australian National University and has also worked at the University of Chiang Mai, Thailand, and the University of Hong Kong, China. Major research interests include terrestrial ecology and biodiversity conservation in tropical East Asia, rain forest biogeography, plant–animal interactions, invasive species, and the impacts of climate change. He is author or coauthor of several books, including *The Ecology of Tropical East Asia* (2009, Oxford University Press) and *Tropical Rain Forests: An Ecological and Biogeographical Comparison* coauthored with Richard Primack (second edition, 2011, Wiley-Blackwell). He is also a lead author for the Asia chapter in the Working Group II contribution to the Fifth Assessment Report of the IPCC, due in 2014.

Wondmagegne Daniel is a PhD fellow in Texas Tech University and is doing his research on civet farming in Ethiopia. His PhD is mainly on Wildlife Management with Geographic Information Systems (GIS) Minor. With a Bachelor's degree in Biology and a Master's in Systematic Zoology, he strives to become part of the solution for wildlife-related problems in Ethiopia. He worked as the Head of the Department of Rangeland and Wildlife Sciences in Mekelle University and handled courses too. While heading the department, he played his leading role in realizing the establishment of a new wildlife and eco-tourism management program. In addition to trying to excel with his teaching activities, he was an active participant of the department's research and community service.

Charbel El-Hani is a Professor at the Federal University of Bahia (UFBA), Brazil. His research focuses on philosophy of biology, science education, ecology, and behavior. He has experience in fostering communities of practice gathering in-service and pre-service high-school biology teachers, graduate and undergraduate students, and researchers. As he also does research on ecology, he has become involved in the experience of building a community of practice among environmental practitioners, researchers, and graduate and undergraduate students.

F.B. Vincent Florens worked for several years with the Mauritian Wildlife Foundation (a conservation nongovernmental organization), and the Mauritius Herbarium before moving to the University of Mauritius where he is currently Associate Professor of Ecology and Head of the Department of Biosciences. He completed his Master's degree in Applied Ecology and Conservation at the University of East Anglia, UK, and received his PhD from the Université de La Réunion, France. He currently serves on several national and regional boards and committees on matters related to biodiversity conservation, and has coauthored various reports on conservation strategy and management principally for the Republic of Mauritius. He coauthored a book on the non-marine mollusks of the Mascarene Islands and has about 40 other publications in regional or international conference proceedings, journals, or books.

Andrés García received his PhD from the University of New Mexico in 2003 and is currently a full-time researcher at Instituto de Biologia at UNAM. He has extensive experience on Mexican herpetofauna, having conducted research on conservation, ecology, and biogeography of tropical dry forest amphibians and reptiles.

Fikirte Gebresenbet is a PhD student in the Department of Zoology, Oklahoma State University, USA. Previously, she taught courses on biodiversity conservation, mammalogy, herpetology, and research methods at Mekelle University, Ethiopia. Her research experience focused on conservation issues with emphasis on reptiles, amphibians, and carnivores. Her PhD research is centered on identifying carnivore conservation priority areas in parts of Ethiopia through habitat modeling and scenario planning. With her good grasp of the conservation activities taking place in Ethiopia, she works to identify and/or solve the problems that are faced by wildlife and conservationists.

Roy E. Gereau is an Assistant Curator at the Missouri Botanical Garden and directs its Tanzania Program. He has been active in African plant taxonomy, floristics, phytogeography, and conservation for 28 years. He has described a number of new species from Tanzania, is a coauthor of the book *Field Guide to the Moist Forest Trees of Tanzania* (Lovett et al., 2006), and first author of the *Lake Nyasa Climatic Region Floristic Checklist* (Gereau et al., 2012). He has received funding from the National Geographic Society for biodiversity inventories, the Critical Ecosystem Partnership Fund for plant conservation assessment in the Eastern Arc Mountains and Coastal Forests of Tanzania and Kenya, and the Liz Claiborne-Art Ortenberg Foundation for studies in biodiversity conservation in the face of climate change.

Luke Gibson is based at the National University of Singapore, where he is studying tropical forest loss in Southeast Asia and its impact on biodiversity. For his PhD, he is recording extinctions of small mammal species from small forest fragments in Chiew Larn reservoir, Thailand, and the persistence – or decline – of other mammalian ungulates and carnivores in the lowland dipterocarp forest surrounding the reservoir. He has worked previously with Phayre's leaf monkeys and long-tailed macaques, studying their sensitivity to habitat fragmentation in Thailand and to biodiversity loss in Singapore, respectively. Before moving to Southeast Asia, he received his Bachelor's degree from Princeton University and his Master's degree from the University of California, San Diego.

Mwangi Githiru is currently the Director of Biodiversity and Social Monitoring at Wildlife Works Sanctuary Ltd., mainly involved in evaluating the environmental and sociological impacts of REDD projects undertaken by Wildlife Works. He has previously worked as a Deputy Director of Research with the Ministry of Higher Education, Science and Technology in Kenya. Dr Githiru has considerable experience in the environment conservation sector at the research, policy, and implementation levels. He received his undergraduate degree in Wildlife Management and his Master's degree in Animal Ecology from Kenyan universities before proceeding to get his doctorate in Zoology from the University of Oxford, UK, as a Rhodes Scholar. He then successfully completed a three-year post-doctorate at the University of Antwerp, Belgium, as a Marie Curie Fellow. Dr Githiru is also an alumnus of the Watson International Scholars of the Environment (WISE) Program at Brown University, USA, and the Archbishop Tutu African Leadership Programme.

Amleset Haile completed her Bachelor's study at Haramaya University, Ethiopia, in Agriculture and her Master's in Management of protected areas at Klagenfurt University, Austria. Previously, she worked in Mekelle University, Ethiopia, as an assistant lecturer in the faculty of Agriculture and Natural Resources Management, responsible for both academic and research activities. At present, she is the assistant national coordinator of the project CASCAPE (Capacity building for scaling up evidence-based best practices in agricultural production in Ethiopia) in collaboration with Wageningen University, Netherlands. Amleset's research interest is in contributing to societal efforts to use natural resources efficiently to assist conservation and livelihood improvement activities. To make her dreams a reality, she also participated in international training including Beekeeping for Poverty Alleviation in Belgium (2008) and Climate Change in Denmark (2009), and in 2010 she was a fellow of the African Women Scientists in Climate Change Award. She also participated in the Environmental Leadership Program at UC, Berkeley, in 2012.

Bila-Isia Inogwabini is the country manager at Christian Aid in the Democratic Republic of Congo. Prior to his appointment at Christian Aid, Inogwabini was a Post-Doctoral Researcher at the Swedish University of Agricultural Sciences, Uppsala, and a Senior Scientific Advisor at the World Wide Fund for Nature. Dr Inogwabini received his PhD in Biodiversity

Management and Master's degree in Conservation Biology, and has 20 years of experience in large mammal conservation. His conservation efforts are internationally recognized: he received the 2007 National Geographic Buffett Award for leadership in conservation in Africa and the UNESCO Young Scholar Award for his research on bonobos.

James D. Jacobi is a biologist with the US Geological Survey's Pacific Island Ecosystems Research Center. He has been conducting research in Hawaii for over forty years with primary emphasis on studying the ecology and status of native plant and bird species, as well the impacts of invasive species on Hawaiian ecosystems. A major focus of his research has also been mapping the distribution of plant species and communities throughout the Islands relative to both current conditions and predicted future climate scenarios.

Richard K.B. Jenkins Richard Jenkins works for the IUCN Global Species Programme and is based in Cambridge, UK. He lived in Madagascar between 2002 and 2012 where he established a biodiversity organization called Madagasikara Voakajy. He obtained his PhD from Cardiff University and his Bachelor's degree from the University of East Anglia.

Michelle Kalamandeen is an environmentalist who has worked extensively for over 12 years in numerous indigenous communities of Guyana. She has a background strongly focused on social justice, ethical implications of conservation actions, and protected area designation and management. Michelle holds a Master's degree in Biodiversity, Conservation and Management from the University of Oxford and has an extensive list of journal and book publications. She currently works as the protected areas coordinator for the Guyana Marine Turtle Conservation Society and is a lecturer at the University of Guyana. She is the coordinating secretary for the National Committee for UNESCO's Man and Biosphere Reserves and an executive board member of the Guyana Human Rights Association.

Gustavo H. Kattan is a biologist with a particular interest in the study and conservation of the biodiversity of the tropical Andes. He received his PhD from the University of Florida and for 14 years was a Conservation Ecologist with the Wildlife Conservation Society in the Colombia Program. Currently he is Associate Professor in the Biology program at Pontificia Universidad Javeriana in Cali, Colombia. He has published over 70 papers in scientific journals and book chapters, and many technical reports, including conservation plans for several species of birds and mammals.

Lian Pin Koh is Assistant Professor of Applied Ecology and Conservation at the Department of Environmental Sciences, Swiss Federal Institute of Technology (ETH Zurich). Lian Pin received his Ph.D. from Princeton University, USA. He has published over 70 articles in peer-reviewed journals including *Science, Nature, Proceedings of the National Academy of Sciences of the USA, Trends in Ecology and Evolution, Ecological Applications,* and *Conservation Biology*. He has served on the editorial board of *Biological Conservation, Animal Conservation, Biotropica, Tropical Conservation Science,* and *Endangered Species Research*. Visit his website (www.lianpinkoh.com) for more information.

Inza Koné holds a PhD in Primate Conservation Biology from the *Université Félix Houphouet-Boigny* (former University of Cocody), Côte d'Ivoire. He is currently a lecturer at the *Université Félix Houphouet-Boigny* and a research associate at the *Centre Suisse de Recherches Scientifiques en Côte d'Ivoire* (CSRS), Abidjan, where he is Head of the Biodiversity and Food Security Department. He is a member of the Scientific Committee of the Great Apes Survival Project of the United Nations Environment Program (GRASP/UNEP), the Advisory Board of the Institute for Breeding Rare and Endangered African Mammals (IBREAM) of the University of Utrecht, the Netherlands, and the Pygmy Hippo subgroup of the IUCN/SSC Hippo Specialists Group. His research focuses on the management of natural resources linking ecology, economics, culture, and the behavioral ecology and conservation of endangered large mammals. For his internationally outstanding achievements in the conservation of species and habitats, Inza Koné was awarded the 2001 WWF award for the conservation of the Taï National Park (a World Heritage site in Côte d'Ivoire), the 2005 Martha J. Galante Award of the International Primatological Society, the 2009 Future For Nature Conservation Award (Netherlands), and the 2012 Whitley Award for Nature Conservation (UK).

Virendra Kumar taught Botany at the University of Delhi for over three and a half decades. He received his PhD from the University of Delhi and has authored a number of books/monographs, including *Chromosome Numbers of the Flowering Plants of the Indian Subcontinent* (1985, Botanical Survey of India, Kolkata) and *Environmental Sensitivity of the Himalayan River*

Basins (2000, CISMHE Monograph Series). He was awarded Homi Bhabha Fellowhip in 1975 and, as a recipient of UNESCO and Royal Society grants, Virendra carried out research on rhododendrons at the Royal Botanic Gardens, Kew; Harvard University; and the University of Virginia. As chairman, he was responsible for providing ecological orientation to the world famous Chipko movement by recommending a ban on deforestation in the Himalaya in the mid-1970s. He is currently engaged in completing his two books: *North American Rhododendrons* and *Evolutionary Divergence in Rhododendrons: An SEM Study of Leaf Surfaces.*

William F. Laurance is Distinguished Research Professor at James Cook University in Cairns, Queensland, Australia, and Prince Bernhard Chair in International Nature Conservation at the Unversity of Utrecht, the Netherlands. He studies how environmental threats, such as habitat fragmentation, logging, hunting, wildfires, and climatic change affect tropical forests and their species. He works in the Asia-Pacific, Amazon, Africa, and tropical Australia, and has published five books and over 300 scientific and popular articles. He is an Australian Laureate, former president of the Association for Tropical Biology and Conservation, and co-winner of the BBVA Frontiers in Ecology and Conservation Biology prize, among other honors.

Nigel Leader-Williams is Director of Conservation Leadership at the University of Cambridge and a Fellow of Churchill College. A trained veterinary surgeon, he completed a PhD on the ecology of introduced reindeer on South Georgia and a post-doctorate on the conservation of rhinos and elephants in Zambia. As previous Director of the Durrell Institute of Conservation and Ecology, and in his present role, Nigel builds capacity in conservation through research and teaching that sits within both natural and social sciences, focusing on large mammals that conflict with peoples' interests. His most recent book, *Trade-offs in Conservation*, was published in 2010.

Jianguo (Jack) Liu holds the Rachel Carson Chair in Sustainability, is a University Distinguished Professor of Fisheries and Wildlife at Michigan State University, and also serves as Director of the Center for Systems Integration and Sustainability. He has been a guest professor at the Chinese Academy of Sciences and a visiting scholar at Stanford University (2001–2002), Harvard University (2008) and Princeton University (2009). Liu takes a holistic approach to addressing complex human–environmental challenges through systems integration, which means he integrates multiple disciplines such as ecology and social sciences. His work has been published in journals such as *Nature* and *Science*, and has been widely covered by the international news media. In recognition of his efforts and achievements in research, teaching, and service, Liu was named a Fellow of the American Association for the Advancement of Science (AAAS) and has received many awards, including the Guggenheim Fellowship Award, the CAREER Award from the National Science Foundation, the Distinguished Service Award from US-IALE and the Aldo Leopold Leadership Fellowship from the Ecological Society of America.

Narong Mahannop (formerly Khao Yai National Park superintendent), is currently the director of Wildlife Conservation Office, Department of National Parks, Wildlife and Plant Conservation, Thailand. He began his career in the Royal Forest Department in 1981, serving as head of the Reforestation Project. During this period, Narong participated in trainings and study trips of many international and regional organizations, including Mississippi State University, USA; the 2nd Conference on ASEAN Heritage Park (AHP), Malaysia; the 4th Regional Conference on Area (PA) in Southeast Asia, Malaysia; the 31st Session of the World Heritage Committee, New Zealand.

Dino J. Martins is a Kenyan conservationist and entomologist. He holds a PhD from the Department of Organismic and Evolutionary Biology at Harvard University. He has published widely on pollinators and insects in scientific, natural history, and environmental journals. His work has been featured in the *Smithsonian Magazine*, on the BBC as well as in *National Geographic*. He is the current Chair of the Insect Committee of Nature Kenya, the East African Natural History Society. Among his awards and fellowships are the Ashford Fellowship in the Natural Sciences, the Graduate School of Arts and Sciences (GSAS), Harvard University, 2004 Smithsonian Institution SIWC–MRC Fellowship, and 2002 and 2003 Peter Jenkins Award for Excellence in African Environmental Journalism. In 2009, he won the Whitley Award for his work with pollinators in East Africa. He was named a *National Geographic* "Emerging Explorer" in 2011. Dino is currently a post-doctoral fellow with the Stony Brook – Turkana Basin Institute. His work continues to explore the intricate connections between insects and human life and livelihoods.

Sittichai Mudsri (formerly Budo-Sungai Padi National Park superintendent), is now a director of Forest Fire Control Division, Department of National Parks, Wildlife and Plant Conservation, Thailand. He also serves as a research collaborator and consultant of "Hornbill Research and Community-based Conservation Project" in Budo-Sungai Padi National Park. He received the Honorary Award in "Creative Solution for Land-Use Problem by Community-based Forest Management" from UNESCO in 2009.

Carolina Murcia is the Science Director for the Organization for Tropical Studies and Adjunct Faculty in the Biology Department at the University of Florida. Previously she was the Sub-regional Coordinator for the Northern Andes at the Wildlife Conservation Society, and for over a decade a Board Member of the Society for Ecological Restoration. She is a biologist with a PhD in tropical conservation, with over 20 years of experience in ecology and conservation of Andean forests. She has published over 30 scientific papers, coauthored one book, and authored a number of technical documents in conservation policy in Colombia.

Maharaj K. Pandit is a Professor at the University of Delhi. A conservation biologist, Raj received his PhD from the University of Delhi. He was elected a Fellow of Linnaean Society of London in 2001 and is a recipient of Raffles Biodiversity Fellowship award from National University of Singapore (NUS). Raj subsequently taught at NUS as Visiting Senior Fellow. His work focuses on the Himalayan ecosystems and the endangered taxa, and his research interests include understanding the ecological, genetic, and genomic causes of plant rarity and invasion, impact of developmental projects on the Himalayan biodiversity, and species' response to climate change in the Himalaya.

Flavia Pardini has been working as a journalist for 20 years in Brazil, the US, and Australia. She has reported on financial and economic matters for Gazeta Mercantil, Reuters and CartaCapital. In 2005, she co-founded Página 22 (www.pagina22.com.br), the first Brazilian magazine dedicated exclusively to sustainability issues.

Renata Pardini is a Professor at the University of São Paulo (USP), Brazil. Her research program focuses on mastozoology and landscape ecology, especially on how structural thresholds affect biodiversity in human-modified landscapes. She has also acted as a consultant in management plans for conservation units, and in discussions on designs for monitoring programs associated with the payment for ecosystem services, as well as on how to improve the ecological components of environmental impact assessments. She participated as a teacher in the short-term course at the Federal University of Bahia (UFBA), from which some results are presented in chapter 10.

Kelvin S.-H. Peh has completed a three-year European Union Marie Curie EST fellowship and received his PhD from Leeds University. Currently an AXA Post-Doctoral Fellow and a Post-Doctoral Researcher at St. John's College (Cambridge), Kelvin's work involves close collaborations with various environmental organizations such as the Global Secretariat of BirdLife International, the Royal Society for the Protection of Birds, and the United Nations Environment Programme's World Conservation Monitoring Centre. Kelvin also has extensive ecological fieldwork experience in Central Africa and Southeast Asia.

Carlos A. Peres, born and raised in Belém, Brazil, was exposed to Amazonian ecology and conservation from age six and his father's 5200-ha landholding in eastern Pará, consisting largely of undisturbed primary forest, became a natural history playground. For the past 25 years, he has been studying wildlife community ecology in Amazonian forests and the biological criteria for designing large nature reserves. He currently co-directs four conservation science programs in Neotropical forests. In 1995, he received a Biodiversity Conservation Leadership Award from the Bay Foundation, and in 2000 he was elected an Environmentalist Leader for the New Millennium by *Time Magazine* and CNN. He is currently a Professor in Tropical Conservation Biology at the University of East Anglia, UK, and divides his time between Norwich and fieldwork at multiple field sites in lowland Amazonia.

Rohan Pethiyagoda has contributed extensively to the scientific literature on Sri Lanka's biodiversity, having worked primarily on the taxonomy and conservation of freshwater fishes and amphibians. The Wildlife Heritage Trust, established in 1990 and since managed by him, is the country's premier publisher of biodiversity-related literature, having contributed also to exploration and research that led to the discovery of hundreds of new species on the island. He is presently a research associate in fishes at the Australian Museum, Sydney, serves as editor for Asian Freshwater Fishes of *Zootaxa*, and is a trustee of the International Trust for Zoological Nomenclature. He has previously served as Environment Adviser to the government of Sri Lanka, as a member – and deputy chair – of the IUCN Species Survival Commission, and as a committee member of the World Commission on National Parks and the Global Amphibian Specialist Group. He is a fellow of the National Academy of Sciences of Sri Lanka and has been recognized through several international awards and prizes including the Rolex Awards for Enterprise.

Patrick Pikacha works with the Solomon Islands Community Conservation Partnership foundation. He has obtained several competitive grants and conducted biodiversity research in the Solomon Islands with a focus on frogs and small mammals; he is lead author of the book *Frogs of the Solomon Islands*. Patrick is currently undertaking a PhD at the University of Queensland, Australia, and has extensive experience of coordinating conservation efforts and working with communities in different island groups to achieve conservation outcomes via working through the local land tenure systems. His passion for the Melanesian environment led him to found the Pacific Island grassroots conservation magazine *Melanesian Geo* in 2005.

Pilai Poonswad is a Professor of Biology in the Department of Microbiology, Faculty of Science, Mahidol University, Thailand. Professor Poonswad's main activities are teaching graduate students in fundamental and advanced parasitology, pre-medical students in medical parasitology, and biology undergraduates in avian biology and conservation. She supervises graduate students in various aspects of microbiology and biology. Her research focuses on hornbill biology and ecology, and her most innovative techniques have been instrumental in a series of practical conservation and awareness initiatives.

Thane K. Pratt is an avian ecologist with a focus on the conservation biology of birds in Hawaii and the tropical Pacific. He is retired from the US Geological Survey, Pacific Island Ecosystems Research Center, where he worked for twenty years as a wildlife biologist engaged in research on endangered Hawaiian forest birds. He is now writing an all-new second edition of *Birds of New Guinea*, with coauthor Bruce Beehler.

Dewi M. Prawiradilaga is a principal scientist at the Research Centre for Biology – Indonesian Institute of Sciences (LIPI). Dewi has been a long term editor of *Treubia*, zoological journal of Indo-Australia archipelago. She received her PhD in Ecology, Evolution and Systematics from the Australian National University, Canberra. In the last fifteen years Dewi has been involved in the ecological research and conservation activities of Indonesian endangered bird species, specifically eagles.

Hajanirina Rakotomanana obtained his PhD from Kyoto University, Japan. He is currently a Professor at the University of Antananarivo, Madagascar, and Vice-President of the steering committee of the Malagasy Birds Group (Asity Madagascar). He has written several scientific papers and books on Malagasy birds. Since 2004, he has become a part of the scientific committee of Pan African Ornithological Congress. In 2005, he received Environmental Leadership courses at Beahrs, University of California, at Berkeley, USA. Currently, he is a member of the Teachers of Tropical Biology Association (a program closely related to Cambridge University).

Jonah Ratsimbazafy received his PhD in Physical Anthropology from the State University of New York at Stony Brook. He is currently the Training and Conservation Coordinator of the Durrell Wildlife Conservation Trust in Madagascar. He is also an Adjunct Professor in the Department of Paleontology and Anthropology, and a Department of Medicine veterinary at the University of Antananarivo. He coauthored the second and third editions of the *Field Guide Series: Lemurs of Madagascar*. From 2006 to 2008, he was the Vice-President of the International Primatological Society for Conservation. He is the Secretary General of a Malagasy Primate Group (GERP). He is the co Vice-Chair of the IUCN/SSC Primate Specialist for Madagascar.

Peter H. Raven has become an influential voice in systematics, ecology, and evolution worldwide over the past 50 years. He served as President of the American Association for the Advancement of Science and other organizations, Home Secretary of the US National Academy of Sciences, and is a member of a number of other academies worldwide. During his 39-year tenure as President of the Missouri Botanical Garden, he guided the Garden to a position of global leadership in conservation, with centers of activity in the tropics of Latin America, Africa, and Asia. From its initiation, he served as co-chair of the editorial committee of the *Flora of China*, a 50-volume work completed in 2013 after 20 years of effort. He is coauthor of the leading textbook in botany, *The Biology of Plants*, and coauthored leading texts in biology and the environment.

With Paul Ehrlich, he originated the important concept of coevolution.

Pedro L.B. da Rocha is a Professor at the Federal University of Bahia (UFBA), Brazil. His experience in ecological research, coordination of the Ecology and Biomonitoring Graduate Studies at UFBA, federal evaluation of graduate courses in Ecology (CAPES), and scientific boards of funding agencies (FAPESB) has contributed to his activities focusing on bridging the research–implementation gap.

Phil Shearman is currently Director of the University of Papua New Guinea Remote Sensing Centre and is a visiting fellow of the School of Biology at the Australian National University College of Medicine, Biology and Environment. Phil received his PhD from the Australian National University, Canberra. His research interests include tropical forest ecology and management, applied remote sensing, and biogeography.

Navjot S. Sodhi (1962–2011) of the National University of Singapore was one of the great minds of conservation biology. A native of Punjab, India, Professor Sodhi graduated from the University of Saskatchewan and then moved to an incredibly fruitful 15 years documenting rain forest loss and degradation in Southeast Asia and their effects on populations of animals and plants. He was best known as a conservationist, someone who cared passionately about those rich lands and the people who lived in the region, and who strove, with a large group of colleagues and students, to devise ways to improve the sustainability of the area while pressures on the forest mounted rapidly.

Herwasono Soedjito is a senior scientist at the Research Centre for Biology – Indonesian Institute of Sciences (LIPI), a former Director of Man and the Biosphere (MAB) Indonesia Program – UNESCO and received his PhD on Forest Ecology from Rutgers University, USA. He has been undertaking research activities since 1980 in the fields of ecology, management of tropical rainforests, and later on expanding his interest in cultural and biodiversity conservation as well as monitoring of indigenous knowledge. He spent years travelling extensively in Indonesian forests and living with local communities in Kalimantan. He was also counterpart of WWF Kayan Mentarang (1991–1995) and Project Manager of WWF Betung Kerihun in West Kalimantan (1995–1998), Scientist and Site Manager of Bulungan Research Forest (BRF)-CIFOR in Malinau – East Kalimantan (1999–2002), Director of Indonesian

Man and the Biosphere Program – LIPI (2003–2007), Director Terrestrial Program of Conservation International Indonesia (2008–2009), and was Fire, Research, and Monitoring Manager of Kalimantan Forests and Climate Partnership (KFCP) – Australia Forest Carbon Partnership (IAFCP) a demonstration area for REDD+ on the Ex-Mega Rice Project in Kapuas District of Central Kalimantan (2010–2011).

Dimitrina Spencer obtained her PhD from the Institute of Social and Cultural Anthropology and Linacre College, University of Oxford. She was the project anthropologist on the DFID funded Cross River State Community Forestry Project in south east Nigeria from 2000 to 2002. She conducts research and teaches anthropology of international development at the University of Oxford. She has co-edited: *Anthropological Fieldwork: a Relational Process* (CSP 2010) and *Emotions in the Field: the Psychology and Anthropology of Fieldwork Experience* (Stanford UP, 2010).

Robert Spencer is a Business Line Director for sustainability services at URS Corporation. A Fellow of the Royal Geographic Society and Chartered Geographer, he has a Masters in forestry and land use from the University of Oxford and Linacre College. He was the Project Manager of DFID's multi-million pound Cross River State Community Forestry Project in south east Nigeria from 2000 to 2003. He went on to deliver the major Yunnan Environment Development Programme (YEDP), also funded by DFID and focussing on the environment-poverty nexus in south west PR China. Robert Spencer has since offered forestry and land use advice in many contexts and countries, most recently Gabon, and is currently focussed on woodland carbon sequestration schemes in the UK. He is Chairman of URS' Infrastructure & Environment division Sustainability Committee for Europe and a member of the UK's Living with Environmental Change (LWEC) Business Advisory Board.

Flora I. Tibazarwa is a lecturer in the Botany Department at the University of Dar es Salaam. She has a Master's degree in Forest Ecology and a PhD in the ecophysiology of marine plants using molecular approaches from the Radboud University in the Netherlands. Her research interests are conservation biology, abiotic stress, and modern biotechnology. She has over 10 years' experience conducting environmental impact assessments and natural resource management. She is the current chairperson of the Tanzania

Biodiversity Information Facility Governing Board, a national facility for coordinating biodiversity informatics. She has won several research grants that have supported both Master's and PhD students.

Marika Tuiwawa has been the Curator of the South Pacific Regional Herbarium since 1998. An internationally acknowledged expert on Pacific Island plants and vegetation, Marika is a much in demand regional botanist and experienced conservation biologist. She is also a founding trustee of Fiji's highly respected conservation organization NatureFiji/MareqetiViti, the only local non-government organization in Fiji focused on the conservation of threatened species. Marika received his Master's degree from the University of the South Pacific in 1996 and has been involved with numerous regional biodiversity and conservation projects since, producing not only many technical reports and major publications in international journals but also mentoring some of the regions' most impressive up-and-coming conservation biologists.

Bethany L. Woodworth is an avian ecologist specializing in the demography and ecology of forest birds. She worked as a Research Wildlife Biologist with US Geological Survey, Pacific Island Ecosystems Research Center, for ten years, focusing on developing conservation and recovery strategies for threatened and endangered Hawaiian forest birds. Bethany now teaches environmental sciences in the Department of Environmental Studies, University of New England.

Alberto Yanosky received his Bachelor's and doctoral degrees from the University of Mar del Plata and his Master's degree from the University of Entre Rios, both in Argentina. He has been working on species conservation, population, and community and landscape ecology, as well as environmental and social safeguards, civil society organizations, and the role of society in conservation. He has experience mainly in the Americas, but has also worked in Africa and Central Asia. His working experience in the environmental and training sectors began in 1985, and now, with more than 25 years of experience, he has published more than 100 scientific papers and books. In 2013, he won the National Geographic Society/Buffett Award for Leadership in Conservation for his efforts to protect threatened habitats and species in Paraguay.

ACKNOWLEDGMENTS

We must acknowledge many people for their assistance in this project, including several anonymous reviewers who helped with polishing particular chapters. We are especially grateful for the support of Wiley-Blackwell, particularly to Ward Cooper, Kelvin Matthews, Delia Sandford, Carys Williams, Izzy Canning, and Ken Chow. We must also thank Anne Abel Smith for copy-editing the chapters, Beth Dufour for checking permissions, and Caroline Jones for arranging the index. Finally, we thank Kathy Syplywczak for her tireless efforts while managing the manuscript through final stages of production.

REMEMBERING NAVJOT SODHI: AN INSPIRING MENTOR, SCHOLAR, AND FRIEND

Maharaj K. Pandit

Department of Environmental Studies, University of Delhi, Delhi, India

Figure 0.1 Navjot S. Sodhi (in front) with the author Raj Pandit, having a roadside lunch at Lachung village in Sikkim Himalaya in June, 2005. Although he always wore his cap backwards, Navjot was always looking forward for new ways to alleviate the biodiversity crisis. Photo courtesy of Tommy Tan ©

In mid-March 2011, when Navjot did not respond to my most recent emails, I felt something amiss. Mails to him rarely went unanswered; he always replied with usual one or two liners or just a word – often his favorite, "Great!" An SOS I sent two weeks later elicited a reply, but it was his graduate student, Luke Gibson, confirming my worst fears – Navjot was in the hospital for a prolonged high fever and the reasons were undi-

agnosed. A month or so later I spoke to Navjot on the telephone to be greeted with a normal welcome – "Hi, Raj." He assured me that he was feeling better, but the fever did not relent. It worried me no end. Sadly, on June 12, 2011 an email from a colleague at the National University of Singapore (NUS) came as a bolt from the blue – Navjot had passed away after undergoing surgery. His illustrious career and a young life of remarkable achievements had been snapped short by a merciless and aggressive lymphoma. Less than a month after the diagnosis, Navjot died from cancer.

THE ÉMIGRÉ

Navjot S. Sodhi was born on March 18, 1962 at Nabha in the Indian State of Punjab. He lived as a proud Sikh, but never displayed a parochial attitude in his interaction with students, colleagues, and friends. He took great pride in calling himself a world citizen, which is also reflected by his strong associations within the academic arena spanning continents. He had his early education at Punjab University and moved to Canada's University of Saskatchewan for his PhD during a politically volatile time for India and Punjab. I recall once, in 2006, having a post-lunch discussion with Navjot and trying to convince him of buying himself a place in India – the usual back-to-the-roots stuff. Initially, my assertion met with his well-known guffaw, but as we neared his "ivory tower" (he was fond of jokingly calling his lab so), Navjot invited me for a cup of tea (Navjot mostly drank coffee, but was aware of my preference). We continued the chat on the famous couch in his lab while the tea brewed. As I had my first sip, Navjot felt a lump in his throat while talking; he narrated his woes of 1980s Punjab and recounted the unfortunate handling of young students at the hands of the security forces. His was a forced migration. I was ashamed of having pushed him this far. He simply ended the conversation by saying, "No thanks. Not that blessed place." I left the lab teary eyed and stupefied. Despite the unpleasant experience as a young student, Navjot by no means had any rancor; he in fact went out of his way to accommodate requests of colleagues from India and I myself have benefited from his magnanimity on more than one occasion. Navjot was deeply interested in seeing the progress of Indian ecologists and conservationists and his recipe was to train more and more youngsters in quantitative skills and modeling techniques.

THE BIG TICKET SCIENCE

Navjot started his academic career by studying birds, with a passion that he maintained to his very last days. He gradually moved from working on micro- to macro-problems in ecology and conservation science. He felt convinced that researching microscopic processes and patterns forms a vital foundation for broader understanding in science, but it was the big ticket issues like landscape transformations and human impacts on biodiversity that interested him more. He attributed this proclivity towards bigger ideas in conservation science to his immense respect for two towering personalities of biology – Ernst Mayr and Edward Wilson. In order to reach where he did by dint of sheer hard work and perseverance, he emulated his heroes that include Alfred Wallace, Charles Darwin, Paul Ehrlich, Peter Raven, Stuart Pimm, and many others by looking at bigger problems facing ecology and biological diversity. Navjot wore his baseball cap backwards not to make a fashion statement (dressing wasn't his forte), but to demonstrate an irreverence for the status quo. He took pleasure in challenging the solidity of paradigms and encouraged his students to do so in order to push the frontiers of scientific enterprise. Navjot almost single-handedly brought the problems of Southeast Asian biodiversity crisis to world attention (Sodhi et al., 2004). His belief in the adage – "think globally, act locally" – was demonstrably more resolute than most of his contemporaries in the tropical Asia. He, along with his collaborators, produced a seminal work on extinctions in Singapore following extensive deforestation, showing the great loss of biodiversity in the small city-state and projecting similarly large losses in the surrounding region using estimated deforestation rates (Brook, Sodhi and Ng, 2003). In yet another important contribution written with his students and colleagues, he highlighted the problem of species co-extinctions driving the biodiversity crisis (Koh et al., 2004). One of the outstanding characteristics of Navjot was his eye for spotting not so fashionable areas of research and bringing those into sharp focus. One such study topic was Southeast Asia's limestone karsts, which Navjot and his students described as "arks" of biodiversity containing high levels of endemism but threatened with overexploitation (Clements et al., 2006). A fitting tribute to the legacy of Navjot's commitment to conservation science is a recent paper (published shortly after his death) in which he and his colleagues led by Luke Gibson provide a global assessment of the impact of disturbance and land conversion on biodiversity in tropical forests (Gibson et al., 2011). They aptly conclude that to maintain tropical biodiversity primary forests are irreplaceable. In the summer of 2005,

Navjot persuaded me to take him to some of the Himalayan sites I was working in. The visit was greatly rewarding, personally and professionally, for me. Seeing the large-scale landscape changes, Navjot asked me to bring this unprecedented story to the attention of the scientific community. The work that emerged (Pandit *et al.*, 2007) was to the best of my information the first one on the Himalayan deforestation and impending species extinctions.

THE FIERCE COMPETITOR AND HUMANIST

Navjot displayed an unwavering dedication to conservation science through a large number of high impact papers and books that he either wrote or edited with leaders in the field. One of his latest books – *Conservation Biology for All* – which he edited along with Paul Ehrlich (Sodhi and Ehrlich, 2010), is a scholarly work that every conservation biologist must have on the shelf; the two editors actually ensured that it happened that way with an agreement with the publisher that made it freely available online one year after print. A few months back, when I received a parcel from William Sutherland of the University of Cambridge containing this book, I was deeply touched. The accompanying letter mentioned that Navjot had bequeathed me a copy of the book (as one of 200 identified beneficiaries worldwide). This gesture not only reaffirmed Navjot's commitment to widening the ambit of conservation scientists, but also ensured access to current and high-quality literature by practicing teachers and researchers. Though a fierce competitor, Navjot had a remarkable heart (I lately learned that he had an interest in poetry). He did not like carrying out poor-quality science and he made it clear to everyone without mincing words; he was critical of wishy-washy science and as an editor of many a journal (he served as editor for 11 journals during his career) he ensured that "only the fittest survived" the review process. Having said that, I also found there was a sensitive person inside him – more encouraging and humane than critical. As an editor, he occasionally asked me to review manuscripts, particularly those dealing with Asia and the Himalaya. I recall that, on at least two occasions when I rejected the submissions (knowing Navjot's penchant for critical appraisal), he accepted my recommendations, but later in private conversation mentioned that

"we may also like to take into account the lack of access to literature and resources these authors may have while expecting rigorous science from them." On most occasions, however, he would use the choicest expressions if there was a hint that the work (published or otherwise) was relatively watered-down stuff; "attack" was his simple instruction.

THE CONCERN FOR HUMAN LIVELIHOODS

Navjot did not remain an aloof conservation scholar, but made conscious effort to take conservation science to social scientists and engaged with them in order to achieve justice for the imperiled ecosystems in Southeast Asia that ranged from oceans to highlands, from swamps to caves and forests. In particular, one may mention the book *Biodiversity and Human Livelihoods in Protected Areas: Case Studies from the Malay Archipelago* (Sodhi *et al.*, 2007) that brought to focus the linkages between human livelihoods of rural communities and species extirpations in two of the world's 25 biodiversity hotspots. The book highlighted the importance of local communities as stakeholders in conservation and suggested that the success of conservation depended on sustaining their economic aspirations. Respect for human dignity and life remained an abiding faith with Navjot. On one occasion while travelling in the interior of Sikkim Himalaya, Navjot spotted a procession of school children carrying placards on environmental protection. He remarked that it was a great sight and asked me to take some pictures, which I did. "I am delighted and hopeful that these children would care more for Earth than we did," said Navjot, as I returned after taking some shots. On another day during the same journey, he spotted graffiti painted on a mountain rock that quoted Mahatma Gandhi: "Earth provides enough for our need, but not for our greed." A picture of the graffiti was again taken, which he later used in one of his books (Sodhi *et al.*, 2007) and the quote appeared in a commentary he coauthored (Brook and Sodhi, 2006). Some of these incidences clearly bring out Navjot's concern for the welfare of Earth and the continued survival of other living organisms. Even as he was conscious that it remained an idealistic goal, he believed in the words of Aldo Leopold: "We shall never achieve harmony with land, any more than we shall achieve absolute justice or liberty for people. In these higher aspirations the important thing is not to achieve, but to strive." Navjot certainly strived more than most.

THE LAST LUNCH

I last met Navjot in June 2010 on a lecture visit to NUS. This luncheon meeting was unusual in the sense that it was intensely personal. While I feel constrained to elaborate on the discussion, I am willing to share that Navjot seemed deeply unhappy with some developments back home in India. He was concerned for his ailing old father and the related distress. I felt greatly humbled by his trust. Towards the end of our meeting, Navjot spoke of the future, having recently bought a new house in his adopted country of Canada where he could start a new life with his family (Navjot had been planning to move to Canada, and received the official offer for a professorship at the University of Toronto the week before he died). Also looking to the future, he finally ended the conversation by inviting me to write a chapter on the Himalaya in his forthcoming book with Peter Raven, which you are holding in your hands. "Write all the stuff on the Himalayan biodiversity crisis you keep telling me," were the last words Navjot said to me. That was the way he encouraged and inspired his colleagues and friends. Navjot passed away on June 12, 2011 leaving his family, colleagues, collaborators, and friends devastated. Although Navjot at the time was wearing his cap backwards, as always, he was always looking forward for new ways to bring attention to the tropical biodiversity crisis and ensure the preservation of the birds and other species that he had come to love.

REFERENCES

Brook, B. W. and Sodhi, N. S. (2006) Conservation biology: rarity bites. *Nature*, **444**, 555–556.

Brook, B. W., Sodhi, N. S. and Ng., P. K. L. (2003) Catastrophic extinctions follow deforestation in Singapore. *Nature*, **424**, 420–423.

Clements, R., Sodhi, N. S., Ng., P. K. L. and Schilthuizen, M. (2006) Limestone karsts of Southeast Asia: imperiled arks of biodiversity. *BioScience*, **56**, 733–742.

Gibson, L., Lee, T. M., Koh, L. P., Brook, B. W., Gardner, T. A., Barlow, J., Peres, C. A., Bradshaw, C. J. A., Laurance, W. F., Lovejoy, T. E. and Sodhi, N. S. (2011) Primary forests are irreplaceable for sustaining tropical biodiversity. *Nature*, **478**, 378–381.

Koh, L. P., Dunn, R. R., Sodhi, N. S., Colwell, R. K., Proctor, H. C. and Smith, V. S. (2004) Species coextinctions and the biodiversity crisis. *Science*, **305**, 1632–1634.

Pandit M. K., Sodhi, N. S., Koh, L. P., Bhaskar, A. and Brook, B. W. (2007) Unreported yet massive deforestation driving loss of endemic biodiversity in Indian Himalaya. *Biodiversity Conservation*, **16**, 153–163.

Sodhi, N. S. and Ehrlich, P. R. (2010) *Conservation Biology for All*. Oxford University Press, Oxford.

Sodhi, N. S., Acciaioli, G., Erb, M. and Tan, A. K.-J. (2007) *Biodiversity and Human Livelihoods in Protected Areas: Case Studies from the Malay Archipelago*. Cambridge University Press, Cambridge.

Sodhi, N. S., Koh, L. P., Brook, B. W. and Ng., P. K. L. (2004) Southeast Asian biodiversity: an impending disaster. *Trends in Ecology and Evolution*, **19**, 654–660.

CHAPTER 1

INTRODUCTION: GIVING A VOICE TO THE TROPICS

Luke Gibson[1] and Peter H. Raven[2]

[1]Department of Biological Sciences, National University of Singapore, Singapore
[2]Missouri Botanical Garden, St. Louis, MO, USA

In this book we deal with an apparent paradox: the apparently lush ecosystems of the world's tropics, teeming with life, are exceedingly vulnerable to disturbance and disintegration. The voices of authors from many tropical countries provide examples of the successes and failures that they have met in pursuing conservation objectives in their countries. Although each person's experience has been unique, they add up to deliver a number of common themes that we shall review in the concluding chapter of this book. The enormous projected population growth in the tropics, coupled with the desire for much higher levels of consumption by a majority of people in a world whose resources are already being used more rapidly than they can be replenished, forms a situation that makes necessary the pooling of our intellectual and financial resources to try to find our way back to sustainability in a socially just world.

Knowledge about the tropics was slow in reaching Europe, although occasional spectacular animals from tropical Africa were brought to the courts of Europe and China even in Classical times. Although the lure of the Indies and their riches of spices were great, it took a massive effort on the part of Portuguese explorers to round the southern tip of Africa and travel on to the East Indies in the half century before Columbus. Subsequently, the Portuguese, Dutch, Spanish, and ultimately the English reached far-flung localities in

tropical lands and established settlements there. These settlements were coastal, however, and mostly lay in regions that had similar coastal and beach plants and animals over wide areas. As a result, the Encyclopedists, who attempted to catalog life from the late seventeenth century onward, including Carl Linnaeus, who founded our modern taxonomic system in the mid-eighteenth century, had no real idea of how biologically rich the tropics actually were. In fact, it was not until the extensive travels of Alexander von Humboldt (1799–1804), who traveled thousands of kilometers through Latin America cataloging and observing its abundant life, that the world began to understand the biological riches of the tropics in their true glory.

Subsequently, Charles Darwin's observations and collections from the voyage of the Beagle (1831–1836) began to reveal the intricacy of tropical ecosystems and to detect the thread of evolution in them. His writings (Darwin, 1845) and those of Alfred Russel Wallace (Wallace, 1869, 1889), who explored the Amazon from 1848 to 1852 and the Malay Archipelago from 1854 to 1862, brought these hitherto mystical regions to life for the general public, stimulating continuous scientific exploration for the subsequent century and a half that has continued to the present day. From the specimens that entered the great museums of the world, it ultimately came to be understood that the majority of the world's species, two-thirds or more of them, occurred in the tropics. The vast majority of them, perhaps 19 out of 20 (mostly small to very small animals and fungi) have yet to be discovered or named scientifically! It is no wonder, then, that our understanding of tropical ecosystems is so limited and our ability to convert them to sustainable agricultural systems often so limited.

Europeans reached the tropics with a colonial mentality, bent on extracting their natural resources as efficiently as possible and thus enriching the countries from which they had come. It was during the course of this effort that many tropical systems were seen to be fragile – incapable of replenishing the resources that had been extracted – and the disintegration of ecosystems that were highly productive initially often caused them to be devastated rapidly.

The knowledge available about tropical ecosystems and the biodiversity that occurs in them is poorer than that concerning temperate ecosystems. Part of the reason for this deficiency lies in the colonial history of most of the regions; those who studied them were often members of expeditions from abroad, and not resident scholars who confronted the special problems of the tropics on a daily basis. As the institutions of tropical countries and their scholarly communities have built up slowly and unevenly, the destruction of their ecosystems has accelerated much more rapidly and uniformly. One of the reasons for this disparity has been rapid population growth, especially in the tropics. At the time of Alexander von Humboldt's travels just over 200 years ago, there were not quite a billion people in the world, a minority of them living in the tropics or what are now considered developing countries. Today, there are more than 7 billion people on Earth, five-sixths of them living in developing countries, including the tropics. One billion people are projected to be added to this number during the next 12 years (Population Reference Bureau, 2012). In Africa, for example, where so many people are malnourished or starving at present, the 950 million people who live in sub-Saharan countries today will be joined by another 500 million people by 2025! The combination of explosive population growth, demand for increased consumption that is growing still more rapidly, and the continued use of often antiquated or damaging extractive technologies are destroying natural communities all over the world, nowhere more rapidly than in the tropics. By the middle of this century, which is as soon as we could begin to hope for global population stability, 2–2.5 billion people will have joined our current numbers,

the great majority of them poor, and almost all of them living in developing countries. Putting the matter another way, each night when the world can be said to sit down to the dinner table, there are 200,000 more people needing to be fed. Against this background, the need for strengthening our knowledge of the tropics is evident, and the need for resident scientists and practitioners who can work to build sustainable systems for the countries with our help is absolutely urgent. This knowledge must then be put to use for the benefit of the people who live in tropical countries so that there can be a hope of attaining global sustainability.

Unfortunately for the achievement of sustainability in tropical regions, the extractive mentality that began with colonialism has persisted in many nations. Moreover, the numbers of resident scientists in most areas are small, especially in the face of the incredible biological diversity that occurs there (Barnard, 1995). A recent review of articles published in two of the leading international tropical ecology journals found that only 62% of tropical countries were represented, and 62% of all those articles came from just 10 countries (Stocks *et al.*, 2008). Furthermore, 62% of the articles were written by lead authors based at institutions outside the country where the research was carried out.

This geographical bias has a historical foundation; much research by Europeans has been and still is conducted in their former colonies in the Old World tropics (Clark, 1985). In the Western Hemisphere, European scientists also began the period of exploration, but they were followed by many scientists from the US. These scientists chose to conduct research in the Neotropics because of their proximity and richness and, more recently, because two of the largest international scientific organizations – the Organization for Tropical Studies and the Smithsonian Tropical Research Institute – are based in Latin America, fostering collaborations with American researchers (Clark, 1985). Regardless of historical patterns and their underlying reasons, we have more to learn about the biodiversity and ecology of the tropics than that of any other part of Earth. Knowledge about tropical biological systems can clearly be accumulated more efficiently by residents than by occasional visitors. There is an important role in these investigations for the major systematic institutions of the world, which are mostly located outside of the tropics, in exploring their biology in the future, but clearly partnerships between them and tropical institutions hold the key for the best and most solid progress. Overall, there is a definite need for greatly accelerated research programs in the tropics, programs that can be conducted best by people who live there. When they are the ones most clearly involved in the research, it can be applied most easily to problems of conservation and sustainability in the regions where these scientists are working.

In this book, we present diverse examples of the ways that scientific principles have been applied to the conservation of tropical ecosystems, and the success of these efforts. We do this by giving a voice to people from these tropical regions, who live there or who have worked there for major portions of their lives, thus gaining practical experience in the application of what we collectively are learning. We invited some of the leading conservation biologists from a variety of tropical countries to share their perspectives on important conservation issues in their countries. The following chapters provide stories of success and loss, of small communities in Thailand working together to protect hornbills, of disordered national governments failing to protect forest habitats and their resident species within, and of many other practical experiences, both successes and failures. Our book is organized by geography, with the chapters arranged into sections corresponding to the four major tropical regions: Africa, the Americas, Asia, and Oceania. At the end, we also include a Diaspora section, with chapters written by people from the tropics who now work on tropical conservation issues at institutions outside of the region.

We deliberately chose the integrative discipline of conservation, implying also the building of sustainable ecosystems, for emphasis here: the needs for conservation are urgent. Conservation, however, rests on a foundation of knowledge built by ecologists, systematists, soil scientists, and especially by social scientists. In many parts of the tropics, the base on which conservation can be carried is notably deficient. Steps must be taken urgently on the basis of what we do know, however, because opportunities are slipping through our fingers with increasing rapidity. At present, the world's people are estimated to be using on a continuous basis about 150% of what the world can produce sustainably (http://footprintnetwork.org), which means that every natural system we know will be simplified, depleted, and made less beautiful until we find a way to live within our means in a world where social justice becomes an important theme.

There are some significant gaps in the coverage in this book. Although many tropical countries are represented, gaps remain particularly in Central America, Andean and northern South America, central and southern Africa, and large parts of Southeast Asia. For one particular country, we invited two people to contribute a chapter, but both ultimately declined to submit a chapter critiquing the conservation measures in their country, fearing backlash by the government and the potential loss of their jobs. This country is regretfully not included in this book, the voices of its potential authors silenced. The limit of freedom of expression is a fundamental concern in many tropical regions, where particular governments can be intolerant of criticism, constructive or not, and sometimes even repressive to those who choose to offer it. The people of the world share a common interest in survival, and the challenge will be met only by finding sustainability for our planet. We hope that one of the benefits of this book will not only be to help empower those who live and study in tropical countries to be able to do an even better job, but also to help us all understand the need for common action based on mutual understanding and some of the means pertinent to taking such actions effectively. We sincerely hope that our book will help to highlight the many problems – and some of the potential solutions – facing countries in the tropics, and ultimately every one of us as well.

REFERENCES

Barnard, P. (1995) Scientific research traditions and collaboration in tropical ecology. *Trends in Ecology and Evolution*, **10**, 38–39.

Clark, D. B. (1985) Ecological field studies in the tropics: geographical origin of reports. *Bulletin of the Ecological Society of America*, **66**, 6–9.

Darwin, C. R. (1845) *The Voyage of the Beagle*. John Murray, London.

Stocks, G., Seales, L., Paniagua, F. *et al.* (2008) The geographical and institutional distribution of ecological research in the tropics. *Biotropica*, **40**, 397–404.

Wallace, A. R. (1869) *The Malay Archipelago*. Oxford University Press, Oxford.

Wallace, A. R. (1889) *A Narrative of Travels on the Amazon and Rio Negro*. Ward, Lock and Co., London.

PART 1

FROM WITHIN THE REGION

Section 1: Africa

CHAPTER 2

CONSERVATION PARADIGMS SEEN THROUGH THE EYES OF BONOBOS IN THE DEMOCRATIC REPUBLIC OF CONGO

Bila-Isia Inogwabini and Nigel Leader-Williams

Durrell Institute of Conservation and Ecology, University of Kent, Canterbury, UK

SUMMARY

Bonobos are the most recently discovered species of great ape, and are endemic to an area within a large convex bend of the Congo River, in the Democratic Republic of Congo (DRC). In this chapter, we highlight issues related to the discovery for science of a significant population of bonobos to the west of their range (Inogwabini *et al.*, 2007a, b), lying some 100 km outside the boundary of the nearest statutory protected area. We use this discovery to compare two opposing conservation paradigms: protected areas as a backbone of species and habitat conservation versus more inclusive conservation models that embrace community- managed conservation areas. The key conservation dilemma that this chapter addresses is why bonobos occur at higher densities in unprotected areas that have long been considered marginal habitats, while habitats previously thought to have been optimal, including areas that are formally protected, have failed to meet their conservation goals for bonobos.

CONSERVATION PARADIGMS

Four centuries before the birth of Christ, the Greek philosopher Plato wrote that removing trees from Attica had led to ongoing loss of soil from high ground and left the landscape looking like a human skeleton, a body wasted by disease (Thirgood, 1981; Plato, 360 BC). This and similar writings suggest that humankind has long sensed the importance of preserving wild nature and all its interrelated components, including forests, waters, and wildlife. Such concerns have been prompted by various underlying motives, including those of a religious, recreational, or economic nature, or through a wish to ensure intergenerational equity and leave the Earth as beautiful and resilient as we found it for our children.

A defining publication for modern-day conservationists worldwide was *Limits to Growth* (Meadows, Randers and Meadows, 1972). This volume predicted a gloomy future for humankind, prompted by our over-extraction of natural resources. Globally, humankind has gained an increasing interest in exploring the role that our species has played in shaping the ecosystems in which we live (Des Jardins, 2001). Since then, conservation and environmental issues have become a persistent theme in wider debates over human development and well-being (Ghimire and Pimbert, 2000). The establishment of numerous international, national, and local multilateral, governmental, and non-governmental organizations and conventions provides evidence of an emerging consensus and a more active willingness by societies worldwide to care for their environment. Indeed, the Convention on Biological Diversity, signed in 1992, now has over 190 state signatories, more than any other United Nations convention.

Some degree of international consensus has emerged over the importance of caring about wild nature and preserving natural resources and the associated wealth of global biodiversity, while a wide range of paradigms to achieve conservation objectives has been developed and tested in different parts of the world. These paradigms range from traditional *resource conservation through use* to *religiously immaculate protection* of untouchable biota. Regardless of where they fall within this wide spectrum, two distinct approaches to conservation have emerged (Ghimire and Pimbert, 2000; Karanth, 2001; Madhusudan and Raman, 2003; Bajracharya and Dahal, 2008). First, many conservationists have sought to protect biodiversity in priority areas for conservation, including for scientific research and for recreation (Ghimire and Pimbert, 2000). This paradigm has been termed an "elitist approach" to conservation, whose main aim is to separate people from wild nature, and to protect biodiversity from direct human use and influence. The origin of this approach has a long history (Dixon and Sherman, 1991), stretching back to decrees promulgated to establish royal reserves in Assyria in 700 BC, followed by similar measures in China in 300 BC (Waley, 1939). Some have come to believe that the "fences-and-fines approach" (Carruthers, 2004) espoused for conservation by modern elites all over the world will not work. In contrast to this traditional approach, there has been a growing sense that the best approach instead may be to integrate humankind as a key part of the ecosystems in which they live (Hutton and Leader-Williams, 2003). This constituency believes that biodiversity and environmental conservation can only be achieved by making multiple arrangements with local communities to manage ecosystems in a sustainable way (Rosser and Leader-Williams, 2010), including through common

property resource regimes (Ashenafi and Leader-Williams, 2005). Hence, many in conservation believe that modern fences-and-fines approaches have to give way to a conservation approach in which protected areas better accommodate human aspirations and needs over the long term (Leader-Williams, Harrison and Green, 1990). This paradigm has been termed the "populist approach," and has been linked to an early definition of conservation attributed to Pinchot, one of the first proponents of modern resource conservation, at the turn of the twentieth century (Ponder, 1987; Smith, 1998). Pinchot defined conservation as prudent use of nature's bounty, as opposed to the unrestricted extraction of natural resources (Eckersley, 1992). Under this paradigm, conservation espouses three important principles: (1) sustained extraction of natural resources based on the principles of scientific management; (2) reduced wastage of natural resources; and (3) equitable sharing of the benefits deriving from those resources by all segments of society. Essentially anthropocentric (Des Jardins, 2001), this approach engenders the concepts of "multiple use" (Kennedy and Quigley, 1994) and "sustainable development" (IUCN, 1991). This approach greatly influenced governments and international agencies to propose resolutions at the 1992 United Nations Conference on Environment and Development that were included as key objectives in the 1992 Convention on Biological Diversity (Glowka, Burhenne-Guilmin and Synge, 1996). As a result, what is increasingly known as "human welfare ecology" (Blaikie and Jeanrenaud, 2000) is starting to emerge at this end of this spectrum, by placing humanity at the center of conservation, essentially accounting for "environmental quality," "social justice," "democratic rights and duties," "equitable access to natural resources," "recreation," and "spiritual" needs (Eckersley, 1992; Des Jardins, 2001). Human welfare ecology is based on the concept of ecosystem management and, recently, on the delivery of ecosystem services (MEA, 2005). This approach has been critical both of excessive economic growth and of the abilities of science and technology to solve all environmental problems.

PARADIGMS IN WILDLIFE SPECIES CONSERVATION

Rates of species loss have reached unprecedented levels (Lawton and May, 1995), and the Earth is now experiencing its sixth mass extinction (Groombridge and Jenkins, 2002). However, in contrast to the five previous mass extinctions, this extinction crisis has largely been caused by one species – our own. To prevent as many additional species as possible from going extinct, conservation practitioners have mainly followed the elitist or fences-and-fines approach, creating large networks of exclusive protected areas. Indeed, conservation has enjoyed great success in increasing the numbers and coverage of terrestrial protected areas all over the world (Ghimire and Pimbert, 2000; Chape, Spalding and Jenkins, 2008). Exclusive protected areas, mainly in high categories of protection, IUCN Categories I and II (IUCN, 2008), have been widely regarded as essential for the long-term maintenance of biodiversity (Terborgh, 2004; Evans *et al.*, 2006). Nevertheless, evidence for their effectiveness remains somewhat equivocal (Putz *et al.*, 2000; Gaveau, Wandono and Setiabudi, 2007). Despite decades of investing significant funds and effort in maintaining highly protected areas from which humans are preferably excluded, global biological diversity has continued to decline rapidly (Bawa, Seidler and Raven, 2004; Moritz and Hammond, 2005; Butchart *et al.*, 2010), even in exclusive protected areas. Furthermore, the fences-and-fines approach is often very narrowly focused on particular species threatened with extinction, and at the same time often uses conservation models that fail to take into account the social and cultural landscapes within which these conservation efforts are embedded (Bawa, Seidler and Raven, 2004). As a result, the long-term conditions under which conservation might succeed may not be put in place, and exclusive protected areas remain tools of uncertain value on which to promote long-term conservation success. Furthermore, in the scramble to conserve charismatic flagship species (Leader-Williams and Dublin, 2000), many protected areas were created without consideration of biological diversity overall (Leader-Williams, Harrison and Green, 1990; Margules and Pressey, 2000). The end result is that many protected areas have failed to provide suitable refuges for the species that were their conservation targets, and at the same time often not included the most critical samples of a particular area's biodiversity within their borders (Rodrigues *et al.*, 2004).

Several key problems are evident while efforts are being made to manage declining species. The available local knowledge of species abundance, distribution, and ecology has often not been taken into account while creating new protected areas. Limited knowledge

available of the areas in which protected sites might best be placed over the past 150 years led to a less than ideal delimitation of some protected areas with regard to the distributions of species they were meant to protect (Pressey, 1997). Furthermore, local social and philosophical beliefs were often ignored. If people were physically displaced from the lands of their ancestors, lands where they had long lived among species that the international conservation community was trying to protect, situations that were by no means ideal for conservation often arose. The imposition by governments of new and non-negotiated regulations upon local communities led to tensions (Colchester, 2000) and conflicts of interests that jeopardized efforts to preserve the biodiversity that was the target for those protected areas.

Efforts to conserve charismatic or flagship species within protected areas have, nevertheless, sometimes been effective in saving critical populations of those species. In some respects, the conservation balance sheet would have looked much worse without the establishment of protected areas (Terborgh, 2004; Oates 2006). Equally, others have pointed out that, even with protected areas, the populations of many charismatic species have continued to decline generally. For example, elephant, rhinoceros, and tiger populations have fallen in many areas across Africa and Asia, despite great efforts and resources devoted to their conservation (Karanth, Sunquist and Chinnappa, 1999; Leader-Williams 2002; Madhusudan and Raman, 2003; Blake and Hedges, 2004). Great apes also face extreme threats throughout their range. They are clearly a critical global conservation priority, as we discuss next.

GREAT APES AND CONSERVATION PARADIGMS

Great apes are our closest living relatives. However, populations of the three surviving genera of non-human great apes have shown alarming declines in recent years, and have become the subject of considerable conservation concern (Butynski, 2001). Populations of both lowland gorillas (*Gorilla gorilla*) and mountain gorillas (*G. beringei*), chimpanzees (*Pan troglodytes*), bonobos (*Pan paniscus*) and orangutans (*Pongo pygmaeus, P. abelii*) are thought to be declining at an increasing rate worldwide. Threats include continuing deforestation through clearing land for agriculture

(March, 1957; Harcourt and Fossey, 1981; Anderson, Williamson and Carter, 1983; Aveling and Harcourt, 1984; Carroll, 1986; Harcourt, Stewart and Inahoro, 1988; Hall *et al.*, 1998; Plumptre and Williamson, 2001), including more recently for plantations of commercial crops such as palm oil (Koh *et al.*, 2011); increased logging activities (Bowen-Jones and Pendry, 1999; Blom *et al.*, 2004); growth in the commercial bushmeat trade (Harcourt, 1996; Bowen-Jones and Pendry, 1999; Kormos *et al.*, 2003); and viral epidemic diseases such as Ebola (Remis, 2000; Vogel, 2000; Formenty *et al.*, 2003; Walsh *et al.*, 2003, 2007; Leroy *et al.*, 2004; Rouquet *et al.*, 2005). A further threat to most great ape populations is the uncontrolled flow of automatic weapons and ammunition resulting from armed conflicts in different range states (Vogel, 2000; Reinartz and Inogwabini, 2001; Draulans and Van Krunkelsven 2002; Kalpers *et al.*, 2003).

Ongoing population declines have led scientists to suggest that most great ape populations will disappear over the course of the next three decades (Miles, Caldecott and Nellemann, 2005). Indeed, most extant populations of great apes are small and fragmented in isolated clusters, posing severe implications for their long-term viability. To decrease the pressure imposed by these threats upon great apes, the international conservation community has put together a suite of measures intended to curb their decline. These measures include the listing of indicative priority great ape populations and sites to ensure the protection of viable populations that represent the full genetic and ecological diversity of great apes. They also include proposals for concrete conservation and political actions, such as the Great Ape Survival Partnership (GRASP) of the United Nations Environment Programme (UNEP) and the United Nations Educational, Science and Culture Organization (UNESCO).

All available evidence from the great ape species that occur in Africa shows that most populations of these species have continued to decline at alarming rates over the last four decades (Hall *et al.*, 1998; Plumptre and Williamson, 2001; Walsh *et al.*, 2003, 2007; Leroy *et al.*, 2004; Rouquet *et al.*, 2005). This trend might well result in some of their populations becoming extinct if appropriate measures are not taken urgently. For instance, large numbers of mountain gorillas were estimated to have occurred in eastern DRC in the early 1960s (Schaller, 1963), but have since declined severely (March, 1957; Harcourt and Fossey, 1981; Anderson, Williamson and Carter, 1983; Aveling and Harcourt 1984; Carroll, 1986; Harcourt, Stewart and Inahoro, 1988; Schaller and Nichols, 1989; Hall *et al.*, 1998). Furthermore, chimpanzees were estimated to number about 2 million individuals across their range a century ago (Oates, 2006); today, an estimated 200,000 chimpanzees, just a tenth of the original population, remain in the wild (Butynski, 2001; Bow, 2004).

Sound applied research on the ecology and abundance of these threatened species is of vital importance to the future of great ape populations. All species of great apes face ever-greater threats, even in once inaccessible forested areas that were long thought safe from exploitation (Hall *et al.*, 1998; Bowen-Jones and Pendry, 1999). Development needs and the search for profits have changed traditional landownership rights and land-use practices in all countries where great apes occur. On a national scale, this has brought about mechanization of resource extraction (e.g., logging, mining), which is changing wild habitats, human social structures, human activities (Anderson, Williamson and Carter, 1983; Harcourt, Stewart and Inahoro, 1988; Gadsby, 1990; Butynski and Koster, 1994; Plumptre and Williamson, 2001), and people's fundamental perception of their immediate environment and ways of life. The effects of these new practices on large mammals, and particularly on great apes, have to be documented thoroughly if the different species of great apes are to be efficiently conserved in their natural habitats. The most recently "discovered" species of great ape is the bonobo or pygmy chimpanzee, whose conservation remains the sole responsibility of DRC, to which the species is endemic.

CONSERVATION AND BONOBOS IN DRC

The Belgian Congo (the former colonial name of DRC) was the first country to create an exclusive national park in Africa, as early as 1925. Subsequently, many of the ecosystems in DRC have altered drastically as the country increasingly opened itself up to the modern world during the early twentieth century. These changes include massive land clearances to extract minerals and other natural resources that, coupled with increasing human populations, have placed great pressure on wild nature in DRC. In 2012, the human population of DRC was 69 million people, of whom about 80% lived in extreme poverty. Furthermore, the population is projected to grow to more than 194 million people by 2050 (Population Reference Bureau, 2012). Struggles over land and natural resources have resulted in crises that have severely impacted ecological processes in fragile regions and depleted populations of large mammals across DRC. The losses of large carnivores, and of elephants and rhinos, illustrate the impact that such changes have wrought on megafaunal groups that play key ecological roles in their habitats. In the Miombo woodland ecosystem, extractive industries have brought the cheetah (*Acinonyx jubatus*) to local extinction throughout its former area in DRC in the early 1970s , while African elephant (*Loxodonta africana*) numbers decreased by 90% during the 60 years from 1945 to 2005. Within the northern wooded savannah ecosystems of DRC, the northern white rhinoceros (*Ceratotherium simum cottoni*) population has declined to a few individuals in Garamba National Park, while savannah elephants in Garamba crashed from 7000 to 1200 individuals, a decrease of 83% during the last decade of the twentieth century (De Merode *et al.*, 2007). Changes have also occurred to the ecosystem processes in the waters of the Congo River. Around Lake Tumba, the rainfall regime has changed drastically, in turn affecting the biophysical composition of the water over the last 34 years, thereby inducing perceptible changes in the biodiversity of the lake. Fish biomass has decreased sharply, and the species composition in Lake Tumba has also been altered markedly (Inogwabini and Lingopa, 2006).

Even though there are no baseline data with which to compare current population estimates, the available evidence indicates that bonobo populations have also declined throughout their range (Kortlandt, 1995; Thompson-Handler, Malenky and Reinartz, 1995; Kempf and Wilson, 1997; Van Krunkelsven, Inogwabini and Draulans, 2000; Bow, 2004). As with other species of great apes, these assumed losses of bonobos have occurred because of habitat destruction, the commercial bushmeat trade (Bennett and Robinson, 2000), increasing human populations in habitats that were formerly considered marginal, logging activities, and epidemics such as sleeping sickness (Kortlandt, 1995). There will almost certainly be further losses of bonobos as risks emerge of epidemic disease outbreaks such as Ebola (Vogel, 2000; Formenty *et al.*, 2003; Walsh *et al.*, 2003, 2007) and anthrax (Leendertz *et al.*, 2006). Other threats to bonobos include increased levels of poverty and armed conflicts (Maisels *et al.*, 2001; Draulans and Van Krunkelsven, 2002; Inogwabini, Omari and Mbayma, 2005). All these threats, especially in the

face of the rapidly increasing human population, call urgently for action to save mankind's evolutionarily closest relative from increasing risk of extinction. However, strategies to conserve bonobos should be based on experience, sound science, and reliable evidence (Blake and Hedges, 2004). In the next section, we address the links between the bonobo populations and conservation strategies adopted by the government of DRC, particularly through the creation of protected areas (PAs) that are supposed to protect the species against events that threaten its long-term survival.

BONOBOS: PART OF A FAILING CONSERVATION PARADIGM?

The bonobo was recognized as a full species only in 1929, and the $36,560 km^2$ Salonga National Park (SNP) was created in 1969 as the first PA to preserve the species (IUCN, 1992). To create SNP, local people were removed and borders were aligned randomly along rivers. However, the populations of bonobos residing within the park boundaries were not documented nor thoroughly assessed until the late 1990s (Van Krunkelsven, Inogwabini and Draulans, 2000; Reinartz *et al.*, 2006). Meanwhile, people continued to make claims on their rights to use resources within SNP, and attempts at serious management of SNP encountered major problems. Almost all data collected subsequently on bonobo distribution and abundance across its range indicate that SNP may not have been the most appropriate site to protect bonobos. Indeed, large zones of SNP have no resident bonobos, suggesting that the borders of the park could have been better sited, and that its area might have been much smaller, thereby reducing conflicts between displaced people and the management of SNP. While SNP was classified as an IUCN Category II protected area, poachers came from long distances and simply ignored the presence of conservation agencies, whether governmental or nongovernmental. By contrast, Lake Tumba became the focus of conservation attention only following a recent study (Inogwabini *et al.*, 2007a, b; Inogwabini, 2010) that documented the presence of what is probably the first or second largest population of bonobos in nature. With an estimated population size of 1880–3550 individuals (Inogwabini *et al.*, 2007a; Inogwabini, 2010), this population represents 9–18% of the estimated global population of wild bonobos. Even more interesting, this population was discovered living among local communities in the Lake Tumba landscape, well outside any formal PA.

The recent discovery of such an important population of bonobos in the vicinity of such a large transport route such as the Congo River, and only a 45-minute flight from the capital city of Kinshasa, highlights the reality that there are large gaps in the current scientific knowledge of bonobos, which are probably the least well documented of any great ape species (Thompson-Handler, Malenky and Reinartz, 1995). Indeed, much greater scientific effort is still required to gain a real understanding of the basic distribution and ecology of bonobos in order to provide the basis for preparing an adequate conservation plan for the species.

IMPLICATIONS FOR CONSERVATION OF BONOBOS

Two major factors explain most of the variation observed in bonobo density across the Lake Tumba landscape: (1) the density of plants of the family Marantaceae, which are the main food for the bonobos; and (2) local willingness to protect bonobos outside protected areas (Inogwabini et al., 2007a, b; Inogwabini and Matungila, 2009; Inogwabini, 2010). To ensure an adequate food supply for bonobos, a better understanding is required of the biology of the Marantaceae, and how their populations might be increased or maintained.

To ensure adequate protection, local communities should be involved in, rather than removed away from, any program intended to conserve bonobos. Where taboos against hunting bonobos are relatively strong, these should be viewed as the key part of the foundation upon which a sound conservation program can be built. Further from Kinshasa, the strength of state authority lessens, while the more local traditional leadership prevails. In turn, this makes it difficult to bring both state and traditional authorities to bear on bonobo conservation in remote areas. Consequently, the strong and organized traditional system of authority in Lake Tumba (Figure 2.1) should become the body designated to enforce conservation law across the landscape. This could best be achieved using available systems of devolution, which clearly indicate the role that traditional authority could play. This recommendation is very timely, given that DRC is currently struggling to implement provisions of its new constitution that favor decentralizing of power as the human population grows explosively over the next few decades. Among major sectors in which issues associated with devolution need to be addressed is the conservation of natural resources and their associated biodiversity. As has been argued elsewhere (Bawa, Seidler and Raven, 2004; Inogwabini, 2007), success in conservation will not be achieved unless there is institutional capacity to manage the natural resources within DRC democratically. In turn, this will have to be reconciled and coordinated with people's legitimate use of the forest and its products to support themselves. In villages where there were no enforceable taboos against hunting bonobos, people simply shied away from the issue of killing bonobos during interviews and focus group sessions (Inogwabini, 2010). Local people may not have discussed the issue because they were afraid of the potential legal consequences of killing bonobos or because they wanted to show that they cared about bonobos in expectation of seeing their villages become part of the World Wide Fund for Nature (WWF)'s Bonobo Community Conservation Project. If the second interpretation is correct, conservationists may have a window in which to act.

BONOBOS AND PARKS AS A CONSERVATION PARADIGM

Apart from advice on technical considerations, a recent study (Inogwabini, 2010) raised a set of important philosophical conservation questions, including one about parks as a conservation paradigm for bonobos. As discussed earlier, conservationists have historically seen protected areas as the last refuge for charismatic and endangered species, such as bonobos, on the brink of extinction (Karanth, Sunquist and Chinnappa, 1999; Madhusudan and Raman, 2003; Bawa, Seidler and Raven, 2004; Inogwabini, Omari and Mbayma, 2005; Moritz and Hammond, 2005; Evans et al., 2006; Butchart et al., 2010). The findings that bonobo numbers in the Lake Tumba landscape appear stable and were clearly protected by local communities in a region that had not previously been considered a conservation priority raised vexing questions over whether protected areas should be delineated at all, over the categories in which they might be designated, and over whether this would have a positive or a negative effect on species conservation. If species have been effectively protected under traditional land-use practices, it is legitimate to re-examine the role of PAs. In setting pri-

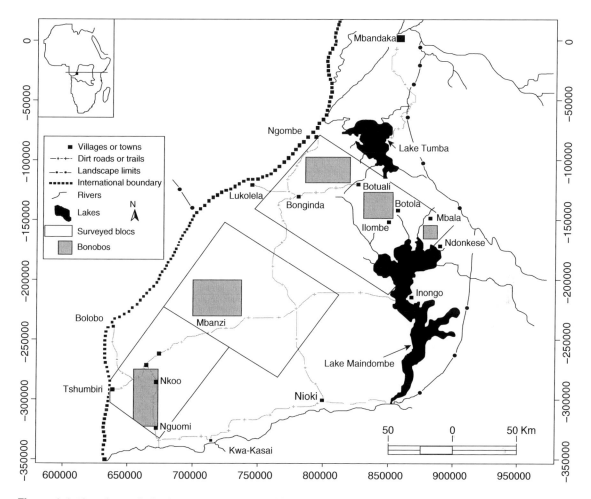

Figure 2.1 The Lake Tumba landscape, Democratic Republic of Congo.

orities for conservation efforts, a recent study (Inogwa-bini and Omari, 2005) suggests that an approach should be adopted in which resources for conservation are allocated to critically endangered species on the basis of *known* self-sustaining populations, rather than on the basis of the protection status of the areas in which they live. In the case of bonobos, this approach could be applied because their overall distribution consists of local populations that can be evaluated individually (Reinartz, 1997; Reinartz *et al.*, 2000; Eriksson *et al.*, 2004; Bruford *et al.*, in press). In turn, ongoing conservation efforts could be assessed and funding

allocated in relation to the success or otherwise of conservation efforts in particular populations. People living in or near the areas could be trained to track animals, document important demographic parameters including group size, birth rates, mortality rates, and epidemiological signs; and to focus on pro-conservation local drivers such as existing local authority, the willingness to pay for conservation of a specific area and/or a specific species, and local community attitudes.

The findings at Lake Tumba strongly suggest that alternative approaches to simply setting up PAs to try and protect bonobos might prove equally or more effective, at least under the circumstances that bonobos currently encounter. As an operational conservation concept, the landscape approach includes not only the biophysical components of an area, but also the social, political, psychological, and other components of that system (Sayer et al., 2006). Since critical numbers of many mammals of conservation concern are located outside of existing PA networks, appropriate strategies should be developed for their conservation (Fay and Agnagna, 1992; Hall et al., 1998; Dupain et al., 2004; Ashenafi and Leader-Williams, 2005). This is especially true for bonobos, since most individuals occur outside PAs. Population trends should be measured both inside and outside PAs, and the lessons applied to the survival of the species as a whole, while paying special attention to the conservation of populations living outside PAs.

The very high densities of bonobos found in Lake Tumba suggest that a range-wide survey of bonobos should also be carried out to set future conservation priorities appropriately, as proposed at the first intergovernmental meeting of GRASP. Knowing that an estimated 9–18% of the world population of bonobos occurs in a single region outside any PA, and that this only became known to science within the last decade, suggests that other populations of substantial size might exist under similar circumstances. Certainly the ongoing protection of the bonobos at Lake Tumba (Figure 2.1) deserves attention in the light of the involvement of local communities. Ways need to be found in which local communities would benefit from conservation activities, a strategy that will be critical to success.

The relationships between the conservation situation of populations of bonobos at Lake Tumba (Figure 2.1) with those in PAs suggest that current conservation law in DRC should be revised. Where local traditional authorities and traditions are respected, these authorities should be empowered as part of the effort to enforce conservation law, as currently practiced in some areas. The other aspect of conservation law that should be revisited is land tenure, which until now has stipulated that all land and anything found therein belongs to the state (Leisz, 1998). This provision prohibits adopting incentives for private conservation action and contradicts the willingness to create community-based natural resource managed (CBNRM) areas, which are envisaged in the new forestry code of DRC. CBNRM should be understood in this context as community conservation areas that loosely resemble

IUCN Category VI, but in which managerial authority is entirely devolved to the local community, rather than to any centralized authority. Clearly, for those areas where critical populations of bonobos have long survived without any official centralized conservation action having been implemented, such as in the case of the Malebo area (Figure 2.1), CBNRM appears to be the optimal route to conservation success. Of course, CBNRMs are not and will not be a panacea for all the conservation problems in DRC. Where necessary, CBNRM approaches should always be complemented by traditional PAs, particularly in areas where traditional practices are at odds with conservation priorities. In this way, we hope that DRC will be able to discharge its international responsibility to conserve the nearest living relative of people for posterity.

ACKNOWLEDGMENTS

Funding to support field studies in Lake Tumba were received from the Beinecke African Scholars Program, through the Wildlife Conservation Society, the British Ecological Society through an Early Career Grant, the African Wildlife Foundation (AWF) through its Charlotte Conservation Fellowship Program, the Young Scientist Award of UNESCO and the National Geographic Buffet Award for conservation leadership in Africa that were awarded to the first author. Field data were collected through the WWF Lac Tumba Program funded by the Central African Regional Program for Environment of the US Agency for International Development. Manuscript writing was supported by the postdoctoral grant received from the Department of Aquatic Science and Assessment of Swedish University of Agricultural Sciences, Uppsala, Sweden.

REFERENCES

Anderson, J. R., Williamson, E. A. and Carter, J. (1983) Chimpanzees of Sapo Forest, Liberia: density, nests, tools and meat eating. *Primates*, **24**, 594–601.

Ashenafi, Z. T. and Leader-Williams N. (2005) Indigenous common property resource management in the Central Highlands of Ethiopia. *Human Ecology*, **33**, 539–563.

Aveling, C. and Harcourt, A. (1984) A census of the Virunga gorillas. *Oryx*, **18**, 8–13.

Bajracharya, S. B. and Dahal, N. (2008) *Shifting Paradigms in Protected Area Management*. National Trust for Nature Conservation, Kathmandu, Nepal.

Bawa, K. S., Seidler, B. and Raven, P. H. (2004) Reconciling conservation paradigms. *Conservation Biology*, **18**, 859–860.

Bennett, E. L. and Robinson, J. G. (2000) *Hunting of Wildlife in Tropical Forests: Implications for Biodiversity and Forest Peoples*. Environment Department, Paper no. 76 (Biodiversity Series – Impact Studies), The World Bank, Washington D.C.

Blaikie, P. and Jeanrenaud, S. (2000) Biodiversity and human welfare, in *Social Change and Conservation* (eds K. B. Ghimire and M. P. Pimbert), Earthscan, London, pp. 46–70.

Blake, S. and Hedges, S. (2004) Sinking flagships: the case of forest elephants in Asia and Africa. *Conservation Biology*, **18**, 1191–1202.

Blom, A., Van Zalinge, R., Mbea, E., Heitkönig, I. M. A. and Prins, H. H. T. (2004) Human impact on the wildlife population within a protected Central African forest. *African Journal of Ecology*, **42**, 23–31.

Bow, P. (2004) *Changing the Future for Endangered Wildlife: Chimpanzee Rescue*. Firefly Books, Altona, Canada.

Bowen-Jones, E. and Pendry, S. (1999) The threat to primates and other mammals from bushmeat trade in Africa, and how this could be diminished. *Oryx*, **33**, 233–246.

Bruford, M. W., Meredith, C., Inogwabini, B. I. and Morgan, E. (in press). Genetic status of the Lac Tumba bonobo (*Pan paniscus*) population. *Folia Primatologica*.

Butchart, S. H. M., Walpole, M., Collen, B., Strien, A. V., Scharlemann, J. P. W., Almond, R. E. A., Baillie, J. E. M., Bomhard, B., Brown, C., Bruno, J., Carpenter, K. E., Carr, G. M., Chanson, J., Chenery, A. M., Csirke, J., Davidson, N. C., Dentener, F., Foster, M., Galli, A., Galloway, J. N., Genovesi, P., Gregory, R. D., Hockings, M., Kapos, V., Lamarque, J.-F., Leverington, F., Loh, J., McGeoch, M. A., McRae, L., Minasyan, A., Morcillo, M. H., Oldfield, T. E. E., Pauly, D., Quader, S., Revenga, C., Sauer, J. R., Skolnik, B., Spear, D., Stanwell-Smith, D., Stuart, S. N., Symes, A., Tierney, M., Tyrrell, T. D., Vié, J.-C. and Watson, R. (2010) Global biodiversity: indicators of recent declines. *Science*, **328**, 1164–1168.

Butynski, T. M. (2001) *Africa's Great Apes: An Overview of Current Taxonomy, Distribution, Numbers, Conservation Status, and Threats*. Smithsonian Institution Press, Washington, D.C.

Butynski, T. M. and Koster, H. (1994) Distribution and conservation status of primates in Bioko Island, Equatorial Guinea. *Biodiversity and Conservation*, **3**, 893–909.

Carroll, R. (1986) Status of the lowland gorilla and other wildlife in the Dzanga-Sangha region of Southwestern Central African Republic. *Primate Conservation*, **7**, 38–41.

Carruthers, J. (2004) Africa: histories, ecologies and societies. *Environment and History*, **10**, 379–406.

Chape S., Spalding, M. D. and Jenkins M. D. (eds) (2008) *The World's Protected Areas: Status, Values, and Prospects in the Twenty-First Century*. University of California Press, Berkeley, CA.

Colchester, M. (2000) Salvaging nature: indigenous peoples, protected areas and biodiversity conservation, in *Social*

Changes and Conservation (eds K. B. Ghimire and P. M. Pimbert), Earthscan, London, pp. 97–130.

De Merode, E., Inogwabini, B. I., Telo, J. and Panziama, G. (2007) Status of elephant populations in Garamba National Park, Democratic Republic of Congo. *Pachyderm*, **42**, 52–57.

Des Jardins, J. R. (2001) *Environmental Ethics: An Introduction to Environmental Philosophy*, 3rd edn. Wadsworth Publishing, CA.

Dixon, J. A. and Sherman, P. B. (1991) *Economics of Protected Areas: A New Look at Costs and Benefits*. Earthscan, London.

Draulans, D. and Van Krunkelsven, E. (2002) The impact of war on forest areas in the Democratic Republic of Congo. *Oryx*, **36**, 35–40.

Dupain, J., Guislain, P., Nguenang, G. M., De Vleeschouwer, K. and Van Elsacker, L. (2004) High chimpanzee and gorilla densities in a non-protected area on the northern periphery of the Dja Faunal Reserve, Cameroon. *Oryx*, **38**, 209–216.

Eckersley, R. (1992) *Environmentalism and Political Theory: Toward an Eco-centric Approach*. University College London Press, London.

Eriksson, J., Hohmann, G., Boesch, C. and Vigilant, L. (2004) Rivers influence the population genetic structure of bonobos (*Pan paniscus*). *Molecular Ecology*, **13**, 3425–3435.

Evans, K. L., Rodrigues, A. S. L., Chown, S. L. and Gaston, K. J. (2006) Protected areas and regional avian species richness in South Africa. *Biology Letters*, **2**, 184–188.

Fay, M. J. and Agnagna, M. (1992) Census of gorillas in northern Republic of Congo. *American Journal of Primatology*, **27**, 275–284.

Formenty, P., Karesh, W., Froment, J. M. and Wallis, J. (2003) Infectious diseases in west Africa: a common threat to chimpanzees and humans, in *West African Chimpanzees: Status Survey and Conservation Action Plan* (eds R. Kormos, C. Boesch, M. A. Bakkar and T. M. Butynski), IUCN, Gland, Switzerland, pp. 169–174.

Gadsby, E. T. (1990) *The Status and Distribution of the Drill (Mandrillus leucophaeus) in Nigeria*. Report to Wildlife Conservation International, WWF (United States of America) and WWF (United Kingdom) and the Nigerian Government.

Gaveau, D. L. A., Wandono, H. and Setiabudi, F. (2007) Three decades of deforestation in southwest Sumatra: have protected areas halted forest loss and logging, and promoted regrowth? *Biological Conservation*, **134**, 495–504.

Ghimire, K. B. and Pimbert, P. M. (2000) Social changes and conservation: an overview of issues and concepts, in *Social Changes and Conservation* (eds K. B. Ghimire and P. M. Pimbert), Earthscan, London, pp. 1–45.

Glowka, L., Burhenne-Guilmin, F. and Synge, H. (1996) *A Guide to the Convention on Biological Diversity*. IUCN Environmental Policy and Law Paper no. 54171.

Groombridge, B. and Jenkins, M. D. (2002) *World Atlas of Biodiversity: Earth's Living Resources in the 21st Century*. University of California Press, Berkeley, CA.

Hall, J. S., White, L., Inogwabini, B.-I., Omari, I., Morland, H. S., Williamson, E. A., Walsh, P., Sikubwabo, C., Saltonstall,

K., Dumbo, B., Kiswele, K., Vedder, A. and Freeman, K. (1998) A survey of the Grauer's gorilla (*Gorilla gorilla graueri*) and eastern chimpanzee (*Pan troglodytes schweinfurthi*) in the Kahuzi-Biega National lowland sector and adjacent forests in Eastern Democratic Republic of Congo. *International Journal of Primatology*, **19**, 207–235.

Harcourt, A. (1996) Is the gorilla a threatened species? How can we judge? *Biological Conservation*, **75**, 165–176.

Harcourt, A. and Fossey, D. (1981) The Virunga gorillas: decline of an island population. *African Journal of Ecology*, **19**, 83–97.

Harcourt, A., Stewart, K. and Inahoro, I. (1988) Conservation status of gorillas in Nigeria. *Primate Eye*, **35**, 26–27.

Hutton, J. M. and Leader-Williams, N. (2003) Sustainable use and incentive-driven conservation: realigning human and conservation interests. *Oryx*, **37**, 215–226.

Inogwabini, B. I. (2007) Can biodiversity conservation be reconciled with development? *Oryx*, **41**, 135–139.

Inogwabini, B. I. (2010) Conserving great apes living outside protected areas: the distribution of bonobos in the Lake Tumba landscape, Democratic Republic of Congo. PhD thesis, University of Kent, Canterbury, UK.

Inogwabini, B. I. and Lingopa, Z. N. (2006) *Inventaire des poisons au Lac Tumba, le fleuve Congo et la Ngiri, République Démocratique du Congo*. WWF-DRC and USAID.

Inogwabini, B. I. and Matungila, B. (2009) Bonobo food items, food availability and bonobo distribution in the Lake Tumba Swampy forests, Democratic Republic of Congo. *The Open Conservation Biology Journal*, **3**, 1–10.

Inogwabini, B. I. and Omari, I. (2005) A landscape-wide distribution of *Pan paniscus* in the Salonga National Park, Democratic Republic of Congo. *Endangered Species Update*, **22**, 116–123.

Inogwabini, B. I., Omari, I. and Mbayma, A. G. (2005) Protected areas of the Democratic Republic of Congo. *Conservation Biology*, **19**, 15–22.

Inogwabini, B. I., Matungila, B., Mbende, L. Abokome, M. and Tshimanga, W. T. (2007a) The great apes in the Lac Tumba landscape, Democratic Republic of Congo: newly described populations. *Oryx*, **41**, 532–538.

Inogwabini, B. I., Matungila, B., Mbende, L., Abokome, M. and Miezi, V. (2007b) The Bonobos of the Lake Tumba-Lake Maindombe hinterland: threats and opportunities for population conservation, in *The Bonobos: Behavior, Ecology, and Conservation* (eds T. Furuichi and J. Thompson), Springer, New York, NY, pp. 273–290.

IUCN (International Union for the Conservation of Nature) (1991) *Caring for the Earth: A Strategy for Sustainable Living*. IUCN, Gland, Switzerland.

IUCN (International Union for the Conservation of Nature) (1992) *Protected Areas of the World: Review of National Systems. Vol. 3: Afrotropical*. IUCN, Gland, Switzerland.

IUCN (International Union for the Conservation of Nature) (2008) *Protected Area Categories*. http://www.environment.gov.au/parks/iucn.html (accessed 12 July 2010).

Kalpers, J., Williamson, E. A., Robins, M. M., McNeilage, A., Lola, N. N. and Muguri, G. (2003) Gorillas in the crossfire: population dynamics of the Virunga mountain gorillas over the past three decades. *Oryx*, **37**, 326–337.

Karanth, K. U. (2001) Debating conservation as if reality matters. *Conservation and Society*, **1**, 64–66.

Karanth K. U., Sunquist, M. E. and Chinnappa, K. M. (1999) Long-term monitoring of tigers: lessons from Nagarahole, in *Riding the Tiger: Tiger Conservation in Human Dominated Landscapes* (eds J. Seidensticker, S. Christie and P. Jackson), Cambridge University Press, Cambridge, pp. 114–122.

Kempf, E. and Wilson, A. (1997) *Les Grands Singes dans la Nature – Rapport du World Wide Fund for Nature sur le Statut des Espèces*. World Wide Fund for Nature, Gland, Switzerland.

Kennedy, J. J. and Quigley, T. M. (1994) Evolution of Forest Service organizational culture and adaptation issues in embracing ecosystem management, in *Ecosystem Management: Principles and Applications (Vol. 2)* (eds M. E. Jensen and P. S. Bougeron), United States Department of Agriculture, Forest Service, Portland, OR, pp. 16–26.

Koh, L. P., Miettinen, J., Liew, S. C. and Ghazoul, J. (2011) Remotely sensed evidence of tropical peatland conversion to oil palm. *Proceedings of National Academy of Sciences of the United States of America*, **108**, 5127–5132.

Kormos, R., Bakkar, M. I., Bonnéhin, L. and Hanson-Alp, R. (2003) Bushmeat as a threat to chimpanzees in West Africa, in *West African Chimpanzees: Status Survey and Conservation Action Plan: West African Chimpanzees* (eds R. Kormos, C. Boesch, M. I. Bakkar and T. M. Butynski), IUCN, Cambridge, pp. 151–155.

Kortlandt, A. (1995) A survey of the geographical range, habitats and conservation of the pygmy chimpanzee (*Pan paniscus*): ecological perspective. *Primate Conservation*, **16**, 21–36.

Lawton, J. H. and May, R. M. (1995) *Extinction Rates*. Oxford University Press, Oxford.

Leader-Williams, N. (2002) Regulation and protection: successes and failures in rhinoceros conservation, in *The Trade in Wildlife: Regulation for Conservation* (ed. S. Oldfield), Earthscan, London, 89–99.

Leader-Williams, N. and Dublin, H. T. (2000) Charismatic megafauna as "flagship species," in *Priorities for the Conservation of Mammalian Diversity: Has the Panda had its Day?* (eds A. Entwistle and N. Dunstone), Cambridge University Press, Cambridge, pp. 53–81.

Leader-Williams, N., Harrison, J. and Green, M. J. B. (1990) Designing protected areas to conserve natural resources. *Science Progress*, **74**, 189–204.

Leendertz, F. H., Lankester, F., Guislain, P., Néel, C., Drori, O., Dupain, Speede, S., Reed, P., Wolf, N., Loul, S., Ngole, M., Peeters, M., Boesch, C., Pauli, G., Ellerbrok, H., Leroy, E. M. (2006) Anthrax in Western and Central African great apes. *American Journal of Primatology*, **68**, 928–933.

Leisz, S. (1998) Zaire country profile. *Country Profiles of Land Tenure: Africa in 1996*, 131–136. Land Tenure Center, University of Wisconsin, Madison, WI.

Leroy, E. M., Telfer, P., Kumulungui, B., Yaba, P., Rouquet, P., Roques, P. Gonzalez, J. P. Ksiazek, T. G., Rollin, P. E. and Nerrienet, E. (2004) A serological survey of Ebola virus infection in central African nonhuman primates. *The Journal of Infectious Diseases*, **190**, 1895–1899.

Madhusudan, M. D. and Raman, T. R. S. (2003) Conservation as if biological diversity matters: preservation versus sustainable use in India. *Conservation and Society*, **1**, 49–59.

Maisels, F., Keming, E., Kemei, M. and Toh, C. (2001) The extirpation of large mammals and implications for montane forest conservation: the case of the Kilum-Ijim Forest, North-west Province, Cameroon. *Oryx*, **35**, 322–331.

March, E. W. (1957) Gorillas of eastern Nigeria. *Oryx*, **4**, 30–34.

Margules, C., and R. Pressey (2000) Systematic conservation planning. *Nature*, **405**, 243–253.

MEA (Millennium Ecosystem Assessment) (2005) *Ecosystem Conditions and Human Well-Being: Current State and Trends.* Island Press, Washington, D.C.

Meadows, D., Randers, J. and Meadows, D. (1972) *Limits to Growth: The 30-Year Update.* Earthscan, London.

Miles, L., Caldecott, J. and Nellemann, C. (2005) Challenges to great ape survival, in *World Atlas of Great Apes and their Conservation* (eds J. Caldecott and L. Miles), University of California Press, Berkeley, CA, pp. 217–241.

Moritz, T. D. and Hammond, T. (2005) *The Conservation Commons: Advocating a New Knowledge Management Paradigm in the Conservation Community.* Millennium Ecosystem Assessment, 2005. IUCN, Gland, Switzerland.

Oates, J. F. (2006) Is the chimpanzee, *Pan troglodytes*, an endangered species? It depends on what "endangered" means. *Primates*, **47**, 102–112.

Plato (360 BC) *Timaeus and Critias.* Translated by Lee, D. (1972). Penguin Classics.

Plumptre, A. J. and Williamson, E. A. (2001) Conservation-oriented research in the Virunga region, in *Mountain Gorillas: Three Decades of Research at Karisoke* (eds M. M. Robbins, P. Sicotte and K. J. Stewart), Cambridge University Press, Cambridge, 362–389.

Ponder, S. (1987) Gifford Pinchot, press agent for forestry. *Journal of Forest History*, **31**, 26–35.

Pressey, R. L. (1997) Priority conservation areas: towards an operational definition for regional assessments, in *National Parks and Protected Areas: Selection, Delimitation and Management* (eds J. J. Pigram and R. C. Sundell), University of New England, Centre for Water Policy Research, Armidale, New South Wales, Australia, pp. 337–357.

Putz, F. E., Redford, K. H., Robinson, J. G., Fimbel, R. and Blate, G. M. (2000) *Biodiversity Conservation in the Context of Tropical Forest Management.* Environment Department, Paper no. 75 (Biodiversity Series – Impact Studies), The World Bank, Washington, D.C.

Reinartz, G. and Inogwabini, B. I. (2001) Bonobo survival and the wartime mandate, in *The Great Apes: Challenges for the 21st Century*, 52–56. Brookfield Zoo, Chicago, IL.

Reinartz, G., Inogwabini, B. I., Mafuta, N. and Lisalama, W. W. (2006) Effects of forest type and human presence on bonobo (*Pan paniscus*) density in the Salonga National Park. *International Journal of Primatology*, **27**, 603–634.

Reinartz, G. E. (1997) Patterns of genetic biodiversity in the bonobo (*Pan paniscus*). PhD thesis, University of Wisconsin, Milwaukee, WI.

Reinartz, G. E., Karron, J. D., Phillips, R. B. and Weber, J. L. (2000) Patterns of microsatellite polymorphism in the range-restricted bonobo (*Pan paniscus*): considerations for interspecific comparison with chimpanzees (*P. troglodytes*). *Molecular Ecology*, **9**, 315–328.

Remis, M. J. (2000) Preliminary assessment of the impacts of human activities on gorillas *gorilla gorilla* and other wildlife at Dzanga-Sangha Reserve, Central African Republic. *Oryx*, **34**, 56–65.

Rodrigues, A. S. L., Akçakaya, H. R., Andelman, S. J., Bakkar, M. I., Boitani, L., Brooks, T. M., Chanson, J. S., Fishpool, L. D. C., Da Fonseca, G. A. B., Gaston, K. J., Hoffmann, M., Marquet, P. A., Pilgrim, J. D., Pressey, R. L., Schipper, J., Sechrst, W., Stuart, S. N., Underhill, L. G., Waller, R. W., Watts, M. E. J. and Yan, X. (2004) Global gap analysis: priority regions for expanding the global protected-area network. *BioScience*, **54**, 1092–1100.

Rosser, A. M. and Leader-Williams, N. (2010) Protection or use: a case of nuanced trade-offs? In *Trade-offs in Conservation: Deciding What to Save*, (eds N. Leader-Williams, W. M. Adams and R. J. Smith), Wiley-Blackwell, Oxford, pp. 135–156.

Rouquet, P., Froment, J. M., Bermejo, M., Kilbourn, A., Karesh, W., Reed, P., Kumulungui, B., Yaba, P., Délicat, A., Rollin, P. E. and Leroy, E. M. (2005) Wild animal mortality monitoring and human Ebola outbreaks, Gabon and Republic of Congo, 2001–2003. *Emerging Infectious Diseases*, **11**, 283–290.

Sayer, J., Campbell, B., Petheram, L., Aldrich, M., Perez, M. R., Endamana, D., Dongmo, Z. L. N., Defo, L., Mariki, S., Doggart, N. and Burgess, N. (2006) Assessing environment and development outcomes in conservation landscapes. *Biodiversity Conservation*, **16**, 2677–2694.

Schaller, G. B. (1963) *The Mountain Gorilla: Ecology and Behaviour.* Chicago University Press, Chicago, IL.

Schaller, G. B. and Nichols, M. (1989) *Gorilla: Struggle for Survival in the Virungas.* Aperture Foundation, New York, NY.

Smith, M. B. (1998) The value of a tree: public debates of John Muir and Gifford Pinchot. *Historian*, **60**, 757–778.

Terborgh, J. (2004) Reflections of a scientist on the World Parks Congress. *Conservation Biology*, **18**, 619–620.

Thirgood, J. V. (1981) Man's impact on the forests of Europe. *Journal of World Forest Resource Management*, **4**, 127–167.

Thompson-Handler, N., Malenky, R. K. and Reinartz, G. (1995) *Action Plan for Pan Paniscus: Report on Free Ranging Populations and Proposals for Their Preservation.* Zoological Society of Milwaukee County, Milwaukee, WI.

Van Krunkelsven, E., Inogwabini, B. I. and Draulans, D. (2000) A survey of bonobos and other large mammals in the Salonga National Park, Democratic Republic of Congo. *Oryx*, **34**, 180–187.

Vogel, G. (2000) Conflict in the Congo threatens bonobos and rare gorillas. *Science*, **287**, 2386–2387.

Waley, A. (1939) *Three Ways of Thought in Ancient China*. George Allen and Unwin, London.

Walsh, P. D., Breuer, T., Sanz, C., Morgan, D. and Doran-Sheehy, D. (2007) Potential for Ebola transmission between gorilla and chimpanzee social groups. *American Naturalist*, **169**, 684–689.

Walsh, P. D. Abernethy, K. A., Bermejo, M., Beyers, R., De Wachter, P., Akou, M. E., Huijbregts, B., Mambounga, D. I., Toham, A. K., Kilbourn, A. M., Lahm, S. A., Latour, S., Maisels, F., Mbina, C., Mihindou, Y., Obiang, S. N., Effa, E. N., Starkey, M. P., Telfer, P., Thibault, M., Tutin, C. E. G., White, L. J. T. and Wilkie, D. S. (2003) Catastrophic ape decline in western equatorial Africa. *Nature*, **422**, 611–614.

CHAPTER 3

GOVERNANCE FOR EFFECTIVE AND EFFICIENT CONSERVATION IN ETHIOPIA

Fikirte Gebresenbet[1], Wondmagegne Daniel[2], Amleset Haile[3] and Hans Bauer[4]

[1]Department of Zoology, Oklahoma State University, Stillwater, OK, USA
[2]Department of Natural Resources Management, Texas Tech University, Lubbock, TX, USA.
[3]Wageningen University and Research, The Netherlands; CASCAPE project, Addis Ababa, Ethiopia
[4]WildCRU, University of Oxford, Tubney, UK

SUMMARY

Governance plays a crucial role in the wildlife conservation sector. Conservation success depends to a large degree on governance efficiency and effectiveness. Ethiopia was one of the first countries to start conservation acts, centuries back. In the past few decades, the sector has been going through different governing regimes that have ultimately meant different policies and degrees of attention towards wildlife conservation. Currently, the Ethiopian Wildlife Conservation Authority (EWCA) is the leading federal authority on conservation issues in Ethiopia. In this opinion piece, we first describe some examples of how different governance systems encourage and/or thwart wildlife conservation in Ethiopia. This is followed by a synthesis of findings and presentation of recommendations for further conservation efforts in Ethiopia. The overall recommendation for conservation in Ethiopia is to enhance governance in terms of using effective conservation systems and implementing a way of accountability for different governing authorities at different levels.

INTRODUCTION

Ethiopia is a country rich in history and full of ancient mysteries. Like elsewhere in the world, the great civilizations of Ethiopia grew at the expense of the country's natural resources. The Axumite Dynasty (AD 150–1150), which was the oldest monarchy in Northeast Africa, established Ethiopia as an important center for trade, including wildlife and their products. The Axumites designated a hunting area in which only the royals and nobles were welcomed (Kobishanov, 1981). In these and other ways, Ethiopia exploited its natural resources and wildlife from an early period. However, Ethiopia later took steps towards conservation. In the fifteenth century, King Zera' Ya'ekob (1434–1468) designated Menagesha-Suba Forest Area[1] as one of the "crown forests" of the country. He ordered the area to

[1]Menagesha-Suba Forest Area was established centuries before the creation of Addis Ababa (the current national capital), but is now located in the outskirts of Addis Ababa. The park is referred to as the "oldest park" in Africa.

be planted with seedlings of indigenous junipers from Wef-Washa Forest, located between Ankober and Debre Sina, and established Menagesha-Suba Park as the country's first protected area (PA).

The first recognized legislation on wildlife conservation in Ethiopia was passed in October 1908 by Emperor Menelik II (1888–1912), who decreed that elephant hunting should be regulated. Further legislation was passed in 1944 to regulate hunting of wildlife by ensuring that certain species were not overhunted. This early legislation demonstrates awareness of the limits of wildlife resources and the dangers people posed to them. Hence, concern for the protection of wildlife was translated into legal acts. Apart from subsistence killings, there was also a widespread culture of prestige killings. Killing animals, especially large carnivores and other big mammals, was something people would do as a way to become recognized by society as heroes. But the early regulation established by the king was a first and crucial step to abate these animal killings.

As in the early examples just described, governance can play a major role in a country's sustainable development that ensures the preservation of its natural resources, or not. Governance, simply, is the process by which a country's authorities make decisions and allocate resources to the citizens. Governance can greatly alter the activities and behaviors that affect wildlife conservation. While most power is often wielded by top government authorities, different formal or informal social and political institutions can also possess the authority to monitor and regulate biodiversity conservation. Institutions and groups can facilitate, or hold back, the activities of individuals or groups. This can be evidently supported by looking at successful or unsuccessful stories of conservation trials in Ethiopia. In this chapter, we first describe some examples of the effectiveness of governance in biodiversity conservation, and then synthesize these examples and present recommendations for further conservation efforts in Ethiopia. With a current population estimated at 87 million, expected to nearly double by 2050, and rising expectations for standards of living, the problems confronted by Ethiopia now will certainly be magnified in the future.

BABILE ELEPHANT SANCTUARY

Babile Elephant Sanctuary is located inside the Somali-Massai center of endemism in eastern Ethiopia; it was established mainly for the conservation of elephants. However, the sanctuary became imperiled in 2006, when a German biofuel company received a permit to start cultivation of castor beans (*Ricinus communis*), a biofuel crop, on 8000 ha of land, more than 70% of which was within the sanctuary boundaries. This situation led to a heated argument between the federal and regional governments, both of which favored the biofuel investment, and local and international conservationists who opposed the planned development with its profound effect on elephant habitat. Thanks to pressure from local and international conservationists, civil society organizations, and the public, the project was terminated in 2008, and elephants continue to occupy the original area of the sanctuary. According to the report of the African Elephants database (Blanc *et al.*, 2007), the number of elephants in Babile in 2006 was estimated to be 264.

NECH SAR NATIONAL PARK

Three different local communities live inside and around Nech Sar National Park in southern Ethiopia. They exploit the park resources in different ways, including cutting trees to obtain firewood, livestock grazing, and crop production (Figure 3.1). Settlements, overfishing, and poaching of wildlife are also common. To protect the park from pressures from the resident

Figure 3.1 Cattle grazing with wildlife inside Nech Sar National Park.
Photo courtesy of © Derek Clark (2007)

and surrounding communities, African Parks PLC (AP), an international wildlife conservation organization, signed a 25-year management agreement with the government to manage the park and its natural resources. However, in 2008, AP terminated the contract, primarily because the government did not accept the core area boundaries that AP established after negotiations with the Guji Oromo and Kore settlers. Additionally, the government had, after three years, still not addressed its contractual obligation to tackle important community issues (Derek Clark, personal communication). The major problem originated with the Guji Oromo, a pastoralist community that raises a large number of cattle, putting pressure on the park and causing conflict with the wild carnivores. Despite the threats they present, the Guji Oromos and Kore people still live inside and on the borders of the park (Figure 3.2).

This conflict between resident communities and wildlife presents a difficult problem, especially in countries like Ethiopia where different ethnic groups with different needs live together. Even if conservation authorities have the will to move resident people or deny them access to the park for its natural resources, they should be able to provide the communities with alternative livelihoods. However, this requires skilled manpower and financial capability, limiting ingredients in these developing countries, especially in the face of rapidly growing populations and expectations of the people.

Figure 3.2 A local bi-weekly market inside Nech Sar National Park.
Photo courtesty of © Mihret Te'amir (2009)

KAFTA SHERARO NATIONAL PARK

Kafta Sheraro National Park is located on the Ethiopia–Eritrea border. On a visit to the park in December 2009, we found it impossible to locate the park boundaries and even local people could not confidently determine if an area was inside or outside the park. There are other problems as well: guides are afraid to visit some areas in the park because of threats by local settlers and lack of support from law enforcement agencies in case of escalation. There is also a general problem of access to the area due to insecurity related to the presence of Eritrean guerrilla fighters crossing the border. Forests inside the park are also burned by locals for habitat encroachment, but also by Ethiopian authorities wishing to clear habitat in patches with high risk of Eritrean fighters creating ambush. These varied problems make conservation in Kafta Sheraro National Park a very challenging task.

AGRICULTURAL INVESTMENT IN SOUTHERN ETHIOPIA

Land-use change in southern Ethiopia is causing the extensive destruction of natural resources, as highlighted by the following two examples. Gambella National Park in southwestern Ethiopia is the country's largest national park (5061 km^2), with a unique ecosystem and wildlife composition. Many recognize great potential for wildlife conservation in this park, but this potential has not yet been tapped; indeed, the resources have been left to deteriorate at the hands of local people (owing to their tradition and livelihood strategies) and various government agencies. The state and federal governments carve up the land for "small" and big investments. Recently, the government established huge tracts of commercial farms within the park. Two large Indian and Saudi commercial plantations, Karuturi (300,000 ha) and Saudi Star (100,000 ha), are currently clearing land and conducting trials within the park (Figure 3.3). As a result of these activities, EWCA and the Park Administration organized a workshop in December 2010 with the objective of re-demarcating the borders of the park. But the re-demarcation has not yet been completed and it is not certain if it would even prevent the plantations from clearing more natural areas.

A second example of the impact of agricultural land change is the proposal to establish a large sugarcane

Figure 3.3 Farming inside what used to be Gambella National Park (it has now been redemarcated and the area is out of the National Park.
Photo courtesy of © Hans Bauer (2011)

Figure 3.4 Settlements inside Bale Mountains National Park.
Photo courtesy of © Fikirte Gebresenbet (2011)

plantation in the corridor between Omo and Mago National Parks. These two parks are among the most remote PAs in Ethiopia and they encompass the country's only viable Lion Conservation Unit (LCU). The corridor is a crucial piece of habitat that links the ranges of east African lions and central African lions, and so the management of the proposed plantation is certain to affect lion populations throughout east and central Africa.

In both examples, the agricultural companies receive licenses from the federal government in Addis Ababa without any consultation with the regional government, the federal conservation authorities, or the Park Administration. Although these agricultural investments are important for supporting the national economy, there are other considerations that should also be weighed, including the development of local communities, local food security, and the protection of wildlife. Although the federal conservation authority plans to promote a wildlife-integrated land-use policy, no attempts have been made so far to avoid the harmful impacts of agricultural developments.

SYNTHESIS

Additional problems in other PAs include encroachment in Bale Mountains National Park. In this park, the only requirement to own a piece of land property is to just fence it (Figure 3.4). Then the person can start

to plough it and also start paying tax to the appropriate office in the local administration. Conflicts in Awash National Park are also a major problem. These conflicts are not only of human–wildlife types, but also conflicts between local people and park staff, which recently caused two deaths.

All these put the existence of EWCA under a bold question mark – what good is it if it cannot at least stop the destruction of its own PAs?

Another problem is the lack of commitment among EWCA staff in putting different agreements into effect. For many, if not most, EWCA staff, having a job represents their top priority, above that of conservation. Otherwise, how can a country sign treaties and conventions like the Convention on International Trade in Endangered Species (CITES), the Convention on Biological Diversity (CBD), and the Convention on Wetlands of International Importance (the Ramsar Convention) and prepare national action plans for the conservation of species but not strive for their implementation?

Obviously, conservation is a complex task. But a country cannot have it both ways: one of the tough jobs of governing bodies is making choices. So, Ethiopian conservation authorities should decide what to do, how to go about it best, and act as fast as possible. It is easier and far cheaper to avoid problems of settlement in PAs than to re-demarcate the park boundaries and/or resettle people later.

Another problem is the process of establishing and gazetting PAs. The establishment of PAs in Ethiopia commenced in the late 1960s and early 1970s. However, starting from their early stages of establishment, all PAs share similar problems including evicting the local people, poorly defined landownership, and conflict between the local people and park offices (Wakjira, 2004). Despite efforts by different stakeholders to improve the management of PAs, wildlife populations are decreasing inside and outside PAs. The major problems that hinder the conservation process in the country include lack of public awareness and political will, improper land use, and socioeconomic problems (Demeke, 1997) that are, directly or indirectly, linked to governance. The problems in governance can be attributed to either limited political will or low capacity (financial or skilled manpower).

Before gazetting a PA, it must undergo a development process for some defined time period. Unfortunately, at the time of writing, 2010, of the 16 national parks in Ethiopia, only two are gazetted: Awash and Simien Mountains National Parks. Despite their being under development, most of the PAs have problems of deterioration that usually result from high human pressure in and around the PAs, pressures that are certain to increase in the future. This has resulted in cases of huge wildlife loss and shrinkage of PAs, due to re-demarcation or re-delineation processes. The re-demarcation happens as a result of different factors, the major ones being settlements inside and/or around PAs, investments inside PAs, and low awareness about wildlife conservation. The worrying question is: How far are the conservation authorities willing to go with shrinking PAs by giving away more land for "development"?

Another problem is the lack of skilled manpower to execute effective conservation. EWCA has gone through a number of changes in different regimes. It was established in 1965 as a chartered wildlife conservation authority, then became an organization, and still later a department within the Ministry of Agriculture. But currently (since 2007) EWCA is under the Ministry of Culture and Tourism. Another absurdity is that, although EWCA is a federal authority, it does not have control over all PAs in Ethiopia because some PAs are managed by regional states. This sometimes means that the federal authority might not have the mandate to decide on some matters in PAs managed by regional offices. Currently, the authority is taking over the management of most PAs in the country, but slowly and partially. These changes in management also create a problem as different governing systems try to operate in their own "new" ways, learning little from previous governing systems.

Lower levels of environmental governance also play a role in the overall conservation of biological diversity.

This includes issues like weak law enforcement by PA staff members (and local judiciaries); weak law enforcement is common in most PAs in Ethiopia. Many park scouts and wardens do not want to oppose the desires and activities of local people because of their fear of being hated and stigmatized. The social system makes their job very difficult because the very people who illegally exploit resources from PAs could be neighbors, relatives, or close family friends, or those employed to enforce the laws. This sometimes makes enforcement both complicated and unpleasant. Many argue by asking: Why should scouts confront locals and be socially shunned in return for their low salaries?

This would not be a valid argument among scouts who have a good level of awareness about conservation. A good example of this could be taken from the time when Nech Sar National Park was being managed by AP. The scouts were doing their job at the level required from them, with better management, awareness, and payment. An anonymous scout from Nech Sar confided to the authors that "when AP was managing the park, we felt like we have somewhere to turn to if we get into a mess with local youngsters or with the police". He also mentioned that in his 25 years of service he witnessed more court cases in the era of AP. The relationship between conservation authorities and the police and judiciary often is not clearly defined. People caught killing an animal, or fishing in a restricted area, or cutting trees or grasses, or grazing their livestock inside a PA often do not face a punishment greater than a talk with the local police, if the case even gets there. Otherwise the park scout or warden will simply talk to the people committing wrong deeds and let them go, only to catch them doing the same thing sometime later. Rare cases make it to the court, but, even in such cases, it does not necessarily mean the guilty parties will receive appropriate punishments. Gaps in the legislation make it difficult for the judges to convict people.

THE WAY FORWARD

In Ethiopia, a system of strong governance that learns from past mistakes and takes future gaps into account will serve best for conserving biodiversity. Federal and regional conservation authorities should push towards strengthening the governance system. Currently, the Ethiopian government is not giving enough attention to conservation. This can be exemplified by the fact that conservation was not part of the Plan for Accelerated and Sustained Development to End Poverty (PASDEP)[2] and is hardly mentioned in the Growth and Transformation Plan (GTP).[3] However, there are some good signs. Wildlife conservation is now being offered as a major in universities, with the intention of creating professionals who will enter careers as conservationists to put their knowledge into practice. Until now, people working for conservation agencies are all either biologists or foresters, and they might lack the needed knowledge and passion about conservation to be effective in their jobs. This change at the universities will produce skilled conservationists who will contribute to efficient conservation in Ethiopia's future. Another positive step forward is the resettlement program in Simien Mountains National Park. Together with an Austrian development project, this project has resettled around 10,000 people from the core area to the buffer zone in the last 10 years.

EWCA should gear to a system that encourages research in wildlife so that the decisions they make are well informed and gaps in the legislation can be filled appropriately. Ethiopian conservationists should become better at problem solving instead of simply identifying problems. Conservation authorities should know what exists in which areas before embarking on a conservation task. This baseline knowledge is necessary to assess any changes in biodiversity and natural resources in response to human pressure. It also contributes in making decisions and evaluation processes including the status of species, land-use patterns, pollution, the effect of global climate change, and the economic value of natural resources. Ethiopia has not yet produced a well-documented, overall survey of its biodiversity, which is surprising for a country that started conservation activities in earnest as early as the fifteenth century. A system should be developed that would facilitate the development and dissemination of such baseline information. Where appropriate, conservation authorities should look more closely at the biosphere reserve model for PAs, a model in which both people and wildlife can coexist sustainably.

Gazetting PAs should represent the first priority because it helps to avoid most problems, especially those regarding settlements and investments, so that further re-demarcation/re-delineation can be avoided. Ethiopia also needs general park regulations with fines, and these regulations need to be enforced consistently. But importantly, park regulations can be enforced only within park boundaries, so there need to be definite and clear boundaries proclaimed in the government gazette. Only doing so would make possible the enforcement of rules and regulations. If current trends are allowed to continue, all of Ethiopia's parks will quickly become nothing more than "paper parks" (Bruner *et al.*, 2001).

Only creating a sense of ownership among local communities will allow the governance system for local PAs to function properly. This can be achieved through increasing the level of conservation awareness in local people and park staff. An equitable benefit sharing that is generated from conservation activities and ecotourism clearly will be helpful in achieving this effect.

EWCA is appointed by the government and its mandate is to conserve wildlife. If wildlife numbers are falling both inside and outside PAs, and the number of people is increasing within them, then EWCA is clearly not fulfilling its responsibilities. Accountability needs to be established. If the Ethiopian conservation effort is failing, then ultimately everyone on the ladder, from government ministers to EWCA's director to park managers to chief scouts, must take responsibility. Hence, although it can be difficult for government appointees, EWCA should promote its own conservation goals and challenge activities that jeopardize wildlife conservation. The continued existence of Ethiopia's natural resources depends on the decisions and actions of the agencies governing Ethiopia's PAs and other natural areas. The protection of these natural resources will not only provide inspiration and enjoyment but will also be a fundamental part of the continued existence of local communities and their traditions. As such, good governance and growing knowledge among Ethiopia's conservationists will both help to preserve Ethiopia's important natural resources for the benefit of today's and future generations.

REFERENCES

Blanc, J. J., Barnes, R. F. W., Craig, G. C., Dublin, H. T., Thouless, C. R., Douglas-Hamilton, I. and Hart, J. A. (2007) *African Elephant Status Report 2007: An Update from the African Elephant Database*. Occasional Paper Series of the IUCN Species Survival Commission, no. 33. IUCN/SSC

---

[2] PASDEP was Ethiopia's 5-year strategic development plan from September 2005 to September 2010.

[3] GTP is Ethiopia's current strategic development plan, which became effective from September 2010 for the following five years.

African Elephant Specialist Group. IUCN, Gland, Switzerland.

Bruner, A. G., Gullison, R. E., Rice, R. E. and da Fonseca, G. A. B. (2001) Effectiveness of parks in protecting tropical bio-diversity. *Science*, **291**, 125–128.

Demeke, Y. (1997) The status of the African elephant (*Loxodonta africana*) in Ethiopia. *Walia*, **15**, 23–32.

Kobishanov, Y. M. (1981) Aksum: political system, economics, and culture, first to fourth century, in *General History of Africa, Vol. II: Ancient Civilization of Africa* (ed. G. Mokhtar), University of California Press, Berkeley, CA, pp. 214–223.

Wakjira, D. (2004) *Strategies for Sustainable Management of Biodiversity in the Nechisar National Park, Southern Ethiopia.* A research report submitted to the Organization for Social Science Research in Eastern and Southern Africa (OSSREA), Addis Ababa, Ethiopia.

CHAPTER 4

WILDLIFE IN JEOPARDY INSIDE AND OUTSIDE PROTECTED AREAS IN CÔTE D'IVOIRE: THE COMBINED EFFECTS OF DISORGANIZATION, LACK OF AWARENESS, AND INSTITUTIONAL WEAKNESS

Inza Koné

Centre Suisse de Recherches Scientifiques en Côte d'Ivoire, Abidjan, Côte d'Ivoire;
Laboratory of Zoology, Université Félix Houphouet-Boigny, Abidjan, Côte d'Ivoire

SUMMARY

The policy of nature conservation in Côte d'Ivoire dates from the colonial era from which it derives its legal and institutional foundations. Since then and especially since the acquisition of political independence in 1960, the country passed a series of laws and decrees and ratified most international conventions related to biodiversity conservation. Meanwhile, several state institutions were created with specific missions and have had mixed fortunes. However, despite these measures that reflect a certain political will for nature conservation, the situation is alarming, 50 years after the independence of this country. Indeed, Côte d'Ivoire is one of the tropical countries that recorded the highest rates of deforestation. Since 1960, the country has lost about 67% of its original forest cover. The effects of deforestation and illegal hunting of wildlife in the country have been devastating. Animal populations are becoming scarce in most national parks and forest reserves, many species having been hunted to near extinction or even extinction. If the enormous human pressures on natural resources in Côte d'Ivoire are inherently related to population growth and poverty, the chaotic situation just described shows the ineffectiveness of conservation policies in the country. This chapter demonstrates that the failure of conservation policies in Côte d'Ivoire can be attributed to the lack of synergy between governmental institutions, a lack of awareness about the importance of nature conservation, and a glaring institutional weakness.

INTRODUCTION

Côte d'Ivoire is located in the middle of the Upper Guinean ecoregion in West Africa. This ecoregion is a biodiversity hotspot characterized by high endemism of fauna and flora, and is considered a top priority for biodiversity conservation in West Africa (Mittermeier, Myers and Mittermeier, 1999). The conservation policy

---

in Côte d'Ivoire dates back to the 1950s, when the country was still a French colony. The original policy was based on the creation and management of an important network of protected areas (PAs), which collectively encompass all major ecosystems within the country. By the time of independence in 1960, more than 5 million hectares (ha) of terrestrial ecosystems were under protection (Ibo, 1993). After independence, existing PAs were reinforced and expanded in size, and additional PAs were created. In Côte d'Ivoire today, there are eight national parks (NPs) covering a total area of 18,568 km², six wildlife and nature reserves covering 3396 km², and 16 botanical reserves covering 1984 km². In total, more than 7% of the country's land area is under official protection. In addition, there are 147 classified forests covering another 3626 km². Classified forests are a special category of PAs managed by a governmental agency primarily for sustainable timber production.

Three of the Ivorian PAs are recognized as World Heritage sites: Taï NP (4540 km²) in southeastern Côte d'Ivoire, Comoé NP (11,492 km²) in northeastern Côte d'Ivoire, and Mount Nimba Nature Reserve (50 km²) in northwestern Côte d'Ivoire. Two others are listed as wetlands of international importance: Azagny NP (199 km²) and Ehotilé Islands NP (6 km²) in southern Côte d'Ivoire. Taï NP, Ehotilé Islands NP, and Mount Péko NP (340 km²) are considered as being of exceptional importance for regional integrated conservation of Upper Guinean forests (Conservation International, 2001).

As part of its biodiversity conservation policy, Côte d'Ivoire has passed a series of laws and ratified many of the international conventions relating to biodiversity conservation, including the Convention on Biological Diversity, the Convention on the International Trade of Endangered Species of Wild Fauna and Flora (CITES), the International Agreement on Tropical Woods, the Ramsar Convention on Wetlands of International Importance, the Convention on the Struggle against Desertification, and the Convention on Climate Change. Importantly, hunting has been forbidden throughout Ivorian territory since 1974, with the aim of reducing human pressures on animal species and populations.

However, despite these measures, Ivorian ecosystems have experienced continuous and increased biodiversity erosion. Today, 50 years after independence, the situation is alarming. Côte d'Ivoire has one of the highest rates of deforestation among tropical countries

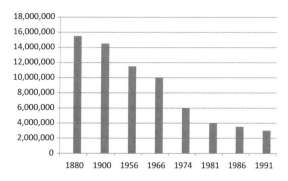

Figure 4.1 Dynamics of the forest cover in Côte d'Ivoire between 1980 and 1991.
Reproduced with permission from Lauginie (2007) Conservation de la Nature et Aires Protégées en Côte d'Ivoire. NSEI/Hachette et Afrique Nature, Abidjan.

(Figure 4.1), and since 1960 the country has lost approximately 67% of its original forest cover (Tockman, 2002). With the population of the country projected to increase from 20.6 million people today to 46.1 million people in 2050, and widespread expectations for increasing standards of living, the situation can only get worse, indicating the need for effective action as soon as it can be initiated.

Today, less than one-quarter of its primary forest remains, totaling approximately 24,000 km². These high deforestation rates were driven by demand for tropical hardwoods and by expansion of commercial and subsistence farming, which were prized by the government as cornerstones of the country's economic development and thus actively encouraged. While this policy has made Côte d'Ivoire the world leader in cocoa production and Africa's largest exporter of coffee, it has come at the expense of extensive forest areas throughout the country. What little forest remains is mostly confined to small fragments on which farmers increasingly encroach. Often, government authorities do little to prevent this encroachment, because it is perceived as a solution to forestall the outbreak of land conflicts between communities.

Like agriculture, logging activities have been characterized by intense development, which was encouraged by low tax rates and high exploitation royalties. Annual timber production increased tenfold between 1950 (228,000 m³) and 1965 (2,560,000 m³), then gradually expanded to over 4,000,000 m³ in 1969 and 5,321,000 m³ in 1977. Given that the production of 10 m³ of timber requires the destruction of two ha of forest (Monnier, 1981), these high timber production rates represent the destruction of millions of ha of native forest. In addition to logging, mining, the construction of major hydroelectric dams, and the development of an impressive network of roads have also had devastating effects on the forest ecosystem (Lauginie, 2007). "Commercial" hunting has also flourished throughout the country, despite the nationwide prohibition against it (Caspary and Momo, 1998; Caspary, 1999). Hunting pressure has reached unsustainable levels for most species even in the best conserved areas of the country (Refisch and Koné, 2005).

These high rates of land clearing and illegal hunting have had devastating effects on the country's wildlife. Populations of many species are now confined to the small areas of remaining forests. Even inside NPs and forest reserves, many animal populations have been hunted to near-extinction or even extinction. For example, the population size of the African elephant (*Loxodonta Africana*), the most emblematic species for the country, has decreased from approximately 100,000 individuals in 1900 to 4000 in 1980 (Merz, 1982, cited by Lauginie, 2007; Pfeffer, 1985). Originally, elephants were present throughout the country; they are now confined to less than 15% of the land area (Direction de la Protection de la Nature, 1991). The pygmy hippopotamus (*Choeropsis liberiensis*) has experienced a similar fate, having been driven to extinction at many sites within its historical range in Côte d'Ivoire because of deforestation. Its population size decreased from 19,000 individuals during the period 1982–1986 (Roth *et al.*, 2004) to 15,000 in 1997, with approximately 12,000 of these in the Taï NP. The most recent population estimates in Taï indicate a maximum of 2000–5000 individuals, with very few surviving at other sites.

Among primates, species sensitive to hunting pressure and habitat disturbance have been extirpated from most areas. This is the case for Miss Waldron's red colobus (*Piliocolobus badius waldronae*) (Oates *et al.*, 2000), and the Diana Roloway guenon (*Cercopithecus diana roloway*) (Koné and Akpatou, 2004; Gonedelé Bi *et al.*, 2008). A recent national survey of chimpanzees revealed that the population size has decreased by more than 90% in the last 20 years, and confirmed that only a few PAs still house viable populations.

If the huge human pressure on natural resources in Côte d'Ivoire is naturally linked with population growth and poverty, the examples just given clearly demonstrate the inefficiency of conservation policies in the country. In this chapter, we argue that the main reasons for this are the disorganization of the conservation sector in Côte d'Ivoire, the lack of awareness among various stakeholders about conservation issues and strategies, and institutional weakness at several levels.

DISORGANIZATION OF THE CONSERVATION SECTOR IN CÔTE D'IVOIRE

Legislative and institutional initiatives implemented in Côte d'Ivoire aimed at conservation, and the control of natural resource exploitation show that these issues have received much attention, but their limitations and the inadequacy of their enforcement also indicate the constant need for improving both the laws and the institutions responsible for their execution (Ibo, 2004). Current Ivorian laws have mainly been inspired by those of the colonial period and have largely proved inadequate for the current situation; the mismatch often effectively constitutes an obstacle to conservation (Ibo, 2004). The observed inefficiency of the many institutions in charge of various aspects of biodiversity exploitation and conservation has partly resulted from the antiquated nature of most key laws, as well as the lack of synergy among the existing laws.

Often the role of institutions is not clearly understood by stakeholders. For example, in 2006, at the launching of the Tanoé program (a pilot community-based management project for the conservation of critically endangered primates in southeastern Côte d'Ivoire), it was not clear which governmental institution should be the liaison with local communities. This resulted in significant delays that could have jeopardized the survival of these primates and other animals. According to the government, the Direction de la Protection de la Nature, is in charge of conservation strategies in the rural domain, and the Direction de la Faune et des Ressources Cynégétiques (DFRC) is in charge of the management of wild animals inside and outside of PAs. Furthermore, the scope of DFRC overlaps with that of the Office National des Parcs et Reserves (OIPR), which is in charge of the management of NPs and forest reserves, and that of the Société de Développement des Forêts (SODEFOR), which is in charge of the management of classified forests. The same issues of collaboration arose when the United Nations Environment Program (UNEP) and partners launched an important Côte d'Ivoire–Liberia corridor project in 2009. The lack of synergy between SODEFOR and DFRC is no doubt one of the major reasons why many classified forests exhibit the "empty forest" syndrome, in which the forest appears relatively intact but animal populations are extremely small or absent (Redford, 1992). SODEFOR focuses on the management of logging activities and surveillance against illegal timber harvesting, and pays little attention to the poaching of wild animals.

The lack of synergy is also apparent between technical institutions from different ministries. For example, while the Ministry of Environment is trying to clear out the people settled in PAs, the Ministry of Education builds schools for these peoples within the PAs arguing that their children must have access to education even if their settlement is illegal.

The logging sector is also disorganized, because logging concessions are not always managed by "certified" companies (Lauginie, 2007). Last but not least, rangers tend to make the management of PAs their exclusive business, despite the crucial importance of contributions by other practitioners such as biologists and social scientists.

The failure of conservation policies is also caused by lax planning and monitoring of the exploitation of natural resources. Indeed, the attribution of logging concessions without respecting certain conditions and the frequent unjustified declassification of classified forests indicate that personal interests often predominate over national interests including nature conservation (Lauginie, 2007). In addition, the frequent changes of ministerial positions and restructuring of these ministries result in regular changes of priorities, leading to lengthy delays or even cancellations of important conservation projects. In addition, there is no clearly defined behavior code for donors, non-governmental organizations (NGOs), and private organizations acting in the field of conservation. These institutions tend to follow their own priorities and develop their own strategies not always in line with those of governmental agencies. Sometimes they impose their own ideas and methods on governmental agencies even if the approach is detrimental to true nature conservation. For example, the GTZ concept of Zones of Controlled Occupation tolerates the exploitation of preexisting plantations within a PA under certain conditions and was imposed on OIPR in Taï NP, although it is fundamentally contrary to all principles of park management.

LACK OF AWARENESS AT MULTIPLE LEVELS

Another deep reason for the inefficiency of conservation policies in Côte d'Ivoire is a lack of knowledge or deliberate violation of laws by foreign immigrants, who are often illiterate and unaware of local laws. Thousands of people have settled in classified forests and even NPs, and the creation of large plantations of cocoa and coffee has led to substantial degradation of the forest habitat. Significant quantities of the cocoa and coffee produced in Côte d'Ivoire were grown in classified forests and PAs. In these rural areas, many of the local residents are unaware of the nationwide prohibition of hunting.

At a higher level, the importance of conservation to local and national economies is not well perceived even by decision makers. Most development strategies are based on the creation of large plantations, domestic animal farms, and fisheries. Wild animals are generally considered to be economically counterproductive because of the damage they cause to plantations or animal farms. However, a simple analysis of the economic value of wild animals could easily prove that in fact they deserve special attention for their potential to generate money through ecotourism. In Kenya, ecotourism generated 320 million euros in 1980, with a single living elephant estimated as generating 13,000 euros per year (Western and Henry, 1979; Brown and Henry, 1989; Whelan, 1991). Even if the Ivorian fauna and ecosystems are not as attractive as those in Kenya or other countries, the potential certainly exists for developing ecotourism as an important source of national income, and thereby providing benefits to the local populations as well (Lauginie, 2007). Until the value for ecotourism is realized, the economic value of wild animals will be limited to their use as bushmeat. Bushmeat is by far the most important source of animal protein in the rural areas of Côte d'Ivoire (Caspary *et al.*, 2001). In 1996, the annual take of wild animals for meat was approximately 120,000 tons, with an estimated commercial value of 118 million euros, representing 1.4% of the national gross domestic product (GDP; Caspary and Momo, 1998). In comparison, the annual production of domestic meat overall was estimated to be fewer than 60,000 tons.

If the importance of natural forests for rainfall is realized and the impact on productivity of agriculture and fisheries is grossly perceived, even by the illiterate, most actions in the field are obviously dictated by the prospective of short-term profit no matter what might happen in the long run. This explains why agro-industries have replaced hundreds of thousands of ha of forest with plantations, without any prior environmental or social impact assessment, even though such assessments have been mandatory since 1996. The indirect economic value of wild animals is even less well perceived; this is one of the major reasons why SODEFOR has not paid the required attention to the protection of wild animals in classified forests under their responsibility. Yet, considering the interdependency of fauna and flora (Clark, Poulsen and Parker, 2001), it is vital to maintain viable populations of wild animals in forest ecosystems, most notably for their importance in maintaining the health and regeneration of these ecosystems (Koné *et al.*, 2008).

INSTITUTIONAL WEAKNESS IN THE FIELD OF CONSERVATION

Most of the governmental agencies in charge of biodiversity conservation in Côte d'Ivoire are faced with a crucial lack of financial, technical, and human means. With very low annual budgets, these agencies have limited means to plan and carry out their own activities independently. Thus, it is not surprising that almost nothing can be done for the effective conservation of PAs without the support of big international non-governmental organizations and cooperation agencies. SODEFOR has tried to initiate collaborations with private logging companies by entrusting them with the expansion and implementation of long-term management of logging concessions, but so far they have obtained poor results because SODEFOR does not have the capacity to monitor the activities of these companies.

In addition, most governmental conservation agencies complain about a lack of personnel. Even when there are enough personnel, they lack the basic equipment needed to be efficient, including vehicles, appropriate arms, and ammunitions. They are also not well trained in modern techniques of forest management such as methods of bio-monitoring, use of geographic information systems (GIS), and participatory approaches.

Conservation NGOs face the same problem. The plethora of local NGOs is particularly ineffective because most cannot raise sufficient funds to achieve

an active presence in the field or to employ and mobilize their members over a sustained period. Because of these problems, only a very limited number of local conservation NGOs have achieved significant results.

In contrast, the support of large international NGOs such as the World Wide Fund for Nature (WWF), Conservation International and Birdlife International has helped conservation efforts in PAs in the 1990s by improving capacity building, infrastructure, and field operations. However, their involvement has generally been limited to brief periods. Following short-lived conservation interventions by these international NGOs, governmental partners have not succeeded in becoming capable of maintaining the level of activities achieved earlier. This difficulty became even more pronounced when many international NGOs left the country suddenly in the face of the military conflicts that occurred in 2002.

In view of the widespread failure of the application of classical conservation policies based on the exclusion of local residents from the use and management of natural resources, the concept of participatory management involving local communities arose in 2000. Unfortunately, local institutions created at the community level have still not been empowered sufficiently to plan and carry out conservation measures. In fact, the governmental authorities have never trusted local communities to manage PAs, as judged by their actions, and so the involvement of such communities has remained limited.

A recently passed law will allow collectives to own and manage PAs, this being an encouraging sign that could support both sustainable development and biodiversity conservation. However, the law has not yet been applied in Côte d'Ivoire, and the associated decentralization policies do not specify how the collectives will be involved.

CONCLUSION AND RECOMMENDATIONS

Wildlife is in jeopardy in Côte d'Ivoire, amid the many signs of an impending ecological disaster. The earlier examples of government inefficiency and failed conservation measures have worsened in the last decade as a result of the sociopolitical crisis that began in Côte d'Ivoire in 2002. The northern half of the country was then occupied by rebel forces, and considerable damage to forest habitats (inside and outside PAs) due to illegal logging, as well as unprecedented levels of poaching, has been reported in these rebel-controlled zones. At the same time, conservation measures in the government-controlled zones have become ever more dependent on the activities of large, international NGOs and on rarely maintained international cooperation. There was a normalization of this situation in Côte d'Ivoire involving an agreement between President Laurent Gbagbo and the rebels, which was signed in Ouagadougou, Burkina Faso. This led to the formation of a unified government in 2007, headed by the leader of the rebellion as prime minister. Since then, the prospect of elections in November 2010 was a source of hope for much-needed stability of the situation and a true revival of development of the country, but unfortunately instead a post-electoral crisis erupted that plunged the country into an even worse disaster until April 2011, when Allassane Ouattara was finally inaugurated as the new president of Côte d'Ivoire. Today there are reasons to believe that the country has finally passed through this time of crisis while the conservation value of many of its ecosystems remains high. A first major action that will be critical following the crisis is an assessment of the impact of the crisis on the natural resources in different regions of the country, with a focus on previously known key biodiversity areas. Once this is completed, governmental authorities and their partners should pay particular attention to long-term conservation of biological resources while they plan the reconstruction of the country and revival of its development. Future development plans must be more coherent and synergy must be established between different ministries and between governmental agencies of the same or different ministries. Laws must be adjusted so that empowerment of local administrative and communitarian institutions is more effective, in particular for the management of their natural resources. Donors must be invited to support the process, but must adhere to the principles of alignment on priorities defined by the country. To secure better commitment from all major stakeholders, national awareness campaigns must be organized using the media, with particular attention paid to building local capacity for sustainable development. In conclusion, a radical change of mentality and behaviors at all levels will be crucial to the protection of Côte d'Ivoire's forests and other natural habitats, and the resident wildlife contained therein.

REFERENCES

Brown, G. and Henry, W. (1989) *The Economic Value of Elephants*. LEEC Discussion Paper. London Environmental Economics Centre, London, UK.

Caspary, H. U. (1999) *Wildlife Utilization in Côte d'Ivoire and West Africa: Potentials and Constraints for Development Cooperation*. GTZ, Eschborn, Germany.

Caspary, H. U. and Momo, J. (1998) *La chasse villageoise en Côte-d'Ivoire – Résultats dans le cadre de l'étude filière viande de brousse (Enquête CHASSEURS)*. Rapport préliminaire No. 1 pour la Direction de la Protection de la Nature et la Banque Mondiale. Abidjan.

Caspary, H. U., Koné, I., Prouot, C. and De Pauw, M. (2001) *La chasse et la filière viande de brousse dans l'espace Taï, Côte-d'Ivoire*. Tropenbos Côte-d'Ivoire séries 2, Tropenbos Côte-d'Ivoire, Abidjan.

Clark, C. J., Poulsen, J. R. and Parker, V. T. (2001) The role of arboreal seed dispersal groups on the seed rain of a lowland tropical forest. *Biotropica*, **33**, 606–620.

Conservation International (2001) *De la Forêt à la Mer: les Liens de Biodiversité de la Guinée au Togo*. Conservation International, Washington, D.C.

Direction de la Protection de la Nature (1991) *Plan de conservation de l'éléphant en Côte d'Ivoire*. Ministère de l'Agriculture et des Ressources Animales, Abidjan.

Gonedelé Bi, S., Koné, I., Béné, J-C. K., Bitty, A. E., Akpatou, B. K., Goné, B. Z., Ouattara, K. and Koffi, D. A. (2008) Tanoé forest, south-eastern Côte-d'Ivoire identified as a high priority site for the conservation of critically endangered primates in West Africa. *Tropical Conservation Science* **1**, 263–276.

Ibo, J. G. (1993) La politique coloniale de protection de la nature en Côte d'Ivoire de 1900 à 1958. *Revue Française d'Histoire d'Outre Mer*, **298**, 83–104.

Ibo, J. G. (2004) L'expérience post-coloniale de protection de la nature en Côte d'Ivoire: quarante-quatre ans de bricolages et d'incertitudes. *Journal des Sciences Sociales*, **1**, 69–89.

Koné, I. and Akpatou, K. B. (2004) Identification des sites abritant encore les singes *Cercopithecus diana roloway*, *Cercocebus atys lunulatus* et *Piliocolobus badius waldronae* en Côte-d'Ivoire. Study Report for CEPA, Abidjan.

Koné, I., Lambert, J. E., Refisch, J. and Bakayoko, A. (2008) Primate seed dispersal and its potential role in maintaining useful tree species in the Taï region, Côte-d'Ivoire: implications for the conservation of forest fragments. *Tropical Conservation Science*, **1**, 291–304.

Lauginie, F. (2007) *Conservation de la Nature et Aires Protégées en Côte d'Ivoire*. NEI/Hachette et Afrique Nature, Abidjan.

Mittermeier, R. A., Myers, N. and Mittermeier, C. G. (1999) *Hotspots: Earth's Biologically Richest and Most Endangered Terrestrial Ecoregions*. CEMEX.

Monnier, Y. (1981) *La Poussière et La Cendre. Paysages, Dynamique des Formations Végétales et Stratégies des Sociétés en Afrique de L'Ouest.* ACCT, Paris.

Oates, J. F., Abedi-Lartey, M., McGraw, W. S., Struhsaker, T. T. and Whitesides, G. H. (2000) Extinction of a West African Red Colobus monkey. *Conservation Biology*, **14**, 1526–1532.

Pfeffer, P. (1985). Elephants en sursis. *Banco*, **3**, 5–11

Redford, K.H. (1992). The empty forest. *BioScience*, **42**, 412–422.

Refisch, J. and Koné, I. (2005) The impact of market hunting on monkey populations in the Taï region, Côte-d'Ivoire. *Biotropica*, **37**, 136–144.

Roth, H. H., Hoppe-Dominik, B., Mühlenberg, M., Steinhauer-Burkart, B. and Fischer, F. (2004) Distribution and status of the hippopotamids in the Ivory Coast. *African Zoology*, **39**, 211–224.

Tockman, J. (2002) Côte d'Ivoire: IMF, Cocoa, coffee, logging and mining. *World Rainforest Movement's Bulletin*, **54** (January).

Western, D. and Henry, W. (1979) Economics and conservation in Third World national parks. *BioScience*, **29**, 414–418.

Whelan, T. (1991) *L'écotourisme. Gérer l'Environnement.* Nouveaux Horizons, Paris.

CHAPTER 5

CONSERVATION CHALLENGES FOR MADAGASCAR IN THE NEXT DECADE

Hajanirina Rakotomanana[1], Richard K.B. Jenkins[2,3,4] and Jonah Ratsimbazafy[5,6,7]

[1]Department of Animal Biology, University of Antananarivo, Antananarivo, Madagascar
[2]Durrell Institute of Conservation and Ecology, University of Kent, Canterbury, UK
[3]School of Environment, Natural Resources and Geography, Bangor University, Gwynedd, UK
[4]Madagasikara Voakajy, Antananarivo, Madagascar
[5]Groupe d'Etude et de Recherche sur les Primates de Madagascar 34, Antananarivo, Madagascar
[6]Department of Paleontology and Biological Anthropology, University of Antananarivo, Antananarivo, Madagascar
[7]Durrell Wildlife Conservation Trust – Madagascar Programme, Antananarivo, Madagascar

SUMMARY

Madagascar is one of the top biodiversity hotspots in the world. Significant progress has been made in conserving its unique flora and fauna through a protected area (PA) network, advancing research techniques (e.g., discovery of new species), and elaborating new legislative environment frameworks (e.g., National Environmental Action Plans, wildlife legislation). These achievements are due to major efforts by scientists, conservation non-governmental organizations (NGOs), and government working in close collaboration. However, extensive loss of native forests to agriculture, illegal logging, poaching, and the spread of invasive species continue to constitute major threats to Madagascar's biodiversity. In order to significantly reduce biodiversity loss, Madagascar faces four main challenges: stopping illegal hunting of wildlife, sustaining the expanded PA network, ensuring that benefits derived from natural resource use are shared equitably, and promoting science as a tool to support conservation. These actions will be crucial to the health of Madagascar's natural resources in the next decade and beyond.

THE STATE OF MADAGASCAR'S BIODIVERSITY

Madagascar, the fourth largest island in the world, lies about 400 km off the eastern coast of mainland Africa. Its isolation from continental land masses (since 143 million years ago; Scotese, Gahagan and Larson, 1989) has made it a living laboratory of evolution and has endowed it with a unique range of species, most of which are endemic (e.g., 85% of plant species, 100% of mammals, 53% of birds, 95% of frogs). Madagascar's high levels of biodiversity and high rates of endangerment make it one of the top biodiversity hotspots in the world (Myers *et al.*, 2000). However, due mainly to human pressures including slash-and-burn agriculture, illegal exploitation of timber, uncontrolled fire, and mining, the island has been subjected to a high rate of deforestation that has caused a phenomenal loss in biodiversity. Since the arrival of humans, many plant and animal species have become extinct and many more are now highly threatened with extinction (Myers *et al.*, 2000). With the current population of 21.9 million projected to grow to 53.6 million by 2050, and considering that the country is one of the poorest in the world, the situation can only become worse. It clearly demands not only the attention of the national and regional authorities, but also of the world community.

CONSERVATION PROGRESS

Efforts to conserve the unique landscapes and natural heritage of Madagascar began in 1927, when a nascent PA network of 10 Réserves Naturelles Intégrales (RNI) covering 560,181 hectares (ha) was established by the French colonial authorities (Andriamampianina, 1987). Thirty years later, two other types of PAs were created: Réserves Spéciales (RS) in 1956 and Parcs Nationaux (PN) in 1958 (Andriamampianina, 1987). These latter two categories became the backbone of Madagascar's PA system during the remainder of the twentieth century and remain largely intact today. National parks are geographically scattered across the island and encompass all major vegetation types. Nevertheless, because of the sheer variety and extent of Malagasy wildlife and plants, many rare and unusual species were not covered by the PA network (Figure 5.1). This gap in coverage, combined with a recognized need to preserve the natural habitats that provide essential services to the Malagasy people, established the need to protect the remaining unprotected wild areas of forest and wetland. Towards this aim, the Malagasy government established the Système des Aires Protégées de Madagascar (SAPM) under the direction of the Ministry of Environment and Forests to implement the "Durban Vision," which aimed to increase Madagascar's PA network from 1.8 to 6.0 million ha from 2003 to 2008 (Randrianandianina *et al.*, 2003).

More recently, conservation planning in Madagascar has begun to encompass the values and needs of local people. From the early 1990s, there was a discernable shift from strict protection to providing more support for the livelihoods of people living near PAs (Ratsim-

Figure 5.1 Ranomafana National Park (southeastern part of Madagascar).
Photo courtesy of © H. Rakotomanana

bazafy *et al.*, 2013). In more recent years, the PA network has aimed to conserve Madagascar's biodiversity and cultural heritage, and to maintain ecological services, as well as promoting the sustainable use of resources to reduce poverty and developing local communities that depend on natural habitats for their livelihoods.

In this chapter, we discuss some of major challenges facing conservation in Madagascar in the next decade. Many species are on the verge of extinction due to extensive habitat loss, and urgent actions are needed to save the island's unique biodiversity. Because the causes of the problems are various, and poverty remains one of the major issues, we acknowledge the need for a multidisciplinary approach involving the inclusion of local communities. However, we concentrate here on the challenges faced in slowing the decline of biodiversity loss in Madagascar.

THREATS TO MADAGASCAR'S BIODIVERSITY

Madagascar qualifies as a global biodiversity hotspot not simply because it has a high diversity of species, most of which are found nowhere else on Earth, but also because an exceptionally high proportion of its endemic fauna and flora is threatened with extinction (Myers *et al.*, 2000). A detailed discussion on the drivers of these threats is beyond the scope of this chapter, but the root cause of environmental degradation is strongly

Figure 5.2. Extensive degradation of the montane tropical rain forest in Ranomafana in 2007.
Photo courtesy of © H. Rakotomanana

linked to the poverty of Madagascar's human population. A large population of rural people living in poverty and facing chronic low food security, with a concomitant high dependency on natural resource use, is a major contributing factor. Corruption, and periodic political instability with its negative repercussions on donor aid, accentuate the challenges faced by conservationists in Madagascar.

The main threat to Madagascar's biodiversity is the loss of native forest (Harper *et al.*, 2007). The slash-and-burn approach used by farmers also leads to uncontrolled bush fires that can be devastating for forests. In many areas, the cleared land is abandoned after only a few years, depleted of its fertility, and new areas of forests must then be cleared for new agriculture (Figure 5.2). On the coast, mangrove forests are cleared for shrimp farms. The loss of trees and other vegetation increases erosion, which can reduce the quality of wetlands and their productivity as fisheries. The selective, illegal logging of commercially valuable trees damages the structure of the forest by opening the canopy and facilitating the creation of new human settlements and roadways. The hunting of protected species for food threatens many mammals, birds, and reptiles in Madagascar (Garcia and Goodman, 2003). Hunting is mainly carried out for subsistence (Jenkins *et al.*, 2011), but some species are also affected by high commercial demand. In general, the removal of endemic animals and plants from nature threatens many species throughout Madagascar because legisla-

tion governing the harvest is rarely implemented or simply ignored. When the rewards for illegal exportation of highly valuable species, like rosewood trees and tortoises, are high, it becomes increasingly difficult to combat the threat, especially when the penalties are lax or simply unenforced.

Invasive plants and animals have extirpated many species on islands and Madagascar is no exception. The introduction of exotic species has caused reductions in the range and diversity of the freshwater fish fauna of Madagascar (Benstead *et al.*, 2003). More recently, concern has been raised about the spread of an exotic crayfish (Jones *et al.*, 2009) and the possible arrival in Madagascar of the chytrid fungus *Batrachochytrium dendrobatidis*, which kills its amphibian hosts and is causing a decline in their populations in many regions throughout the world. Below, we summarize the major challenges to the conservation of Madagascar's unique biota.

CHALLENGES

Stopping the illegal exploitation of animals

Although deforestation and the illegal logging of precious wood grab the headlines, illegal harvesting of endemic animals constitutes another major threat to Malagasy biodiversity. The illegal exploitation of animals must be stopped to prevent local extirpations and species extinctions. In Madagascar, where many species have very small geographic ranges, the extirpation of a single population may equate to the extinction of an entire species.

The collection of ploughshare tortoises (*Astrochelys yniphora*) for illegal export to Asia and the hunting of threatened lemur species for bushmeat within Madagascar's national parks are just two examples of the illegal exploitation occurring within the island's PA system (Garcia and Goodman, 2003). It is critical to stop, or at least reduce, this illegal activity because unchecked it might soon cause the extinction of some species with restricted ranges. This is especially true for tortoises, which have slow generation rates and thus limited biological capacity to withstand collection.

If critically endangered species of animals are to be protected inside PAs and generally, more resources need to be allocated to the personnel working to deter and apprehend illegal poachers. In addition, efforts need to be undertaken so that judges and lawyers will better understand and thus be able and willing to enforce the law. This could be achieved as part of a wider campaign to inform the public about wildlife legislation and the penalties for breaking the law. The high prices being offered on the international market for certain species of animals from Madagascar make these actions imperative if conservation goals are to be met.

Illegal bushmeat is mainly linked to supplying protein and income and is rarely undertaken (except for tenrecs) because of a strong taste preference for the meat (Jenkins *et al.*, 2011). Reducing bushmeat hunting and other illegal exploitation of animals in PAs is mainly a question of law enforcement, while elsewhere it can probably be reduced only if the supply of domestic meat is improved.

Certainly, the sustainable, legal exploitation of game species for food needs to be incorporated into the management plans of Madagascar's new sustainable use PAs. Scientific research into "sustainability" is costly and can be inconclusive. There has been a reported decline in the knowledge and application of traditional methods governing natural resource use, but these should be rejuvenated where possible. In areas where the demand for meat from game species is high, projects that could lead to the production of alternative sources of animal protein could be developed. Careful studies would need to be made of the peoples' preferences for, and likely use of, fish, bushmeat, and other sources of animal protein.

SUSTAINING THE EXPANDED PROTECTED AREA NETWORK

The key to sustaining the expanded PA network is recognizing that the ultimate goal is to utilize natural resources in a sustainable manner. The two main challenges facing this goal are therefore reducing deforestation and improving forest management.

Reducing the destruction of natural resources

Abusive and illegal exploitation of timbers (e.g., rosewood and ebony), along with expanding farmland, are the major causes of deforestation. The integrity of Madagascar's national parks and other strict nature reserves must be maintained through the tireless enforcement of legislation. Through restricting sus-

tainable use in these areas to non-consumptive activities, the opportunity to promote wise harvests in other forest areas is justifiable, as well as essential. Significant blocks of forests have already been allocated for use and the Ministry of Environment and Forest is expected to implement sustainable forest management techniques, including forest plantations, in order to satisfy the demand for forestry products and tree seedlings. The continued expansion of agricultural land therefore not only directly threatens biodiversity through forest loss, but also effectively reduces the available forest areas for sustainable use, which could lead to the spread of illegal or unsustainable practices in the remaining forests. In some areas, the development and use of alternative energy resources such as biofuels (e.g., palm oil, jatropha, soy, and sugar cane) could be promoted following appropriately robust social and environmental surveys. Essentially though, improved forms of land management and agriculture are needed to make more efficient use of the available land.

Strengthening the effectiveness of forest management

The degradation of natural resources has been accentuated during the last two decades and the government and its partners have struggled to respond at an equivalent pace. Since 2003, the government has tried to implement some profound changes in the environmental sector. These changes include series of activities including improving the accountability of all stakeholders involved in the management of natural resources and strengthening the enforcement of forest and environmental regulations. To improve coordination and management, and to disseminate and use research data at the national, regional, and local levels, it is necessary to increase the human, material, and infrastructural capacities of the ministry. In parallel, the forest revenue system and tax laws should be reformed to ensure a taxation system that promotes low environmental impacts.

IMPROVING BENEFITS SHARING AND EQUITABLE USE OF NATURAL RESOURCES

The benefits from conserving biodiversity should be shared by the local communities that depend on these natural resources for their own livelihood. These benefits are often perceived at the international or national level, but it is the local people, often the rural poor struggling to ensure their livelihoods, who experience the highest opportunity costs from conservation actions that restrict the use of natural resources. There may be some existing local benefits from conservation, such as ecological services like watershed protection or employment or revenue sharing from ecotourism. However, not all sites can benefit from these services, and sites with exceptional levels of local endemism like Menabe in western Madagascar might succumb to logging and deforestation pressures because of insufficient local incentives for conservation (J. Durbin, personal communication).

Direct incentives for community-based conservation through contracts and payments are therefore considered in order to ensure the effectiveness of this challenge:

Community management rights and capacity: local community centers must have clear management rights, either legal or customary, and be capable of maintaining the resource, of controlling the behavior of community members, and also of protecting the resources from threats originating outside the community.

Initial inventory and clear monitoring criteria: the investors must be assured that the area concerned maintains sufficient levels of biodiversity.

Monitoring: it must be possible to demonstrate, through independent and transparent evaluation, that biodiversity has been maintained to the quality/quantity specified.

Clear link between resource maintenance and benefits: the provision of direct benefits should be contingent on the results of monitoring that demonstrate maintenance of the resource to the levels specified.

Significant benefits: the benefit should provide a sufficient incentive to the resource managers to maintain the resource (e.g., surveillance) and opportunity costs from not exploiting the resource.

Effective, equitable distribution of benefits: Benefits should be sufficiently perceived by individuals within the community to persuade them to desist from activities that degrade the resource.

Long-term perspective with continued funding prospects: the agreement will be of greater interest to both parties if there are long-term prospects for continuation; this provides greater comfort for the investors that biodiversity will be conserved forever and for the managers that they will have a steady source of benefits.

ENSURING THAT SCIENCE IS BETTER USED TO SUPPORT CONSERVATION

The scientific approach of evaluating hypotheses, statistical testing, and transparent peer-reviewed publication of results is used by conservation biologists. Many international NGOs state that their conservation planning and decision making is based on sound science. Even when scientific data or publications on a subject are unavailable, managers, donors, and governments often seek the opinion of scientists. That science can make an important contribution to conservation in Madagascar, and elsewhere, is not in question. However, scientific studies must be aligned to the needs of conservation in order to be of use to conservation planning and implementation.

Given its unique natural habitats and exceptionally high proportion of endemic species, Madagascar has long attracted natural scientists from overseas, and in recent years these scientists have trained new generations of Malagasy scientists. In the field of conservation, scientists have made extraordinary progress in describing the taxonomy of plants and animals and much of this information was incorporated into the recent national PA planning process (Kremen *et al.*, 2008). Madagascar still attracts a large number of overseas scientists driven by their personal pursuits of puzzling academic questions in their areas of expertise. Their research projects typically include an important contribution to training Malagasy students, but in many cases probably do not contribute directly to conservation.

Madagascar is a fascinating island and will long continue to attract scientists. A challenge in the future is for the overseas institutions and universities to make an improved contribution to the research needed by Madagascar. At the tenth meeting of the Conference of the Parties to the Convention on Biological Diversity, some of the targets that were agreed concerned ecosystem services, improving benefits from biodiversity, avoiding extinction, and improving the sustainable use of natural resources. These aims could be adopted into research in Madagascar to help promote conservation of its threatened natural resources and wildlife. As an example, it might be more useful to focus research on the ecological role of lemurs as seed dispersers, their value to ecotourism, and the drivers of illegal bushmeat hunting that threaten this group of primates instead of studying aspects of their social behavior that are not crucial for conservation planning. Indeed, a recent collaborative publication by a number of scientists active in Madagascar concluded that more research should be orientated towards conservation issues (Irwin *et al.*, 2010).

The other challenge facing scientists concerns the communication of their results. Overseas scientists are expected to publish their research findings in the most influential journals. A high-profile publication in a well-respected scientific journal about conservation, that might be genuinely useful in Madagascar, will have minimal impact unless greater effort is made to communicate the results to the potential beneficiaries within Madagascar. Scientists should identify the potential target audience, whether they are personnel from a PA, a group of fishermen, or government decision makers, and explain their research so that it can be appreciated and adopted by the people living in those areas. It would be surprising if the feedback during meetings was in itself not helpful to the scientists in interpreting the results differently or in designing a new research project. Stakeholders in Madagascar need to welcome such initiatives, even if undertaken through translation, to maximize the benefit from research projects. When compared with halting deforestation and sustaining PAs, getting more from scientists seems less of a pressing priority. In many cases though, the results from research and monitoring are key to deciding when, and how to conserve forests and species. Conservation research has a lot to offer Madagascar but greater effort is needed to harness the resources and efforts of scientists and align them with national priorities and needs.

CONCLUSION

The ongoing political crisis in Madagascar that started in January 2009 has had a substantial, negative impact on Madagascar's conservation efforts. Indeed, conservation is not easy when 70% of the population lives under the poverty threshold, and difficulties are exacerbated during periods of instability. Many Malagasy people are directly dependent on the remaining natural resources that need to be managed sustainably and for the long term – a point that can be overlooked by politicians. In addition, Madagascar is facing a drastic challenge in saving its endangered and unique species while helping its people to alleviate poverty.

There is no doubt that a multidisciplinary approach would be a way to address Madagascar's problems before it is too late. In addition to the challenges mentioned and discussed in this article, it is also promising to see that since 2009 there has been increasing cooperation among conservation practitioners. As an example, the year 2009 saw the creation of the Alliance Voary Gasy, a civil society composed of Malagasy conservation associations and NGOs aimed at defending good governance and protecting Madagascar's resources against corruption and illegal trade. Such groups advance the implementation of a criminal justice court in Madagascar to prosecute violators of environmental legislation. Certainly, more importantly, local communities with whom we have worked for several years still hold out hope, if benefits of conservation can be realized locally. Especially in remote areas where government control is weak, the empowerment of local communities may be for Madagascar the only chance of preserving its endangered species.

Seeking innovative "win-wins" in this context is the necessary challenge for conservationists. Success will require a coordinated effort among the Malagasy government and local communities, lemur biologists, and conservation planners to devise realistic survival strategies for the island's unique biodiversity.

REFERENCES

Andriamampianina, J. (1987) Statut des parcs et réserves de Madagascar in *Priorités en Matière de Conservation des Espèces à Madagascar* (eds R. A. Mittermeier, L. H. Rakotovao, V. Randrianasolo, E. J. Sterling and D. Devitre), IUCN, Gland, Switzerland, pp. 27–30.

Benstead, J. P., de Rham, P., Gattolliat, J-L., Gibon, F. M., Loiselle, P. V., Sartori, M., Sparks, J. S., and Stiassny, M. L. J. (2003) Conserving Madagascar's freshwater biodiversity. *BioScience*, **53**, 1101–1111.

Garcia, G. and Goodman, S. M. (2003) Hunting of protected animals in the Parc National d'Ankarafantsika, northwestern Madagascar. *Oryx*, **37**, 115–118.

Harper, G. J., Steininger, M. K., Tucker, C. J., Juhn, D. and Hawkins, F. (2007) Fifty years of deforestation and forest fragmentation in Madagascar. *Environmental Conservation*, **34**, 325–333.

Irwin, M. T., Wright, P. C., Birkinshaw, C., Fisher, B., Gardner, C. J., Glos, J., Goodman, S. M., Loiselle, P., Rabeson, P., Raharison, J. L., Raherilalao, M. J., Rakotondravony, D., Raselimanana, A., Ratsimbazafy, J., Sparks, J., Wilmé, L. and Ganzhorn, J. U. (2010) Patterns of species change in anthropogenically disturbed forests of Madagascar. *Biological Conservation*, **143**, 2351–2362.

Jenkins, R. K. B., Keane, A., Rakotoarivelo, A. R., Rakotomboavonjy, V., Randrianandrianina, F. H., Razafimanahaka,

H. J., Ralaiarimalala, S. R. and Jones, J. P. G. (2011) Analysis of patterns of bushmeat consumption reveals extensive exploitation of protected species in eastern Madagascar. *PLoS ONE*, **6**, e27570.

Jones, J. P. G., Rasamy, J. R., Harvey, A., Toon, A. and Oidtmann, B. (2009) The perfect invader: a parthenogenic crayfish poses a new threat to Madagascar freshwater biodiversity. *Biological Invasions*, **11**, 1475–1482.

Kremen, C., Cameron, A., Moilanen, A., Phillips, S., Thomas, C., Beentje, H., Dransfield, J., Fisher, B., Glaw, F., Good, T., Harper, G., Hijmans, R., Lees, D., Louis, E., Nussbaum, R., Raxworthy, C., Razafimpahanana, A., Schatz, G., Vences, M., Vieites, D., Wright, P. and Zjhra, M. (2008) Aligning conservation priorities across taxa in Madagascar with high-resolution planning tools. *Science*, **320**, 222–226.

Myers, N., Mittermeier, R. A., Mittermeier, C. G., da Fonseca, G. A. B. and Kent, J. (2000) Biodiversity hotspots for conservation priorities. *Nature*, **203**, 853–858.

Randrianandianina, B. N., Andriamahaly, L. R., Harisoa, R. M. and Nicoll, M. E. (2003) The role of the protected areas in the management of the island's biodiversity, in *The Natural History of Madagascar* (eds S. M. Goodman and J. P. Benstead), University of Chicago Press, Chicago, IL, pp. 1423–1432.

Ratsimbazafy, J. H., Arrigo-Nelson, S. J., Dollar, L. J., Holmes, C. M., Irwin, M. T., Johnson, S. E., Stevens, N. J., and Wright, P. C. (2013) Conservation of Malagasy Prosimians: A View from the Great Red Island, in *Leaping Ahead: Advances in Prosimian Biology* (eds J. Masters, M. Gamba and F. Genin), Springer, New York, NY, pp. 387–396.

Scotese, C. R., Gahagan, L. M. and Larson, R. L. (1989) Plate tectonic reconstructions of the Cretaceous and Cenozoic ocean basins, in *Mesozoic and Cenozoic Plate Reconstructions* (eds C. R. Scotese and W. W. Sager), Elsevier, Amsterdam, Netherlands, pp. 27–48.

CHAPTER 6

CONSERVATION IN MAURITIUS AND RODRIGUES: CHALLENGES AND ACHIEVEMENTS FROM TWO ECOLOGICALLY DEVASTATED OCEANIC ISLANDS

F.B. Vincent Florens

Department of Biosciences, University of Mauritius, Réduit, Mauritius;
Université de la Réunion, La Réunion, France

SUMMARY

Mauritius and Rodrigues are among the last places on earth to have been reached by humans and yet are also among the most ecologically devastated, thus illustrating our great propensity to destroy the environment. The resulting situation, with several species on the brink of extinction, has attracted extensive conservation efforts, mostly from abroad. Some species near extinction, whose situation appeared hopeless, have recovered and represent conservation success stories. Today, conservation approaches and techniques continue to be innovated, developed, and tested on the two islands, which consequently represent a kind of "conservation laboratory" for the tropics. In some ways, the islands could be seen as representing what awaits the rest of the tropical world as the latter catches up in terms of human overpopulation, habitat destruction, and with fragmentation and alien species invasion being accelerated. Despite some notable successes, con-servation problems on the islands persist and are being exacerbated, on the one hand by a low and declining commitment of the government to the conservation of biodiversity, and on the other by an overwhelmingly prominent conservation non-governmental organiza-tion (NGO) that appears to be increasingly drifting away from biodiversity conservation and towards self-preservation. There is a growing need to shift the focus of local conservation efforts away from the current expensive, predominantly species-centric approach towards a more all-encompassing and economically more sustainable ecosystem approach. Some of the missing ingredients for this outcome seem to include involving alternative NGOs to carry out conservation work, an increased capacity of the authorities to take evidence-based decisions, and a reduction of the powers of politicians, who, facilitated by the country's laws, often interfere and pressure conservation profes-sionals and scientists to produce outcomes that are often contrary to the country's stated conservation policies.

Conservation Biology: Voices from the Tropics, First Edition. Navjot S. Sodhi, Luke Gibson, and Peter H. Raven.
© 2013 John Wiley & Sons, Ltd. Published 2013 by John Wiley & Sons, Ltd.

INTRODUCTION

Mauritius (1865 km²) and Rodrigues (109 km²), two volcanic oceanic islands formed 8–10 million years ago in the Mascarene Archipelago in the southwest Indian Ocean, had one of the worst possible starts in relation to biodiversity conservation. Within merely 370 or so years of human presence, these formerly pristine islands, teeming with endemic and often evolutionarily remarkable species, were transformed into two of the worst impacted places on earth ecologically (Cheke and Hume, 2008). Mauritius is most famous among conservation biologists for having provided the world with the very symbol of human-induced species extinction, the remarkable dodo (*Raphus cucullatus*) (Turvey and Cheke, 2008). Rodrigues had its own large flightless bird, the solitaire (*Pezophaps solitaria*), which was also quickly driven extinct by humans. But these two species are only the tip of the iceberg of extinction on each island. Parrots, owls, rails, giant tortoises and lizards, fruit bats, snails, and many other animal and plant species disappeared rapidly after humans first set foot on the islands (Table 6.1). The two islands are unusual in that their original biota was relatively well known from an early stage of human presence, owing to their late discovery. Thus, helped by some relatively complete fossil records (e.g., Rijsdijk *et al.*, 2009), Mauritius and Rodrigues provide us with one of the best records of what was initially present and what was lost, when,

and why, leaving little ambiguity about the heavy-handed role humans played directly and indirectly in rapidly driving so many species to extinction (Cheke and Hume, 2008; Rijsdijk *et al.*, 2011).

DAUNTING CONSERVATION CHALLENGES

Except for certain recent conservation successes, the two islands can be regarded as textbook examples of what should not be done if we are to preserve biodiversity and ecosystem functioning and, ultimately, ourselves. Natural habitats on both islands were destroyed for settlement or agriculture, the rate of destruction jumping to catastrophic levels in the nineteenth century, mainly as sugar cane plantations were established widely on Mauritius. This development continued thereafter, leaving a confetti of fragments of native habitat sprinkled mostly over the steeper slopes and other areas least suitable for agriculture (Figures 6.1, 6.2). Many extant native species that survived this environmental transformation, particularly plants, remain threatened due to extinction debt (Tilman *et al.*, 1994; Vellend *et al.*, 2006). The rarest species in the world is from Mauritius: the palm *Hyophorbe amaricaulis*, down to a single individual. All propagation attempts over at least 50 years have failed (e.g., Sarasan, 2010), making it a top contestant to become the symbol of the living dead. Many other species, particularly

Table 6.1 Native and endemic terrestrial species diversity in selected groups in Mauritius (Mau) and Rodrigues (Rod), with respective total number of extinctions. Percentages are given in brackets

	Total Native		Total Endemic		Total Extinct		Endemic Extinct	
	Mau	Rod	Mau	Rod	Mau	Rod	Mau	Rod
Angiosperms[1]	691	150	273 (39.5%)	47 (31.3%)	61 (8.8%)	17 (11.3%)	30 (11.0%)	10 (21.3%)
Mammals[2]	5	2	1* (20.0%)	0	2 (40.0%)	1 (50.0%)	0	0
Land birds[2,3]	28	14	19 (67.9%)	13 (92.9%)	16 (57.1%)	11 (78.6%)	12 (63.2%)	11 (84.6%)
Reptiles[2]**	17	8	16 (94.1%)	8 (100.0%)	5 (29.4%)	8 (100.0%)	5 (31.3%)	8 (100.0%)
Butterflies[4]	30	10	5 (16.7%)	0	4 (13.3%)	1 (10.0%)	1 (20.0%)	0
Snails[5]	125	30	81 (64.8%)	16 (53.3%)	43 (34.4%)	7 (23.3%)	36 (44.4)%	5 (31.3%)

[1] Baider *et al.* 2010; [2] Cheke and Hume 2008, [3] Hume 2011; [4] Williams 2007; [5] Griffiths and Florens 2006;
*Goodman *et al.* 2008
**one species of gecko survives on Rodrigues but it was first recorded after 1884 and is believed to be cryptogenic

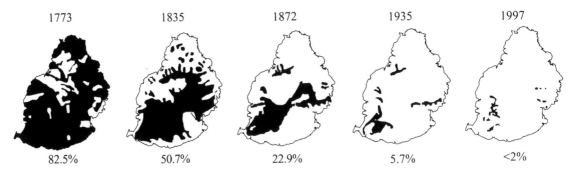

1773	1835	1872	1935	1997
82.5%	50.7%	22.9%	5.7%	<2%

Figure 6.1 The percentage of native habitats (in black) remaining at different dates on Mauritius. Only remnants dominated in their canopy by native species are shown. Those dominated in the canopy by invasive alien plants but with some native relicts accounted for another 3.4% in 1997. The situation on Rodrigues is even worse.
Adapted from Vaughan and Wiehe (1937) and Page and D'Argent (1997).

Figure 6.2 Current habitat destruction on Mauritius. (a) The last remnant of mainland coastal native forest in the Mascarenes being cleared at Roches Noires, Northeast Mauritius, for developing an "environmentally friendly" integrated resort scheme, demonstrating major loopholes in the local environmental impact assessment process (May 2008). (b) Pile of uprooted native and endemic plants on Ilot Gabriel Nature Reserve, North Mauritius shortly after the islet was leased to the friend of a minister. Officially only dead wood was cleared (August 2008). (c) Illegal wetland back-filling in northern Mauritius (March 2008). Most coastal wetlands in Mauritius now suffer from severe disturbances like fragmentation and filling (Laurance et al., 2012) (d) One of the illegal clearings for deer hunting within the dense forest of the Nature Reserve of Cabinet (note shooting platform indicated by the arrow). Those responsible were not fined, despite being identified (August 2006).
Photos © F.B.V. Florens

plants, have each shrunk down to a continuously decreasing handful of individuals. For example, the Mauritian endemic *Badula ovalifolia*, an understory tree discovered in 1821, is today known from only three adults and one sapling with nearly half of its adult population lost since 1997 (Florens, Baider and Bosser, 2008). But apart from the extremely small and fragmented populations left over from habitat destruction, the native biota also faces growing pressures from many invasive alien animals (particularly predators), and plant species that often reach extreme densities in even the best preserved habitat remnants (Strahm, 1993; Florens, 2008).

Overexploitation has also contributed to the overall decay of biodiversity on both islands. Easily hunted animals like the four species of giant tortoises (two on each island) were swiftly driven extinct (Cheke, 1987). Populations of economically useful plants were also decimated. The best examples are some palm species with edible hearts, which went from being extremely abundant (Bernardin de Saint Pierre, 1773) to the brink of extinction within two centuries (Maunder *et al.*, 2002). In addition to all these major classic threats, there is a suite of other biological problems such as agricultural insect pest introduction (Kaiser, Hansen and Müller, 2008), lost interactions like seed dispersal by now extinct fauna (Hansen, 2010), and the hampering of pollination or seed dispersal by alien species (Hansen, Olesen and Jones, 2002; Hansen and Müller, 2009), all of these piling up pressure against the survival of native species of the two islands.

Climate change may also worsen matters, although evidence so far suggests that the pre-human colonization biota of Mauritius has been rather resilient to past climatic stress (Rijsdijk *et al.*, 2009, 2011; van der Plas *et al.*, 2012). Climate change over the last 38,000 years seems to have triggered transitions between vegetation communities that mainly involved species reassortments or changes in vegetation distribution, with little evidence for plant species having gone extinct (de Boer *et al.*, 2013a, b). At any rate, a large portion of the surviving native species is nowadays under threat of extinction according to the IUCN Red List criteria. Among angiosperms, 81.7% and 77.8% of species should classify as currently threatened in Mauritius and Rodrigues, respectively (Baider *et al.*, 2010). The corresponding figures for land mollusks, an often-used indicator group, stand at 80% and 60% for Mauritius and Rodrigues, respectively (Griffiths and Florens, 2006). In addition, all nine extant land bird species

endemic to one or the other island are threatened (Cheke and Hume, 2008).

CONSERVATION AND SUCCESSES

Such a grim situation laid a perfect stage for attracting substantial conservation efforts, mainly from abroad from institutions like the Peregrine Fund and the Jersey Wildlife Preservation Trust (Jones, 2008). These efforts led to spectacular conservation successes starting in the 1970s with the Mauritius Kestrel (*Falco punctatus*), then known from only four individuals and more recently boasting over 800 wild birds (Jones, 2008). Recoveries of the Pink Pigeon (*Columba mayeri*) and Echo Parakeet (*Psittacula eques*) on Mauritius were achieved through intensive management, including captive breeding, and represent some of the other great conservation achievements that followed in the 1980s and 1990s (Jones and Swinnerton, 1997; Swinnerton *et al.*, 2004). In parallel with these species-centric approaches, habitat restoration has been attempted, mainly through the control of invasive alien species on important offshore islets such as Round Island, which holds the highest density of threatened vertebrate species in the world, or Ile aux Aigrettes, home to the last relics of the coastal dry forests of the Mascarenes (Parnell *et al.*, 1989).

Habitat restoration has also been undertaken on Mauritius and Rodrigues mainland, where so-called "conservation management areas" (CMAs) have been set up and managed (Cheke and Hume, 2008). CMAs (totaling <1% of remaining mainland native habitats) are typically located within well-preserved remnants of native vegetation. Ranging from 0.3–19.3 hectares in size, CMAs are usually regularly cleared of invasive introduced plants and are fenced against large alien mammals like deer, feral pigs, and goats (Figures 6.3a, b). The successful eradication of alien species, including rats, goats, and rabbits, from several islets, resulted in dramatic recovery of the native vegetation and fauna (e.g., North *et al.*, 1994). On the mainland, the control of invasive alien plants like the strawberry guava (*Psidium cattleianum*) in the CMAs on Mauritius or the rose apple (*Syzygium jambos*) on Rodrigues had marked beneficial effects on both native plants (Baider and Florens, 2006, 2011; Monty, Florens and Baider, 2013) and animals (Florens and Baider, 2007; Florens *et al.*, 2010; Hugel, 2012a). However, it has been noted more recently that this restoration activity is

Figure 6.3 Some conservation activities on Mauritius and Rodrigues. (a) Fencing forest remnants to try to exclude large alien mammals. This measure generally provides weak benefits relative to its cost, particularly on Mauritius, but continues to be implemented. (b) Weeding of invasive alien plants from native forests. This represents by far the most judicious use of the conservation dollar, but has a low popularity relative to more "exciting" species-centric projects. (c) Supplementary feeding of pink pigeon on Mauritius. This conservation action has proven invaluable in saving the species when it was on the brink of extinction, but now helps maintain the species' dependence on intensive management of questionable sustainability. (d) Innovative use of Aldabra tortoises within a valley undergoing reforestation with native species at the Francois Leguat Tortoise Park, Rodrigues. The animals are used as analogues to replace extinct endemic species and their lost ecological functions. Photos © (a), (c) F.B.V. Florens; (b) Courtesy of D. Florens; (d) Photo courtesy of François Leguat Reserve

unnecessarily being made both more expensive and damaging to native biodiversity merely due to a lack of basic evidence-informed decision (Florens and Baider, 2013). Furthermore, expensive attempts to exclude large hoofed mammals from the CMAs have generally been ineffective (Florens, 2008; Baider and Florens, 2011) and their presence and impact did not prevent a strong recovery of the forest community in the CMAs, particularly on Mauritius, indicating that attempts to exclude large mammals might not be as important as is generally assumed compared with alien plant control, at least locally.

More recently, a number of translocations of threatened species to safer havens have been successful. For example, individuals of the Mauritius Fody (*Foudia rubra*), a species declining in the face of alien predation and habitat degradation in its last strongholds in the Mauritian uplands (Nichols, Woolaver and Jones, 2004), were translocated to the lagoon islet of Ile aux Aigrettes (Cristinacce *et al.*, 2008), where alien predators are absent. Other highly restricted vertebrates that have survived on a single offshore islet have been reintroduced onto other islets following the eradication of their alien predators there. For example, individuals of the endemic skink *Leiolopisma telfairii* have been moved from their last refuge on Round Island to Ile aux Aigrettes, thus establishing a second population that will help reduce the risk of extinction (Thébaud *et al.*, 2009). In other cases, the entire species has gone extinct, and further action is obviously impossible. This has been the case with the four giant tortoise species, which formerly played important ecological functions as herbivores and seed disseminators (Hansen, Kaiser and Müller, 2008). In some cases, alien analogues have been introduced to try to restore the former ecological interactions. Thus, Aldabra tortoises (*Aldabrachelys gigantea*) have been introduced to Ile aux Aigrettes where they now disseminate seeds of endemic ebony (Griffiths *et al.*, 2011). This tortoise, along with a smaller species (*Astrochelys radiata*), have been introduced to Rodrigues to restore ecological function in the François Leguat Giant Tortoise Reserve, where they successfully disseminate seeds of native plants and control invasive alien weeds (Burney, 2011; Figure 6.3d). However, *A. gigantea* has more recently also been shown to enhance the germination of some invasive alien weeds (Waibel *et al.*, 2013). Another limitation with their use as analogue species is that they may not be able to breed in some sites where they have been introduced (Griffiths *et al.*, 2012).

Mauritius and Rodrigues can thus boast a number of conservation achievements and continue to play an important role as a laboratory where new conservation approaches and techniques are being innovated, tested, and developed. The outcomes are useful not only locally but as examples elsewhere, since many places in the world are or will soon be facing similar problems, including habitat destruction, overexploitation, disrupted mutualisms, and species invasion. Indeed, to a substantial extent, Mauritius and Rodrigues already represent situations that await much of the tropical world based on current worldwide trends in human overpopulation (densities of 668 people in Mauritius and 364 in Rodrigues per km^2 [CSO, 2010]), habitat destruction and fragmentation, extinction, and invasion by alien species. These multiple threats combine to produce acute conservation problems but have one advantage: they incite urgent and concrete conservation interventions. By attracting such attention, Mauritius and Rodrigues have played useful roles as conservation laboratories for the world beyond, and so far many such experiments are working effectively. Despite the massive habitat loss and other threats, there is some cause for measured optimism given the

surprisingly large proportion of native species, particularly of plants, that persist (Baider *et al.*, 2010; Florens *et al.*, 2012). Furthermore, new native or endemic plant species still continue to be discovered despite the fact that only tiny habitat remnants survive (Le Péchon *et al.*, 2011; Baider *et al.*, 2012; Baider and Florens, 2013). Even in a few cases where populations have been extirpated from one or the other island, reintroduction might still be possible. Some plants extinct on Rodrigues can be reintroduced from Mauritius. Neighboring Réunion Island might also supply Mauritius with some species for eventual reintroduction, including plants such as *Hernandia mascarenensis* (Bosser *et al.*, 1976 onwards), the butterflies *Antanartia borbonica* and *Salamis augustina* (Williams, 2007; Martiré and Rochat, 2008), the snail *Erepta setiliris* (Griffiths and Florens, 2006), and the bird *Circus maillardi* (Jones, 2008). As in the case of the introduced tortoises, close relatives of other extinct endemic species could be considered as introduction candidates to serve as analogues in natural communities. However, stemming further losses and extinctions is a much higher priority.

REMAINING CHALLENGES

The conservation successes outlined here are heartening, particularly the recoveries of some charismatic vertebrates, but the disturbing fact remains that, over the course of some four decades of sustained and often intense and costly conservation efforts, only a handful of species seem to have been saved from virtually certain extinction. Most of these have been vertebrates (Jones, 2008), although a few plant species have also recovered much, including *Ramosmania heterophylla* (Rubiaceae), a Rodrigues endemic (Strahm, 1989). In contrast, the vast majority of native species, particularly the less charismatic ones such as at least 750 endemic insect species (e.g., Motala *et al.*, 2007; Hugel, 2009, 2010, 2012b), continue to decline, as the overwhelming majority (99%) of native habitats on which most depend continues to decay into impoverished ecosystems dominated by alien invasive plants. The number of species that have been saved for the medium to long run thus represent a mere drop in the ocean of endemic and threatened species, a situation that calls for more decisive and meaningful conservation action. It is true that several species once thought to be extinct, having escaped all surveys for sometimes one or even two centuries, have been relocated in the reduced patches of remaining habitat (Florens, Florens and Sevathian, 2001; Florens and Baider, 2007; Baider and Florens, 2011). But given the evident extinction debt and the degree to which the two islands' natural habitats have been and continue to be impacted (e.g., Figure 6.2), it would be naïve to conclude that enough is being done to stem further extinctions.

A major weakness in the current regional conservation approach, particularly in Mauritius, where the bulk of the two islands' biodiversity survives, is the disproportionate importance being given to a case-by-case species-level approach to conservation, especially for birds and more recently for reptiles. This species-specific approach often takes away efforts and funds from a more comprehensive ecosystem approach that would benefit many more species simultaneously, including the very birds and reptiles of central attention as the habitat as a whole became more functional. Thus, a close examination at some of the hailed conservation successes, such as that of the pink pigeon on Mauritius, reveals that the continued maintenance of the birds depends heavily on the ongoing intensive management of the habitat as a whole, including the provision of supplementary food, the control of alien predators, and the management of disease, which is rife in the "wild" subpopulations (Bunbury et al., 2008). The use of feeding hoppers, made necessary because there is still too little restored native forest to support the recovering bird population – less than 1% of the native vegetation currently left – has been observed to facilitate predation on the birds by alien vertebrates, presumably feral cats (personal observation, 1996). It is also plausible that the feeding hoppers may increase risks of transmission of water-borne diseases, despite the fact that only dry grains are dispensed in the hoppers (Namah, 2010) and that the bird-concentrating effect that feeding hoppers have is likely to favor disease transmission as in the recent outbreak of beak and feather disease among Echo Parakeets (Kundu et al., 2012). If conservation is "The implementation of policies/programs for the long-term retention of natural communities/species under conditions which provide for continuing evolution" (Primack, 1998), then we have not yet succeeded with the pink pigeon but instead have saved it from extinction only to succeed in maintaining it as it were "under drip in an intensive care unit." The heavy dependence of this and other species such as the Echo Parakeet on intensive management seriously questions the viability of these

"conservation successes" over the long run. It is also true, however, that a conservation NGO, like the one currently spearheading the conservation of these species, can itself derive substantial benefits with continued intensive species-centered management; such activities make the NGO effectively indispensable and at the same time help guarantee an influx of funds through management fees, which generally increase with project cost. Incentives to favor species-centered conservation management, which is the most expensive approach but also the one that builds the greatest dependence on continued management for the longest time, can thus be very strong. The virtual monopoly of non-governmental conservation action, which this NGO holds on the two islands, exacerbates this problem. Such a situation can constitute a brake to the promotion of the much more desirable approach of addressing conservation from an ecosystem point of view – for example, by restoring whole habitats as through the control of invasive alien species. While the need for such large-scale restoration has recently been recognized for Mauritius and Rodrigues (NBSAP, 2006), it is still proving difficult to implement with less than 10% of the objective attained at midterm of the National Biodiversity Strategy and Action Plan (NBSAP), much of which is concentrated on islets of importance mainly for a few vertebrate species but which overall comprise a tiny fraction of the total native or threatened biodiversity.

This situation prevails despite an ongoing Protected Area Network (PAN) project funded at US$16 million, mainly by the UNDP. A small fraction of that amount would suffice to exceed the NBSAP (2006) target of larger scale habitat restoration at current proven costs of initial invasive alien plant control (<US$2,000 per ha) (Florens and Baider, 2013). Unfortunately, from the project document (UNDP-GEF 2009), it appears that much of this record funding for the country, entrusted mainly to the same NGO and the National Parks and Conservation Services, is earmarked for consultancies and other activities that might often duplicate existing knowledge, in particular that concerning weed control costs and efficacy, leaving relatively little for on the ground conservation work. On a positive note, Mauritius has set up a Conservation Fund, and some progress has been made in restoring wider habitat areas. However, disproportionally more effort still goes to expensive and relatively low-impact projects. A recent example is the translocation attempt made on one of the populations of an endemic species of reptile

(*Gongylomorphus fontenayi*) that was under threat on Flat Island, a nature reserve north of Mauritius that is highly subject to pressure by the human population. An exaggeration of the taxonomic status of the reptile (from population to full endemic species *contra* published works [Austin, Arnold and Jones, 2009]) seems to have triggered the authorities to contribute substantial funding there while neglecting other much higher impact conservation work elsewhere, like larger scale habitat restoration. In light of such situations, it appears vital to strengthen the capacity of the governmental authorities like the National Parks and Conservation Services, most of whose staff is trained not in ecology or conservation but in agriculture, to take more effective and practical conservation decisions. Breaking the monopoly of the NGO seems vital too.

One other major conservation problem is the weak and declining commitment of government to conservation of biodiversity (e.g., Figure 6.2, Caujapé-Castells *et al.*, 2010; Florens 2012a, b; Florens, 2013), which itself appears to largely reflect the lack of importance that ordinary Mauritians attach to conservation, generally perceived as a luxury. This problem of perception is more acute on the more urbanized Mauritius than it is on Rodrigues. Despite the commitment of many government officers and bodies to addressing conservation challenges, their efforts are often curtailed by interference intense enough to make the officers yield to the whims of politicians and end up backing decisions that are detrimental to conservation. One example is the recent lease of the Ilot Gabriel Nature Reserve to a minister's friend for touristic exploitation, where the promoter not only violated clauses of the lease with impunity, but benefited by a cover-up of the wrongdoings (Figure 6.3b). A similar situation happened on another nature reserve (Flat Island). Currently, the government is taking a disturbingly long time to approve the management plan for the National Park of Islets, a situation that incidentally favors plans to develop hotels on biologically important islets. The declining commitment of the government to conservation is also apparent in the management of the Conservation Fund, which is meant to finance conservation projects but which is instead being used largely for the day-to-day running of the National Parks and Conservation Service, or even to finance litter-picking campaigns and other expenses only remotely related to conservation. This misallocation of funds continues to deprive the National Biodiversity Strategy and Action Plan of funding. Furthermore, while laws and regulations to protect biodiversity on Mauritius and Rodrigues are fairly robust, their implementation often remains a remote dream. For example, the endemic *Pteropus niger*, the only of three original species of frugivorous bats that has escaped extinction on Mauritius, has been legally protected since 1993. After some 20 years, no fines have been levied, despite the fact that many animals are illegally killed every year by fruit growers (Anonymous, 2010) and sometimes for sport. The government is even considering relaxing the protection law (Anonymous, 2010) and adopting a culling program for this endangered bat species under pressure from fruit growers (Florens, 2012a)! A discouraging sign of the times is that even refraining from culling populations of this endangered species would start to be regarded as a conservation victory. Another similar example of *status-quo* "conservation victory" concerns one of the last areas of native forest in southwest Mauritius (at Ferney), which was scheduled to be sliced in half by a new highway but which in the end was not (Cheke and Hume, 2008). As part of the solution, it appears imperative that the discretionary powers that ministers have in law be reduced; there are clearly too many examples of undue pressure placed on officers and bodies to take actions that are often in conflict with the country's official strategies and the international conventions that it has have adopted. A bill has been drafted by an international team of legal experts that should, if enacted, help fill this and other gaps. But the civil society should also take up a more vigorous role in taking authorities to task on the numerous violations that are continuing to erode what is left of the unique biodiversity of the two islands.

CONCLUSION

Mauritius and Rodrigues islands still hold exceptional biodiversity of global significance despite massive habitat destruction, invasion by alien species, and other threats that have led to many extinctions and to a biota that is today among the most threatened in the world. Decisive conservation actions have saved some species from virtually certain extinction and a suite of conservation and restoration management actions continue to be innovated, tested, or improved, often with encouraging results. However, apart from the classic threats of habitat destruction and fragmentation, invasive species and diseases, extinction debt, or broken down mutualisms that are besetting the islands' biodiversity, certain generally little-mentioned threats appear poised to worsen the situation or at least hinder progress. The low and declining commitment of government to biodiversity conservation needs to be addressed and remedied. The country should seek to implement its laws to protect biodiversity rather than seeking to relax or ignore them. There is also much room for improvement in the capacity of practitioners to adopt a more robust, evidence-based approach to conservation and restoration action as well as in prioritization of tasks, particularly concerning restoration of habitats. Finally, it appears essential to encourage NGOs' action to move towards more all-encompassing and sustainable efforts like embracing an ecosystem approach and community involvement in conservation in a more meaningful manner, keeping species-centered action only when it is absolutely necessary.

ACKNOWLEDGMENTS

The National Parks and Conservation Service for providing permission to carry out research within the National Parks. Cláudia Baider and Anthony S. Cheke for their very constructive comments on earlier versions of the manuscript.

REFERENCES

Anonymous (2010) *Fourth National Report on the Convention on Biological Diversity – Republic of Mauritius.* http://www.cbd.int/doc/world/mu/mu-nr-04-en.doc (accessed March 19, 2013).

Austin, J. J., Arnold, E. N. and Jones, C. G. (2009) Interrelationships and history of the slit-eared skinks (*Gongylomorphus, Scincidae*) of the Mascarene Islands based on mitochondrial DNA and nuclear gene sequences. *Zootaxa*, **2153**, 55–68.

Baider, C. and Florens, F. B. V. (2006) Current decline of the "Dodo-tree": a case of broken-down interactions with extinct species or the result of new interactions with alien invaders? In *Emerging Threats to Tropical Forests* (eds W. F. Laurance and C. A. Peres). Chicago University Press, Chicago, IL, pp. 199–214.

Baider, C. and Florens, F. B. V. (2011) Control of invasive alien weeds averts imminent plant extinction. *Biological Invasions*, **13**, 2641–2646.

Baider, C. and Florens, F. B. V. (2013) *Eugenia alletiana* (Myrtaceae), a new critically endangered endemic species to the island of Mauritius. *Phytotaxa*, **94**(1), 1–12.

Baider, C., Florens, F. B. V., Rakotoarivelo, F., Bosser, J. and Pailler, T. (2012) Two new records of *Jumellea* (Orchidaceae)

for Mauritius (Mascarene Islands) and their conservation status. *Phytotaxa*, **52**, 21–28.

Baider, C., Florens, F. B. V., Baret, S., Beaver, K., Matatiken, D., Strasberg, D. and Kueffer, C. (2010) Status of plant conservation in oceanic islands of the Western Indian Ocean. 4th Global Botanic Gardens Congress. Dublin, Ireland. http://www.bgci.org/files/Dublin2010/papers/Baider-Claudia.pdf (accessed March 19, 2013).

Bernardin de Saint Pierre, J.-H. (1773) *Voyage à l'Isle de France, à l'Isle de Bourbon, au Cap de Bonne Espérance; etc. par un officier du Roi.* Société Typographique, Neuchâtel, France.

Bosser, J., Cadet, T., Guého, J. and Marais, W. (1976 onwards) Flore des Mascareignes – La Réunion, Maurice, Rodrigues. MSIRI/ORSTOM-IRD/Kew.

Bunbury, N., Stidworthy, M. F., Greenwood, A. G., Jones, C. G., Sawmy, S., Cole, R. E., Edmunds, K. and Bell, D. J. (2008) Causes of mortality in free-living Mauritian pink pigeons *Columba mayeri*, 2002–2006. *Endangered Species Research*, **9**, 213–220.

Burney, D. A. (2011) Rodrigues island: hope thrives at the François Leguat Giant Tortoise and Cave Reserve. *Madagascar Conservation and Development*, **6**, 3–4.

Caujapé-Castells, J., Tye, A., Crawford, D. J., Santos-Guerra, A., Sakai, A., Beaver, K., Lobin, W., Florens, F. B. V., Moura, M., Jardim, R., Gomes, I. and Kueffer, C. (2010) Conservation of oceanic island floras: present and future global challenges. *Perspectives in Plant Ecology, Evolution and Systematics*, **12**, 107–129.

Cheke, A. S. (1987) An ecological history of the Mascarene Islands, with particular reference to extinctions and introductions of land vertebrates, in *Studies of the Mascarene Islands Birds* (ed. A. W. Diamond), Cambridge University Press, Cambridge, pp. 5–89.

Cheke, A. S. and Hume, J. P. (2008) *Lost Land of the Dodo: An Ecological History of Mauritius, Réunion and Rodrigues.* T & AD Poyser, London.

Cristinacce, A., Ladkoo, A., Switzer, R., Jordan, L., Vencatasamy, V., de Ravel-Koenig, F., Jones, C. G. and Bell, D. (2008) Captive breeding and rearing of critically endangered Mauritius fodies *Foudia rubra* for reintroduction. *Zoo Biology*, **27**, 255–268.

CSO (2010) *Population and Vital Statistics, Republic of Mauritius, Year 2010.* Central Statistics Office, Port Louis, Mauritius.

de Boer E. J., Slaikovska, M., Hooghiemstra, H., Rijsdjik, K. F., Vélez, M. I., Prins, M., Baider, C. and Florens, F. B. V. (2013a) Multi-proxy reconstruction of environmental dynamics and colonization impacts in the Mauritian uplands. *Palaeogeography, Palaeoclimatology, Palaeoecology.*

de Boer, E. J., Hooghiemstra, H., Florens, F. B. V., Baider, C., Engels, S., Dakos, V., Blaauw, M. and Bennett, K. D. (2013b) Rapid succession of plant associations during the glacial-Holocene transition in Mauritius: an alternative mechanism to tracking climate change in a small oceanic island? *Quaternary Science Reviews*, **68**, 114–125.

Florens, F. B. V. (2008) Ecologie des forêts tropicales de l'île Maurice et impact des espèces introduites envahissantes. PhD thesis, Université de La Réunion, Réunion, France.

Florens, F. B. V. (2012a) Going to bat for an endangered species. *Science*, **336**, 1102.

Florens, F. B. V. (2012b) National parks: Mauritius is putting conservation at risk. *Nature*, **481**, 29.

Florens, F. B. V. (2013) Conservation: Mauritius threatens its own biodiversity. *Nature*, **493**, 608–609.

Florens, F. B. V. and Baider, C. (2007) Relocation of *Omphalotropis plicosa* (Pfeiffer, 1852), a Mauritius endemic landsnail believed extinct. *Journal of Molluscan Studies*, **73**, 205–206.

Florens, F. B. V. and Baider, C. (2013) Ecological restoration in a developing island nation: how useful is the science? *Restoration Ecology*, **21**(1), 1–5; DOI: 10.1111/j.1526-100X.2012.00920.x

Florens, F. B. V., Baider, C. and Bosser, J. M. (2008) On the Mauritian origin of *Badula ovalifolia* (Myrsinaceae), hitherto believed extinct, with complementary description. *Kew Bulletin*, **63**, 481–483.

Florens, F. B. V., Florens, D. and Sevathian, J. C. (2001) "Extinct" species rediscovered in Mauritius. *Phelsuma*, **9**, 53–54.

Florens, F. B. V., Baider, C., Martin, G. M. N., and Strasberg, D. (2012) Surviving 370 years of human impact: what remains of tree diversity and structure of the lowland wet forests of oceanic island Mauritius? *Biodiversity and Conservation*, **21**, 2139–2167.

Florens, F. B. V., Mauremootoo, J. R., Fowler, S. V., Winder, L. and Baider, C. (2010) Recovery of indigenous butterfly community following control of invasive alien plants in a tropical island's wet forests. *Biodiversity and Conservation*, **19**, 3835–3848.

Goodman, S. M., Jansen van Vuuren, B., Ratrimomanarivo, F., Probst, J.-M. and Bowie, R. C. K. (2008) Specific status of populations in the Mascarene Islands referred to *Mormopterus acetabulosus* (Chiroptera: Molossidae), with description of a new species. *Journal of Mammalogy*, **89**, 1316–1327.

Griffiths, C. J., Hansen, D. M., Jones, C. G., Zuël, N. and Harris, S. (2011) Resurrecting extinct interactions with extant substitutes. *Current Biology*, **21**, 1–4.

Griffiths, C. J., Zuël, N. Tatayah, V., Jones, C. G., Griffiths, O. and Harris, S. (2012) The welfare implications of using exotic tortoises as ecological replacements. *PLoS ONE*, **7**, e39395.

Griffiths, O. L. and Florens, F. B. V. (2006) *A field guide to the non-marine molluscs of the Mascarene Islands (Mauritius, Rodrigues, Réunion) and the northern dependencies of Mauritius*. Bioculture Press, Mauritius.

Hansen, D. M. (2010) On the use of taxon substitutes in rewilding projects on islands, in *Islands and Evolution* (eds V. Pérez-Mellado and C. Ramon), Institut Menorquí d'Estudis. Recerca, Menorca, Spain, pp. 111–146.

Hansen, D. M. and Müller, C. B. (2009) Invasive ants disrupt gecko pollination and seed dispersal of the endangered plant *Roussea simplex* in Mauritius. *Biotropica*, **41**, 202–208.

Hansen, D. M., Kaiser, C. N. and Müller, C. B. (2008) Seed dispersal and establishment of endangered plants on oceanic islands: the Janzen-Connell model and the use of ecological analogues. *PLoS ONE*, **3**, e2111.

Hansen, D. M., Olesen, J. M. and Jones, C. G. (2002) Trees, birds and bees in Mauritius: exploitative competition between introduced honey bees and endemic nectarivorous birds? *Journal of Biogeography*, **29**, 721–734.

Hugel, S. (2009) New Landrevinae from Mascarene islands and little known Landrevinae from Africa and Comoros (Grylloidea: Landrevinae). *Annales de la Société Entomologique de France*, **45**, 193–215.

Hugel, S. (2010) New and little known predatory katydids from Mascarene islands (Ensifera: Meconematinae and Hexacentrinae). *Zootaxa*, **2543**, 1–30.

Hugel, S. (2012a) Impact of native forest restoration on endemic crickets and katydids density in Rodrigues island. *Journal of Insect Conservation*, **16**, 473–474.

Hugel, S. (2012b) Trigonidiinae crickets from Rodrigues island: from widespread pantropical species to critically endangered endemic species. *Zootaxa*, **3191**, 41–55.

Hume, J. P. (2011) Systematics, morphology, and ecology of pigeons and doves (Aves: Columbidae) of the Mascarene Islands, with three new species. *Zootaxa*, **3124**, 1–62.

Jones, C. G. (2008) Practical conservation on Mauritius and Rodrigues. Steps towards the restoration of devastated ecosystems, in *Lost Land of the Dodo* (eds A. S. Cheke and J. P. Hume), T & AD Poyser, London, pp. 226–259.

Jones, C. G. and Swinnerton, K. (1997) A summary of conservation status and research for the Mauritius kestrel *Falco punctatus*, pink pigeon *Columba mayeri* and echo parakeet *Psittacula eques*. *Dodo*, **33**, 72–75.

Kaiser, C. N., Hansen, D. M., Müller, C. B. (2008) Exotic pest insects: another perspective on coffee and conservation. *Oryx*, **42**, 1 4.

Kundu, S., Faulkes, C. G., Greenwood, A. G., Jones, C. G., Kaiser, P., Lyne, O. D., Black, S. A., Chowrimootoo, A. and Groombridge, J. J. (2012) Tracking viral evolution during a disease outbreak: the rapid and complete selective sweep of a circovirus in the endangered Echo Parakeet. *Journal of Virology*, **86**, 5221–5229.

Laurance, S. G. W., Baider, C., Florens, F. B. V., Ramrekha, S., Sevathian, J-C. and Hammond, D. H. (2012) Human drivers of tropical wetland disturbance and their impacts on biodiversity. *Biological Conservation*, **149**, 136–142.

Le Péchon, T., Baider, C., Gigord, L. D. B., Haevermans, A. and Dubuisson, J.-Y. (2011) *Dombeya sevathianii* (Malvaceae): a new critically endangered species endemic to Mauritius. *Phytotaxa*, **24**, 1–10.

Martiré, D. and Rochat, J. (2008) *Les papillons de La Réunion et leurs chenilles. Biotope, Mèze (Collection Parthénope)*, Muséum national d'Histoire naturelle, Paris, France.

Maunder, M., Page, W., Mauremootoo, J. R., Payendee, R., Mungroo, Y., Maljkovic, A., Vericel, C. and Lyte, B. (2002) The decline and conservation management of the threatened endemic palms of the Mascarene Islands. *Oryx*, **36**, 56–65.

Monty, M. L. F., Florens, F. B. V. and Baider, C. (2013) Invasive alien plants elicit reduced production of flowers and fruits in various native forest species on the tropical island of Mauritius (Mascarenes, Indian Ocean). *Tropical Conservation Science*, **6**, 35–49.

Motala, S. M., Krell, F.-T., Mungroo, Y. and Donovan, S. E. (2007) The terrestrial arthropods of Mauritius: a neglected conservation target. *Biodiversity and Conservation*, **16**, 2867–2881.

Namah, J. (2010) Ecology of birds in managed and non-managed forests of Mauritius. BSc thesis, University of Mauritius, Mauritius.

NBSAP (2006) *Mauritius National Biodiversity Strategy and Action Plan*. Port Louis, Mauritius.

Nichols, R., Woolaver, L. and Jones, C. G. (2004) Continued decline and conservation needs of the Endangered Mauritius olive white-eye *Zosterops chloronothos*. *Oryx*, **38**, 291–296.

North, S. G., Bullock, D. J. and Dulloo, M. E. (1994) Changes in the vegetation and reptile populations on Round Island, Mauritius, following eradication of rabbits. *Biological Conservation*, **67**, 21–28.

Page, W. and D'Argent, G. A. (1997) A vegetation survey of Mauritius (Indian Ocean) to identify priority rainforest areas for conservation management. IUCN/MWF report. Port Louis, Mauritius.

Parnell, J. A. N., Cronk, Q., Wyse Jackson, P. and Strahm, W. (1989) A study of the ecological history, vegetation and conservation management of Ile aux Aigrettes, Mauritius. *Journal of Tropical Ecology*, **5**, 355–374.

Primack, R. B. (1998) *Essentials of Conservation Biology*. Sinauer Associates, Sunderland, MA.

Rijsdijk, K. F., Zinke, J., de Louw, P. G. B., Hume, J. P., van der Plicht, H., Hooghiemstra, H., Hanneke, J. M., Meijer, H. J. M., Vonhof, H., Porch, N., Florens, F. B. V., Baider, C., van Geel, B., Brinkkemper, J., Vernimmen, T. and Janoo, A. (2011) Mid-Holocene (4200 kyr BP) mass mortalities in Mauritius (Mascarenes): insular vertebrates resilient to climatic extremes but vulnerable to human impact. *The Holocene*, **21**, 1179–1194.

Rijsdijk, K. F., Hume, J. P., Bunnik, F., Florens, F. B. V., Baider, C., Shapiro, B., van der Plicht, J., Janoo, A., Griffiths, O. L., van den Hoek Ostende, L. W., Cremer, H., Vernimmen, T., de Louw, P. G. B., Bholah, A., Saumtally, S., Porch, N., Haile, J., Buckley, M., Collins, M. and Gittenberger, E. (2009) Middle-Holocene concentration-Lagerstätte on oceanic island Mauritius provides a window into the ecosystem of the dodo (*Raphus cucullatus*). *Quaternary Science Reviews*, **28**, 14–24.

Sarasan, V. (2010) Importance of in vitro technology to future conservation programmes worldwide. *Kew Bulletin*, **65**, 549–554.

Strahm, W. A. (1989) *Plant Red Data Book for Rodrigues*. Koeltz Scientific Books, Konigstein, Germany.

Strahm, W. A. (1993) The conservation and restoration of the flora of Mauritius and Rodrigues. PhD thesis, University of Reading, Reading.

Swinnerton, K. J., Groombridge, J. J., Jones, C. G., Burn, R. W. and Mungroo, Y. (2004) Inbreeding depression and founder diversity among captive and free-living populations of the endangered pink pigeon *Columba mayeri*. *Animal Conservation*, **7**, 353–354.

Thébaud, C., Warren, B. H., Cheke, A. S. and Strasberg, D. (2009) Mascarene Islands, biology, in *Encyclopedia of Islands* (eds R. G. Gillespie and D. A. Clague), University of California Press, Berkeley, CA, pp. 612–619.

Tilman, D., May, R. M., Lehman, C. L. and Nowak, M. A. (1994) Habitat destruction and the extinction debt. *Nature*, **371**, 65–66.

Turvey, S. T. and Cheke, A. S. (2008) Dead as a dodo: the fortuitous rise to fame of an extinction icon. *Historical Biology*, **20**, 149–163.

UNDP-GEF (2009) Expanding coverage and strengthening management effectiveness of the protected area network on the island of Mauritius. UNDP Project document. PIMS 3749, GEF Project ID 3526. Unpublished report.

van der Plas, G., de Boer, E., Hooghiemstra, H., Florens, F. B. V., Baider, C. and van der Plicht, H. (2012) Mauritius since the last glacial: environmental and climatic reconstruction of the last 38,000 years from Kanaka crater. *Journal of Quaternary Science*, **27**, 159–168.

Vaughan, R. E. and Wiehe, P. O. (1937) Studies on the vegetation of Mauritius I. A preliminary survey of the plant communities. *Journal of Ecology*, **25**, 289–343.

Vellend, M., Verheyen, K., Jacquemyn, H., Kolb, A., Van Calster, H., Peterken, G. and Hermy, M. (2006) Extinction debt of forest plants persists for more than a century following habitat fragmentation. *Ecology*, **87**, 542–548.

Waibel, A., Griffiths, C. J., Zuël, N., Schmid, B., and Albrecht, M. (2013) Does a giant tortoise taxon substitute enhance seed germination of exotic fleshy-fruited plants? *Journal of Plant Ecology*, **6**, 57–63.

Williams, J. R. (2007) *Butterflies of Mauritius*. Bioculture Press, Mauritius.

CHAPTER 7

DESIGN AND OUTCOMES OF COMMUNITY FOREST CONSERVATION INITIATIVES IN CROSS RIVER STATE OF NIGERIA: A FOUNDATION FOR REDD+?

Sylvanus Abua[1], Robert Spencer[2] and Dimitrina Spencer[3]

[1]Calabar, Cross River State, Nigeria
[2]Environmental Consultant, UK
[3]University of Oxford, UK

SUMMARY

This chapter reflects on a series of donor-funded interventions in the forestry sector in Cross River State (CRS), southeast Nigeria. We explore the distinctions between project outputs and impact. We show that today, looking back, certain project outputs could be evaluated as poor but the observable and potential impacts may form a useful basis on which to build the foundations for an equitable REDD+ system from the bottom up. The key impact of the projects under review was the emergence of system builders who continue to lead social and institutional innovation. We conclude the chapter with lessons learned from the successes and failures of community forestry interventions in CRS to inform the design of further interventions in Nigeria as it prepares for a long-term engagement with the REDD+ approach, as well as in other developing tropical countries.

INTRODUCTION

Drawing on Principle 22 of the Rio Declaration on Environment and Development, Joint Forestry Management and Community Forestry (Younis, 1997; Tewari and Isemonger, 1998) have become well-established approaches in conservation and the sustainable management of natural resources worldwide. According to the dominant community conservation paradigm, conservation must be participatory, must treat local people as equal partners, and should be organized in such a way that the conservation effort yields economic and social benefits both to local people

---

and the wider economy. Research has proposed that the success of such conservation initiatives depends significantly on the role of local people (e.g., Cernea, 1993), including their participation in the design and also in the implementation of projects (e.g., M'Gonigle and Parfitt, 1994). Various authors have explored the challenges in community conservation and resource management and have proposed different models to understand and apply them effectively (e.g., Hardin, 1968; Dawes, 1973, 1975; Runge, 1981; Olson, 1935, 1971; Ostrom, 1999). More recently, the social, political, and institutional innovation underlying the ongoing re-valuation of natural resources in the context of REDD+ has posed additional questions about the possibilities for equitable and sustainable resource management. The emerging research has brought useful insights from innovation studies suggesting the importance of bottom-up system building led by champions, and, in this process, the co-construction of new values and network innovations bridging technologies, markets, and organizations (Ramirez, 1999, Hajek *et al.*, 2011, Ventresca and Hajek, 2010).

This chapter reflects on a series of donor-funded interventions in the community forestry sector in CRS, southeast Nigeria, to explore their outputs and impacts. We draw a distinction between project outputs (showing how effective the project was in achieving its goals) and impact (which goes beyond the project goals and may include unforeseen developments that could often be described as aspects of innovative social or institutional practices referred to by Newman (2012) as "outside the project system"). While outputs are expected and measure the effectiveness of a project, the impact points to (often unexpected) innovations. We show that today, looking back, certain project outputs may be evaluated as poor but the observable and potential impacts may form a useful basis on which to build the foundations for an equitable REDD+ system from the bottom up. The key impact was the emergence of system builders who continue to lead social and institutional innovation. As discussed in Hajek *et al.* (2011), "System builders . . . have highly bridged networks and their core attribute, rather than technical or market expertise, appears to be the capability to marshal resources to the challenges they encounter and broker knowledge among different actors." Such system builders and the networks they have created and maintained form the main impact of the projects discussed here. We conclude the chapter with lessons learned from the successes and failures of community forestry interven-

tions in CRS to inform the design of further interventions in CRS, Nigeria, as well as in other developing tropical countries.

CONTEXT AND DRIVERS FOR COMMUNITY FORESTRY IN CRS, NIGERIA

Nigeria is Africa's most populous country with over 162 million people, projected to grow to 433 million people over the next 40 years (Population Reference Bureau, 2011). Any conservation action must be viewed in the context of this growth as well as that of expectations for human welfare based on consumption. CRS, covering 23,074 km^2 in southeastern Nigeria, is home to about 3 million people. Of the remaining tropical high forest in Nigeria, 50% is found in CRS. Over 70% of the population of CRS lives in forested areas, depending on land-use systems that combine farming and forest use. CRS plays a significant role in the national economy, producing large proportions of the country's main staples of yam and cassava. It is estimated that agriculture currently employs about 80% of the state's labor force and contributes about 40% to the state's gross national product (GDP).

Biological research confirms the diverse fauna and flora of these forests and their high degree of endemism (Davis *et al.*, 1994), and they are described as a biodiversity hotspot of global significance (Myers *et al.*, 2000, Oates, Bergl and Linder, 2004, Bergl, Oates and Fotso, 2007). While in 1991 the forest cover of CRS was 7,920 km^2, by 2008 it had been reduced to 6,102 km^2 (the rate of deforestation being estimated at 2.2% per annum). The forests and wildlife within this region are subject to intense and growing utilization from hunting for the commercial bushmeat trade, collection of firewood, clearing for farms, timber extraction, and the conversion of forest into oil palm plantations. Poor enforcement of forest laws and regulations presents further challenges to the sustainable management of forests in CRS.

Some authors have argued that poverty may be related to an increase in natural resource exploitation (Gray and Moseley, 2005). Other factors may include the growth of the cash economy (associated with payments for schooling, health, and higher education), migration, and diversification of household income sources. Various forest uses have become not only a way to survive for poorer people, but also a source of fast cash (by selling timber) or investment (by developing plantations) for those who can afford to engage in such activities. Thus, using forest resources has become both a survival and an accumulation strategy.

The main forest management regimes in CRS consist of national parks (covering roughly 4000 km²), controlled by the federal government; forest reserves (covering about 2,700 km²), controlled by the state government; and community forest estate (estimated to cover 1,632.75 km²).

DONOR INTERVENTIONS AND ACCOMPLISHMENTS

Since the early 1990s, at least five donor-supported projects have focused on forest conservation in the state, latterly with specific objectives targeting poverty (see Table 7.1). Table 7.1 summarizes the key donor-led interventions between 1991 and 2007.

With the exception of the Integrated Conservation and Development Programme (see later), all the donor interventions have aimed to enhance the incomes of forest-dependent communities through improved management of their forest resource.

The DFID Cross River State Community Forestry Project (CRSCFP) is examined more closely as the principal source of understanding of donor-led project challenges and successes, while the other projects noted earlier are referenced when appropriate. Some further details on the aims, objectives, and outcomes of CRSCFP are, therefore, described in the following section.

The *DFID CRS Community Forestry Project (CRSCFP)*: 1999–2002

Consultants Scott Wilson and ERM, contracted by DFID, worked with CRS Forestry Commission (FC) staff to help local communities set up Forest Management Committees (FMCs) and prepare resource maps and community-based regulations for the management of their forest resources. The project included a program of internal reform within the State FC resulting in the formation of a dedicated Community Forestry Support Unit and the creation of task forces committed to tackling illegal logging, staff re-organization, and sustainable revenue systems.

The main goal of CRSCFP was the capacity building of communities and the FC for improved management of the natural resource base. Capacity building of communities was intended to support them to manage their forests and derive livelihood benefits, while capacity building of the FC focused on better support for forest-dependent communities.

Significant project outputs included eight community-based forest management plans with associated maps and local bylaws; the successful introduction of alternative rural livelihood options to target communities (including the domestication of selected non-timber forest products (NTFPs) and forest game, and practical techniques for improved beekeeping); and draft new forestry legislation, including pioneering sections on the recognition and support of community-based forest management by the State FC. These outputs were demonstrated at the project's international lesson-sharing workshop in Calabar in 2002.

Significant challenges included outputs requiring further time and effort to properly embed. For instance, the local forest management plans were only completed as the project was closing, so implementation support was not available. Second, the draft new forest law was left with the government of the state just as it was completed, allowing no time for lobbying or awareness raising about its potential.

ANALYSIS AND DISCUSSION

The results of the approach that donor-led projects pursued in CRS were not always predictable and, consequently, some aspects of these projects could indeed be questioned in terms of their inherent sustainability. Subsequent manifestations of project delivery pose ethical and moral questions about the origin and appropriateness of community-based approaches to natural resource management in varying social and cultural contexts. The following section explores some of the key challenges and opportunities to sustaining outputs and making an impact, particularly with regards to capacity.

We discuss the outcomes of the projects with respect to three main fields of capacity: the CRSFC, the development projects, and the (FMCs). We focus on capacity as the ideal legacy of conservation efforts – that is, capacity to keep conserving and using sustainably the natural resource base today and for future generations.

Table 7.1 Details of sampled projects

Project title and Duration	Funding Agency	Implementing agency	Focus
Project 1. Integrated Conservation and Development Programme (1991–1999)	European Union	Worldwide Fund for Nature (WWF) in partnership with Cross River National Park	Biodiversity conservation complemented by rural development in areas contiguous to the national park to improve the basic conditions of human existence. Logic: "hard" infrastructure such as road construction and establishment of health clinics and schools will motivate local people to reduce levels of exploitation of timber and non-timber forest products (NTFPs) and desist from annual burning to clear additional forest for farms.
Project 2. Cross River State (CRS) Forestry Project (1991–1995)	British Overseas Development Administration (ODA), now UK Department for International Development (DFID)	ODA in collaboration with Forest Development Department (FDD)	Technical assistance addressing forest conservation and management by the FDD: studies focused on timber, NTFPs, socioeconomic factors, farming, and perceptions of forest value by local communities. Main output: Forest Sector Strategy (1994), recommending greater participation of communities in future forest management.
Project 3. CRS Community Forestry Project (CRSCFP) (1999–2002)	UK DFID	Scott Wilson (now URS) and Environmental Resources Management in collaboration with the CRS Forestry Commission (CRSFC) and Forest Management Committees	Development of sustainable community forest management in CRS through capacity building of community-based forestry organizations and the CRSFC together with income generation activities and better marketing of timber and NTFPs.
Project 4. Cross River Environmental Project (2003–2005)	Canadian International Development Agency	One Sky – Canadian Institute for Sustainable Living and a coalition of four Nigerian non-governmental organizations (NGOs) and one environmental network (NGOCE). The NGOs were partnered with Canadian organizations and businesses under Joint Initiative activities.	Strengthening capacity of local NGOs to influence gender-sensitive policy change and environmental improvement in CRS and to build a network of long-term partnerships among Canadian and Nigerian environmental organizations to work together for the benefit of the region. The NGOs supported community-based organizations (CBOs) in the buffer zones of key protected areas: the Guinean Lowland Forest Ecosystem in Cross River National Park and the Montane Ecosystem of the Obudu Plateau.
Project 5. Sustainable Practices in Agriculture for Critical Environments (SPACE) (2004–2007)	United States Agency for International Development (USAID)	Associates in Rural Development (ARD), now TetraTech	To conserve the ecological values and processes of CRS and slow agricultural expansion into tropical forest through sustainable agricultural activity that will also enhance the welfare of communities that border areas of ecological value. Also to address critical and complex governance and livelihood issues transforming the people and natural landscape of CRS.

Project capacity

As noted earlier, the main goals of most development projects in the state have dwelt on the need to improve the livelihoods of forest-dependent people by enhancing management of their natural resource base. The principal challenge for CRSCFP was to provide an alternative model for forest management than the existing one of uncontrolled timber exploitation and conversion to agriculture, including subsistence agriculture and cash crops. A necessary first step was demonstration of the merits of sustainable forest management through compelling and accessible economic arguments based on local markets and prices. Unfortunately, insufficient and inadequate research during project design, together with early implementation, failed to support this because effort was spent on the rolling-out of the FMC model, rather than generating market information on the local forest economy. It was only towards the very end of the project that adequate information and guidelines on key timber and non-timber forest products were promulgated, so there was no time for this to be embedded before project close-out.

Moreover, markets and local cash economies were distorted by graft within the FC, inadequate knowledge of forest regulations on the part of community members and local FC officers, and shady business dealing by timber merchants and processors. Thus, project management for outputs, rather than for awareness and empowerment (Newman 2012), together with short project cycles, reduced project capacity for effectiveness and legacy. At the time, it was indeed more lucrative for local communities to wholesale their timber and clear fell for agriculture, rather than collect NTFPs or maintain forest stands in the institutional context of an outdated legal framework and poor administration (e.g., enforcement) of fees and tariffs connected with legal forest exploitation. In this context, lacking a working economic, political, and social model, it was challenging to reveal the intrinsic forest value, especially as donor-led approaches ground arguments in culturally different perspectives on biodiversity and environmental protection. Some of these key challenges may be overcome through the new system of forest valuation developed by REDD+, provided these guarantee equitable distribution of forest value and adequately explain alien concepts of carbon sequestration.

FMCs' capacity

An FMC is a community-based institution devised by the DFID CRSCFP as a focal point for local forest management decision making. FMCs were originally envisaged as local management institutions to work with the FC to generate value from the forest in a sustainable way though better management and the sale of timber and NTFPs, and to ensure that gains from forest resources were used for effective community development, with a particular focus on the poorest members of the community (i.e., those with little access to farming, plantations, or government work). The following factors limited overall success in improvement in FMCs' capacity:

1. The way the FMC idea was implemented in each village limited some of its capacity. There were two main issues with implementation. First, the FMCs were imposed by the CRSCFP, initially as an institutional construct to the FC and then by the project and FC to the local communities. Second, pre-existing local community institutions (so-called "bush committees," "land-use committees," "allocation committee," or "village forest protection organizations") for forest monitoring and game management were ignored and by-passed due to lack of knowledge about them. The FMCs often duplicated the work of such local organizations with various responsibilities related to the forest (e.g., vigilante groups set up to stop thieves and loggers; or tree spotters to guide loggers). The new institutions (FMCs) had no traditions or historical ties to forest management and were susceptible to hijacking by new elites, thus producing bad feelings and increasing the perception that FMCs were government proxies, not community-based institutions – that is, they were not credible forest stewards in the eyes of the community. The duplication of forest institutions sometimes contributed to internal community conflict, and thwarted effective coordination.

2. The capacity of FMCs was also constrained by contradictions between project aims and local expectations. In some villages, FMCs were associated with building false expectations and failing to deliver – especially when they lacked forest resources. False expectations may have resulted from previous conservation projects – for example, when the EU National Park Programme (1991–1991, see Table 7.1) was implemented, communities enclosed in the park demanded roads in pristine forests as compensation for economic

opportunities lost because they could not get their farm crops to markets (cf. Alashi, 1999). Some villages in the support zone demanded, and, at times, were granted roads close to the park boundary. Despite the road construction, there was much resistance from the youth, contending that the incentives did not match the opportunities lost. Some of these feelings continued on into the CRSCFP and limited the capacity of the FMCs – that is, FMCs or their leaders' credibility were devalued by locals because they were unable to respond to such expectations.

3. The capacity of local FMCs was also disrupted due to their unforeseen social reengineering functions. Although the FMCs were conceived to represent all community interests and groups (including the women and the poorest) and to work closely with traditional institutions of power such as the chiefs, in some cases they were captured by male-dominated youth groups seeking to reverse village power; or by timber dealers, working closely with the major processor; or by local teachers and civil servants who were actively seeking alternative income-earning opportunities but not necessarily in the community interest.

4. Finally, at the time of the project, the FMCs did not have delegated legal authority. Thus, the bylaws that each FMC created were not legally binding and their enforcement depended on good will and local community moral values, which could be superseded by the chief or other important players in the village. Additionally, by-laws raise the issue of plurality of legislation. Identifying and resolving conflicts between the state and local communities with regards to these laws vis-à-vis the supremacy of statutory law is rather challenging. This may be so because there is a tendency for government officials to limit devolution of juridical powers to local community institutions, given that the statutory law supersedes community bylaws.

System builders: a major prerequisite for effective and impactful community conservation projects

The shortcomings of the CRSCFP discussed in the previous section are chiefly associated with the narrow focus of development project design and outputs and with the organizational and temporal challenges in a specific socioeconomic and political context.

"Community capacity" (in terms of resources, human, financial, and ecological; and of commitment to build on local strengths and needs, to seize opportunities, and to address problems) has been proposed by many authors (e.g., Sharma, 1993; McGuire *et al.*, 1994; Aspen Institute, 1996; Litke and Day, 1998; Mehta and Kellert, 1998; Vodden ,1999; Gunter, 2000; Markey *et al.*, 2005) as a key ingredient for sustainable development. The DFID-funded community-based management system only really worked well in relatively smaller and rather more cohesive communities where the power of the village chief and his associates was still relatively strong and effective in mobilizing various forms of engagement and resolving conflicts. In many communities, the power of the traditional chiefs' institution is seen as waning and there are hardly any effective institutions to replace them. Indeed, some of the literature on CBNRM initiatives has argued that "social cohesion" is an important factor in community conservation (e.g., Vodden, 1999; Skutsch, 2000; Lane and McDonald, 2005; Markey *et al.*, 2005).

Today, looking at the project impact from the prism of the REDD+ initiative, the singular theme that is revealed when looking for success stories is the emergence of system builders (Hajek *et al.*, 2011) – that is, local villagers, leaders, and institutions capable of recognizing opportunity, marshalling resources, adapting and acting with flexibility, building networks, and bridging existing ones. Three key categories of system builders emerged during the projects and have continued building technical capability, networks, and a knowledge base. The authors identify system builders at international, national/regional, and state/local levels.

A CASE STUDY ON A LOCAL SYSTEMS BUILDER

Chief Peter Ikwen of Okorshie in the north of CRS was a civil servant and a local politician. His retirement coincided with the establishment of the CRSCFP FMCs. He was elected as the chairman of the Okorshie FMC. Due to his personal qualities and his extensive past experience in local administration and politics – in marshalling resources and building networks across institutions and communities in CRS, he became a champion of the FMC model. A visionary, he had seen some value in the project from his own point of view and that of his community – the exchange of ideas and the intercommunity networks for opportunities to improve local incomes. Aware that these would have ceased with the end of the short-term project cycle, he engaged in institutional innovation. He established the Association of Chairmen of FMCs in the state in order to maintain the connections between them and to act as a new layer of governance of forests serving local interests and being flexible in drawing in resources. This Association played a role in the certification of the FMCs by the state government, who formally recognized 18 FMCs in 2004. While CRSCFP did not aim to create networks between communities, the local leaders saw the value in this. This is an important impact, which now could be employed in the mobilization for REDD+. Chief Peter is now on the National REDD+ Committee for Nigeria and plays an even more active role in building strong links between local communities and the FC and other stakeholders in improving livelihoods. For example, he recently played a key role in the delivery of state-wide training in innovative livelihood options in 2011 including new techniques for bee keeping and snail farming, initially rolled out under the CRSCFP budget. While the CRSCFP did not have, for example, bee keeping as a formal output, it seemed to acquire local value and grew significantly, initially in the Okroshie community and now statewide. The introduction of the new technology for bee keeping locally has formed an important plank in the development of alternative livelihoods in the state and in enhancing biodiversity. Rather than being an effective output, this could be considered a project impact. This shows the key role that the capability of local system builders can play in developing and maintaining local community capacity through institutional innovation, including roles and networks' creation, re-formulation, or maintenance. These local social networks have become the "locus for behavioral change with respect to governance and land-use of the forest asset" (Hajek et al., 2011), which will be essential for project origination and development under the REDD+ initiative.

CONCLUSION AND RECOMMENDATIONS

In CRS, Nigeria, community forestry conservation – a specific form of the application of conservation biology – emerged from the tensions between top-down, institutional restructuring programs capturing significant natural resources on one hand and global international development trends on the other. The projects, which aimed to involve local people in CRS in community conservation and sustainable management, were flawed due to a mismatch between local and international development values attributed to forests amplified by market and institutional failures. The main lesson thus suggests that re-valuation of forest resources and building new value systems bridging together institutions, markets, and technologies (Hajek et al., 2011) are key success factors in future projects. While project outputs were sometimes not met, the projects' legacy seems to have been the impact on local capacity visible in a cohort of local, regional, and national system builders. This capacity is unevenly spread across the forest value chains and geographies. The challenge for REDD+ is to find, draw on, and support these visionaries, institutional innovators, and value builders to continue expanding and strengthening their networks. The specific lessons learnt in this context include:

1. Learn from past projects and allow a period for an extensive project design study, which involves local system builders who invariably know how to innovate institutions and behavioral norms by working bottom up. Local system builders have the know-how to marshal resources and develop economic models for forest value with local participation, and they can play a key role in both design and implementation.

2. Thus, socioeconomic research has to be an integral part of any conservation biology research, bringing the system builders in to help define the questions and parameters of the research. This is an effective way to ensure cultural, technical, and economic value alignments between local, regional, national, and international understandings.

3. A more effective capital-intensive timeline for future projects should focus for at least the first two years on intensive research, design, engagement, and value matching. This intensive period will hand over to partnerships of governmental and non-governmental institutions, private businesses, and community organizations, which will engage in a process of relationship building, role and institutional innovations, and further model development embedding sustainable conservation for the future.

REFERENCES

Alashi, S. (1999) National parks and biodiversity conservation: problems with participatory forestry management. *Review of African Political Economy*, **26**, 140–144.

Aspen Institute (1996) *Measuring community capacity building: a workbook-in-progress for rural communities.* Aspen Institute Rural Economic Policy Program, Queenstown, MD.

Bergl, R. A., Oates, J. F. and Fotso, R. (2007) Distribution and protected area coverage of endemic taxa in West Africa's Biafran forests and highlands. *Biological Conservation*, **134**, 195–208.

Cernea, M. (1993) Strategy options for participatory reforestation: focus on the social actors. *Regional Development Dialogue*, **14**, 3–33.

Davis, S. M., Gunderson, L. H., Park, W. A., Richardson, J. R. and Mattson, J. E. (1994) Landscape dimension, composition, and function in a changing Everglades ecosystem, in *Everglades: The Ecosystem and Its Restoration* (eds S. M. Davis and J. C. Ogden), St. Lucie Press, Delray Beach, FL, pp. 419–444.

Dawes, R. M. (1973) The commons dilemma game: an n-person mixed-motive game with a dominating strategy for defection; ORI Research Bulletin, in *Governing the Commons: The Evolution of Institutions for Collective Action* (ed. E. Ostrom), Cambridge University Press, New York, NY.

Dawes, R. M. (1975) Formal models of dilemmas in social decision making, in *Governing the Commons: The Evolution of Institutions for Collective Action* (ed. E. Ostrom), Cambridge University Press, New York, NY.

Gray, L. C. and Moseley, W. G. (2005) A geographical perspective on poverty-environment interactions. *The Geographical Journal*, **171**.

Gunter, J. (2000) Creating the conditions for sustainable community forestry in BC: a case study of the Kaslo and District Community Forest. Master's thesis. Burnaby: Simon Fraser University.

Hajek, F., Ventresca, M., Scriven, J. and Castro, A. (2011) Regime building for REDD+: evidence from a cluster of local initiatives in south eastern Peru. *Journal of Environmental Science and Policy*, **14**, 201–215.

Hardin, G. (1968) The tragedy of the commons. *Science*, **162**, December, 1243–1248.

Lane, M. B. and McDonald, G. (2005) Community-based environmental planning: operational dilemmas, planning principles and possible remedies. *Journal of Environmental Planning and Management*, **48**, 709–731.

Litke, S. and Day, J. (1998) Building local capacity for stewardship and sustainability: the role of community-based watershed management in Chilliwack, British Columbia. *Environments*, **25**, 91–109.

Markey, S. (ed.) (2005) *Second Growth: Community Economic Development in Rural British Columbia*. University of British Columbia Press, Vancouver, Canada.

McGuire, M., Rubin, B., Agranoff, R. and Richards, C. (1994) Building development capacity in nonmetropolitan communities. *Public Administration Review*, **54**, 426–433.

Mehta, J. and Kellert, S. (1998) Local attitudes toward community-based conservation policy and programs. *Environmental Conservation*, **25**, 320–333.

M'Gonigle, M. and Parfitt, B. (1994) *Forestopia: a practical guide to the New Forest Economy*, Harbour Publishing, Madeira Park, B.C.

Myers, N., Mittermeier, R. A., Mittermeier, C. G., da Fonseca, G. A. B. and Kent, J. (2000) Biodiversity hotspots for conservation priorities. *Nature*, **403**, 853–858.

Newman, S. (2012) *Understanding and maximizing impact in tropical forestry interventions, Oxford Centre for Tropical Forests Annual Members Day*, March 2,m 2012, OUCE, University of Oxford, Oxford. http://www.tropicalforests.ox.ac.uk/sites/tropicalforests.ox.ac.uk/files/Steven%20Newman.pdf (accessed March 19, 2013).

Oates, J. F., Bergl, R. A. and Linder, J. M. (2004) Africa's Gulf of Guinea forests: biodiversity patterns and conservation implications. *Advances in Applied Biodiversity Science*, **6**, 1–90.

Olson, M. (1971) *The Logic of Collective Action: Public Goods and the Theory of Groups*. Harvard University Press, Cambridge, MA.

Ostrom, E. (1999) *Self Governance and Forest Resources*. CIFOR Occasional Paper no. 20.

Population Reference Bureau (2011) *2011 World Population Data Sheet*. http://www.prb.org/pdf11/2011population-data-sheet_eng.pdf (accessed March 19, 2013).

Ramirez, R., (1999) Value co-production: intellectual origins and implications for practice and research. *Strategic Management Journal*, **20**, 49–65.

Runge, C. F. (1981) Common property externalities: isolation, assurance and resource depletion in a traditional grazing context. *American Journal of Agricultural Economics*, **63**, 595–606.

Sharma, R. (1993) The socioeconomic evaluation of social forestry policy in India. *Ambio*, **22**, 219–224.

Skutsch, M. (2000) Conflict management and participation in community forestry. *Agroforestry Systems*, **48**, 189–206.

Tewari, D. D. and Isemonger, A. G. (1998) Joint forestry management in south Gujarat, India: a case of successful community development. *Community Development Journal*, **33**, 22–40.

Ventresca, M. J. and Hajek, F. (2010) Technology-Market-Organization (T-M-O): strategies for study, policy and action in dynamic technology fields. Course note, Saïd Business School.

Vodden, K. (1999) Nanwakola: co-management and sustainable community economic development in a B.C. fishing village. MA dissertation. Burnaby: Department of Geography, Simon Fraser University.

Younis, T. (1997) Bottom-up implementation after Rio: rural community participation in Scottish forestry. *Community Development Journal*, **32**, 299–331.

CHAPTER 8

SHADES OF GREEN: CONSERVATION IN THE DEVELOPING ENVIRONMENT OF TANZANIA

Flora I. Tibazarwa[1] and Roy E. Gereau[2]

[1]Department of Botany, University of Dar es Salaam, Dar es Salaam, Tanzania
[2]Missouri Botanical Garden, St. Louis, MO, USA

SUMMARY

Many developing countries like Tanzania are endowed with natural resources that are needed for sustenance of the poverty-stricken populaces. Balancing the benefits of development that is needed to move the population out of poverty with the threat of loss or degradation of the natural resources on which the communities depend requires a complex and well-informed decision-making process. In this chapter, two conservation areas, Lake Natron and Kitulo Plateau, are used to exemplify the challenges and delineate best practices, with Kitulo presenting a notable example of conservation to be emulated. With the explosive population growth projected for Tanzania over the coming decades, the stakes for human welfare, biological conservation, and the achievement of sustainability are very high.

INTRODUCTION

"Meeting the needs of the present without compromising the ability of future generations to meet their own needs" is commonly referred to as sustainable development, a concept that is constantly being revised. For development to be sustainable, it is argued that it should be equitable and balanced; that is, including the interests of the different stakeholders in the three inter-related areas of economic, environmental, and social concerns, as defined by the World Bank in 2004 (Soubbotina, 2004).

For all countries, sustainable development is an accepted underlying best-practice principle for economic growth, increasingly so as the global community strives towards achieving the UN millennium development goals (IIED, 2000). However, for developing countries such as Tanzania, the promotion of sustainable development in light of the realities of challenges to survival in both rural and urban settings and the dynamics of environmental, social (including political), and economic dimensions presents scenarios for discussion on where the balance between present and future needs lies.

With a population of over 48 million people, projected to grow to more than 138 million by 2050, Tanzania has a booming population, of which an estimated 88% live on less than $2 per day (Population Reference Bureau 2011, 2012). Its economy is almost solely dependent on the environment as a source of the basic human rights of food, energy, and shelter. In this context, Tanzania is ranked 32nd out of the 33 least-developed countries on the African continent and one

Conservation Biology: Voices from the Tropics, First Edition. Navjot S. Sodhi, Luke Gibson, and Peter H. Raven.
© 2013 John Wiley & Sons, Ltd. Published 2013 by John Wiley & Sons, Ltd.

of the poorest in the world by the UN Office of the High Representative for the Least Developed Countries, Landlocked Developing Countries and Small Island Developing States. The economy depends predominantly on agriculture, which accounts for more than 40% of the gross domestic product (GDP), provides 85% of exports, and employs 80% of the work force. The country is heavily dependent on donor assistance, with restructured macroeconomic and developmental policies currently starting to take root with an aim of changing the status quo (Government of Tanzania, 2008).

Tanzania is particularly conscious of the inseparable relationship between economic development and the exploitation of natural resources, and is committed to bringing about balance through strategies that outline specific means to achieve this. In this context, the National Vision 2025 acknowledges the importance of linking economic development and environmental sustainability. Likewise, the National Strategy for Growth and Reduction of Poverty (NSGRP) is outcome focused and based on the achievement of three clusters of poverty reduction outcomes that cover cross-cutting issues, including environmental protection and sustainable utilization of natural resources for development. The strategies are complemented by a comprehensive legal framework for environmental protection as evidenced by the National Environmental Policy (1997), the Environmental Management Act (2004), and the National Environmental Impact Assessment and Audit Regulations (2005). These legal instruments are supported by sector policy and legislation for natural resources (tourism, wildlife, forestry, and beekeeping), agriculture, lands, and rural and urban development, as well as several international agreements.

Tanzania is signatory to several international agreements that have a mandate to ensure the sustainable utilization of natural resources, coupled with the conservation of biodiversity for future generations. The Ramsar Convention on Wetlands is one such agreement that Tanzania has signed. Since 2000, Tanzania has designated four wetlands for protection. These include two freshwater wetlands, Moyowosi and Kilombero; a marine wetland, Rufiji-Mafia-Kilwa; and an inland soda wetland, Lake Natron.

Notwithstanding Tanzania's demonstrated commitment to conservation and the sustainable use of natural resources, the country faces socioeconomic development needs, particularly in the areas where protected areas are designated. In this chapter, we demonstrate the challenges to and opportunities for achieving sustainable development in Tanzania, based on two development projects. One is a soda ash extraction project and the other a dairy farm converted to a national park. In particular, we focus on the gray areas where benefits are considered losses and vice versa and the realities of striving towards a balance between development and conservation.

THE LAKE NATRON SODA ASH PROJECT

Lake Natron is a soda lake situated in the bottom of the Rift Valley basin at an elevation of around 600 m above sea level. It is located in the northern part of Tanzania, 2 degrees south of the equator. The northern shore of the lake touches the national boundary with Kenya, with the site distant from any town in Kenya or Tanzania. The lake's main freshwater inflow is provided by the Ewaso Ngiro River, whose catchment is in Kenya, supplemented by the Peninj and Monik Rivers, which flow off the Eastern Rift escarpment in Tanzania. According to Guest and Steven (1951), there are about 28 hot (30–50°C) alkaline springs constantly flowing into Lake Natron. The springs are postulated to be the main source of brine (diluted with fresh water) and trona (concentrate) that are observed on the lake (Norconsult, 2007b).

Lake Natron was designated a Ramsar site in July 2001 and is known to be one of the few remaining breeding sites for the Lesser Flamingo population in the Rift Valley system, which comprises approximately 1.5 million birds (Brown, 1955; Childress et al., 2004). The position of Lake Natron is self-protecting because access is limited due to the extensive volcanic ash plain that hinders driving and settlement, especially in the dry season (Figure 8.1). Furthermore, the flamingos breed in parts of Lake Natron that are not readily accessible to tourists, but feed in the other soda lakes that are more easily accessible. The water divide inflow into the lake makes up the Ramsar site boundary. Prior to its declaration as a Ramsar site, the lake and its environs were part of the Lake Natron Game Controlled Area, home to resident wildebeest populations and several other wildlife species; the current Ramsar site is surrounded by the Game Controlled Area.

Ramsar recommends that land-use activities in designated sites be those that ensure maintenance of the ecological integrity of the area; that is, the principle of "wise use" is recommended. At Lake Natron, pastoralism with isolated patches of agriculture on the alluvial

Figure 8.1 Overview of Lake Natron, photo taken from the west side of the Lake in Peninj. The mountain in the background is Mt. Gelai, which supports the only forest reserve in the area.
Photo courtesy of D. Parry (2007)

soils of the rivers, ecotourism targeting the dormant volcano Oldonyo Lengai and the Engare Sero waterfalls, and trophy hunting on the eastern side of the lake currently make up the predominant land-use activities (Norconsult, 2007a).

From the 1950s, Tanzania has contemplated the economic potential to extract soda ash from Lake Natron. These possibilities have involved various investors (Guest and Steven, 1951), and especially Toyo Soda Manufacturing in 1972–1976 and Ingenierie in 1993. Soda ash is a basic component in industry: a key component of glass and detergents, and of general use in chemical industries. The global soda ash demand growth was forecast to average 3.8% per year in the period up to 2010, with a growth rate estimated at 7% per year in the "Growth Corridor" by the Chemical Industries Intelligence Services web site (www.icis.com). In 2006, Tanzania, through the National Development Cooperation, solicited an investor, Tata Chemicals Limited, a chemical company with five soda ash mines worldwide. It commissioned a technical feasibility and environmental and social impact assessment (ESIA) study for a plant with a capacity of producing 500,000 to 1,000,000 tons of soda ash per year for a lifetime of 100 to 150 years (Norconsult, 2007a).

In addition to the revenue that would be generated from soda ash production, the plant would provide substantial socioeconomic development to an area with virtually no health centers, roads, schools, or access to safe domestic water. The pastoral Maasai communities that inhabit the Lake Natron area, particularly on the eastern lakeshore, are entirely dependent on livestock, on the lake for soda (a feed supplement), and on the *Acacia* woodlands for edible and medicinal herbs. The soda plant would attract key services, primarily for the soda extraction process and its staff, but in so doing the nearby communities would also benefit. The effects of development would be extended to what is known as the Tanga Corridor, thus revitalizing the transport system in three Regions by reinstating a defunct railway line and raising the profile of a marine port in Tanga.

The ESIA process highlighted several challenges within the current land-use practices to the preservation of the lake, and more importantly pointed to the lack of sufficient data on the dependence of the birds for breeding in the lake to support development of appropriate and adequate mitigation measures. Furthermore, there were minimal institutional capacity and mechanisms to preserve the site in the absence of an integrated management plan, which had not been developed at the time the feasibility study was done. The transboundary nature of Lake Natron also required the decision-making process for its development to be agreed upon using regional agreements in addition to Ramsar and the African Eurasian Water Bird Agreement. The ESIA thus served to raise a number of concerns voiced strongly by the regional and international community; these resulted in the government of Tanzania reconsidering approval of the project in its presented form.

As with most developments with significant economic potential, these concerns have resulted in new efforts to address and research environmental options and to develop mitigation strategies that will serve to facilitate development while maintaining the ecological integrity of the area. At present, a species-specific action plan has been developed for the Lesser Flamingo (Childress, Nagy and Hughes, 2007) and a biodiversity assessment has been conducted, though it is still under review as a draft. The data needs for assessment and planning are so extensive that some argue that it is almost inconceivable that the project can be considered at all.

THE KITULO NATIONAL PARK

Kitulo National Park is the first area in tropical Africa protected specifically for its botanical diversity (Davenport, 2002a, b). Located in the Southern Highlands of

Tanzania, the Park is predominantly montane grassland and montane forest covering an area of over 400 square kilometers. The Park includes the Kitulo Plateau and the adjoining Livingstone Forest Reserve and is administered by Tanzania National Parks (TANAPA). With a well-documented vascular plant flora rich in regional and local endemics (Gereau *et al.*, 2012), the park is an ideal tourist destination for botanists and orchid lovers (Figure 8.2).

Protection of the Kitulo Plateau's unique flora was first proposed to the Tanzanian government by the Wildlife Conservation Society (WCS), in response to documented evidence of the growing international trade in orchid tubers (as a traditional food, "chikanda") and increased hunting of small game and logging activities in the surrounding forests (Davenport and Ndangalasi, 2002). WCS has conducted several surveys and published extensively on the value of and threats

Figure 8.2 Impatiens cribbii (Grey-Wilson) Grey-Wilson, Kitulo Plateau.
Photo courtesy of C. Davidson (2008), www.FloraoftheWorld.com

to the Kitulo Plateau as part of their conservation efforts in the Southern Highlands. Orchids are protected by the Convention on International Trade in Endangered Species (CITES), but over 4 million tubers from 85 different species, including regional endemics, are traded annually over the border from Tanzania to Zambia (Davenport and Bytebier, 2004). The conversion of land to agriculture for pyrethrum and potato farming, and the invasion of natural grasslands by *Pinus patula*, threaten the orchid-rich grasslands.

Prior to its current protection status, the larger portion of the Kitulo Plateau was a national ranch for sheep and cereals (from 1965) that later (from 1977) changed to dairy farming. However, the natural and cultivated pasture species grown on the plateau were not sufficiently palatable to support dairy farming and the farm outputs declined (Davenport and Bytebier, 2004). With increased documentation of the threat from the orchid trade, in 2004 Kitulo was gazetted as a National Park to afford it the protection of a unique resource of exceptional quality.

In addition to its botanical value, Kitulo National Park is one of the only two homes of the Kipunji (*Rungwecebus kipunji*), a recently described genus and species of monkey (Davenport *et al.*, 2006), which is registered as critically endangered (Davenport *et al.*, 2008). Other rare species found in Kitulo include Tanzania's only population of the rare Denham's bustard; a breeding colony of the endangered blue swallow; and range-restricted bird species like the mountain marsh widow (*Njombe cisticola*) and Kipengere seedeater; there are also a number of endemic species of butterflies, chameleons, lizards, and frogs known from the area (Davenport, 2006).

ARGUING FOR SUSTAINABLE DEVELOPMENT AND CONSERVATION

Lake Natron and Kitulo National Park are protected for their biodiversity and conservation value under Tanzanian policies and legislation for natural resource management. Ramsar sites are under the portfolio of the Division of Wildlife in the Ministry of Natural Resources and Tourism and national parks are under TANAPA, a government agency that is semi-autonomous but subscribes to national policies and legislation for natural resource management.

The protection of orchids under CITES and gazetting of the Kitulo National Park enable the enforcement of laws to ensure that agricultural expansion is checked, the trade in orchids is reduced, and strategies for the removal of the rapidly spreading *Pinus patula* are established. The protection of the botanical diversity of Kitulo National Park became possible largely because of the wealth of scientific data available for the area; this information facilitated the discussion on how best to preserve the area in line with its socioeconomic development priorities. The protection of Kitulo, therefore, is based on a relatively straightforward and acceptable argument for conservation. It also must be said, however, that pressure for increasing agricultural land in the Kitulo area is not as intense as in other parts of the country; thus socioeconomic factors did not hinder the declaration of a Park to the same extent as would have been the case elsewhere.

The Lake Natron Ramsar site, on the other hand, has received less scientific attention, with the exception of that from Birdlife International, whose records on the flamingos support arguments for its being the only breeding site in the Eastern Rift Valley (Figure 8.1). With the exception of the project-based ESIA (Norconsult, 2007a), no other study from the area has documented its biodiversity value. ESIA reporting relies largely on existing documented information that is supported by a limited amount of fieldwork to confirm documented information and fill in knowledge gaps. Thus, following the ESIA, the species action plan for the Lesser Flamingo and a biodiversity assessment plan for the Lake Natron area were developed as concerted efforts to guide protection of the site in the absence of an approved integrated management plan.

In Tanzania, conservation is invariably linked to economic benefits to try to meet the "opportunity cost" of conservation. Kitulo National Park thus markets non-consumptive ecotourism that generates revenue to ensure its sustainable protection. Furthermore, TANAPA has a well-established community relations policy targeting communities surrounding its parks, that furthers the agency's ability to ensure protection of resources. The economic benefits from Lake Natron, on the other hand, are largely driven by central-government-projected investment that is not primarily conservation oriented. The anticipated economic returns of establishment of a soda ash facility may heavily impact the commitment of the nation towards conservation. The lack of sufficient scientific information on the area's ecological dynamics limits the arguments for its conservation status and priorities, while the precautionary approach being advocated by some

sectors does not respond to the needs for development. These challenges are further complicated by international obligations where decisions taken at the national level have international consequences – economic, social, and environmental. When these consequences are perceived as negative, the situation is sometimes referred to as "exporting unsustainability" such as in the case of Lake Natron.

There are opportunities to apply best practice from the Kitulo process to Lake Natron; that is, systematic assessment and publication of its biodiversity value. However, this requires commitment of capacitated national and international advocacy and implementing agencies to address matters in a multidisciplinary fashion, whereby equitable balance is attained for both socioeconomic development and biodiversity.

REFERENCES

Brown, L. H. (1955) The Breeding Lesser and Greater Flamingo in East Africa. *Journal of the East African Natural History Society*, **22**, 159–162.

Childress, B., Nagy, S. and Hughes, B. (2007) *International Single Species Action Plan for the Conservation of the Lesser Flamingo*. Phoenicopterus minor. AEWA Technical Series No. 34 Bonn, Germany.

Childress, B., Harper, D., Van den Bossche, W., Berthold, P. and Querner, U. (2004) Satellite tracking lesser flamingo movements in the Rift Valley, east Africa: pilot study report. *Ostrich*, **75**, 57–65.

Davenport, T. R. B. (2002a) Tanzania's new national park to protect orchids. *Oryx*, **36**, 224.

Davenport, T. R. B. (2002b) Garden of the Gods: Kitulo Plateau, a new national park for Tanzania. *Wildlife Conservation*, June, 15.

Davenport, T. R. B. (2006) Plants, primates and people: conservation in the Southern Highlands of Tanzania. *Miombo*, **28**, 7–8.

Davenport, T. R. B. and Bytebier, B. (2004) Kitulo Plateau, Tanzania: a first African park for orchids. *Orchid Review*, **112**, 161–165.

Davenport, T. R. B. and Ndangalasi, H. J. (2002) An escalating trade in orchid tubers across Tanzania's Southern Highlands: assessment, dynamics and conservation implications. *Oryx*, **36**, 55–61.

Davenport, T. R. B., De Luca, D. W., Jones, T., Mpunga, N. E., Machaga, S. J., Kitegile, A. and Picton Phillips, G. P. (2008) The critically endangered kipunji Rungwecebus kipunji of southern Tanzania: first census and conservation status assessment. *Oryx*, **42**, 352–359.

Davenport, T. R. B., Stanley, W. T., Sargis, E. J., De Luca, D. W., Mpunga, N. E., Machaga, S. J. and Olson, L. E. (2006) A new genus of African monkey, Rungwecebus: morphology, ecology, and molecular phylogenetics. *Science*, **312**, 1378–1381.

Gereau, R. E., Kajuni, A. R., Davenport T. R. B. and Ndangalasi, H. J. (2012) Lake Nyasa Climatic Region floristic checklist. Monographs in systematic Botany from the Missouri Botanical Garden, **122**, i–v, 1–118.

Government of Tanzania (2008) Millennium development goals report: mid-way evaluation 2000–2008, Government of Tanzania.

Guest, N. J. and Steven, J. A. (1951) Lake Natron, its springs, rivers, brines, and visible saline reserves. G. S. o. Tanganyika.

IIED (2000) *The Millennium Development Goals and Conservation: Managing Nature's Wealth for Society's Health*. D. Roe. Russell Press, London, p. 176.

Norconsult (2007a) Environmental and Socio-economic impact assessment for the development of a Soda Ash Facility at Lake Natron, Tanzania.

Norconsult (2007b) Hydrological Investigations to support the Extraction of Concentrated Lake Brine from Lake Natron, Tanzania.

Population Reference Bureau (2011) *2011 World Population Data Sheet*. http://www.prb.org/pdf11/2011population-data-sheet_eng.pdf (accessed March 19, 2013).

Population Reference Bureau (2012) World Population Data Sheet 2012. http://www.prb.org/pdf12/2012-population-data-sheet_eng.pdf (accessed March 25, 2012.

Soubbotina, T. (2004) *Beyond Economic Growth: An Introduction to Sustainable Development*. WBI Learning Resource Series. The International Bank for Reconstruction and Development.

CHAPTER 9

SUSTAINABLE CONSERVATION: TIME FOR AFRICA TO RETHINK THE FOUNDATION

Mwangi Githiru

Department of Zoology, National Museums of Kenya, Nairobi, Kenya;
Wildlife Works, Voi, Kenya

SUMMARY

Global environmental policy and negotiations are important. They are the arenas where we seek to distinguish ourselves from other animals by summoning our empathy and benevolence – so-called *humanity* – and work collectively towards a common goal. It is a lofty aspiration certainly worth pursuit, albeit we really have little to show for it, either from the negotiating tables or post-negotiations implementation. In the same vein, some conventional economic development theories, including sustainable development and gross domestic product-measured growth, are increasingly being questioned. One such economic hypothesis is founded around the environmental Kuznets curve, which, for Africa, implies that we are too poor to be worrying about environmental conservation. In line with Maslow's hierarchy of needs, it suggests that we should instead strive to increase consumption and accelerate development in order to increase investment in conservation. Here, I contend that this does not seem to augur well for Africa, and conservation advocacy based on these kinds of rationale shall not stand for the environment or biodiversity in the face of other competing interests. I believe that what we need in lieu, in Africa and perhaps elsewhere too, is a deeper recognition and agreement on why/what to conserve. Then,

for the things we decide to conserve, we can ponder about how to actually go about doing it. One way we can do this is by changing our tactics at the global arena from hedgehogs to foxes, and, without disregarding the future, also by starting to look back rather than forward, which for us in Africa usually means to the West.

SOMETHING IS NOT WORKING

Besides the 4-yearly World Conservation Congress, perhaps the two leading global environmental conservation efforts – the Convention on Biological Diversity (CBD) and United Nations Framework on Convention on Climate Change (UNFCCC) – hold their annual Conference of Parties (CoP) meetings at various exotic destinations around the world. It is difficult for people not directly involved in the actual negotiations, like myself, to keep track. From Bali, to Nagoya, Copenhagen, Durban, Cancún to Rio, these CoPs come and go. I have not attended any of these meetings, but, reading through some real-time blogs and commentaries coming through, not much is achieved, in essence (see e.g., Hulme, 2009). Except for a smattering of potentially exciting seminars (such as the South Korea's GreenGrowth Initiative at UNFCCC's CoP 16 in Cancún in 2010) and potentially interesting initiatives (such as the Reduced Emissions from Deforestation and forest Degradation [REDD] concept first introduced at UNFCCC CoP 13 in Bali in 2007), it is largely business as usual: another plan, another strategy, more promises, more buck passing and blame games culminating in the disturbingly familiar finger pointing.

Paradoxically, in 2010, which was the so-called International Year of Biodiversity, it was made abundantly clear that governments signed up to the CBD had missed their 2010 biodiversity target of achieving a significant reduction in the rate of biodiversity loss (Stokstad, 2010). Leading scientists concluded that, despite some local successes, the rate of biodiversity loss was not slowing (Butchart et al., 2010), declaring that progress towards achieving the 2010 targets had failed (Walpole et al., 2009). Although both biological diversity and cultural heritage (including ways of life) are indeed inherently dynamic, concern is based upon the rates at which they are currently being eroded (Pimm et al., 1995). I am puzzled: how does conservation continue to lose ground in the face of all the ardent, honest, and ever-growing conservation efforts worldwide?

THE USUAL SUSPECTS: HEDGEHOGS OR FOXES

The reasons proffered for the global inertia towards conservation leave me with a heavy sense of déjà vu. Our failure to meet the 2010 global conservation targets, for example, was blamed on the usual suspects, perhaps the commonest two being lack of "solid" evidence for the biodiversity–poverty link (e.g., Mooney and Mace, 2009) and ineffective communication with decision makers. It follows an all too familiar pattern: scientists declare the importance of scientific evidence (e.g., Balmford et al., 2005), then document conservation failure (e.g., based on IUCN Red List status), then convene to develop or adopt a strategic plan or revise some existing plans and targets (Stokstad, 2010). In effect, they basically develop a new plan of action, supported by more targets – in this current example, fresh targets for 2020 (Perrings et al., 2010): to me, doing more of the same, but expecting different results.

My take? First, solid evidence (if that can be found) and better communication will not necessarily shift the focus of the politicians, decision makers, or other involved parties. There is already considerable informa-

tion out there so it is worth asking: Are they so impervious? What if they just do not care and are simply greedy? Or, what if other equally valid, pressing concerns drive their actions? It is easy to see that more evidence or better communication, while usually desirable, may not change much. Moreover, it is all too easy to blame the government and politicians, but they are elected by people and to a great extent reflect the people's needs, and fears (except those flat-out dictators). Clearly, if voters changed significantly enough, politicians would have to change too (e.g., the recent events of the "Arab Spring"). Yet, we continue doing and blaming the same things, and expect different results. Unsurprisingly, this does not happen, so we end up with Bali, Nagoya, Cancún, Copenhagen, many prior, and many others to follow. We seem unwilling or unable to change tactics. To me, our perpetual settling on these usual suspects is a form of displacement behavior (Whitten et al., 2001), whereby we are unable to muster the actions needed to aptly address a problem that increasingly appears overwhelming. A hammer is all we seem to have at our disposal, and all problems start to look like nails. Figuratively, I think the great philosopher Isaiah Berlin would have considered us classic hedgehogs. Hedgehogs simplify life around one great idea and bet on it, more or less disregarding everything else, even shoehorning facts into something that will support their ideology. In contrast, foxes embrace uncertainty and know they are never fully in control, changing their minds when they realize they are wrong about something, or something better exists out there (Sunter and Illbury, 2001). Though both strategies do have their merits, the greatest challenge is figuring out when to switch between them, and having the ability to do so. I feel it is time conservationists altered tactics, became foxier when seeking areas where answers to the conservation problems may lie. We need to stop walking north on this southbound train (Orr, 2003); exit the train and try something fresh. Besides lack of evidence and better communication, sustainable development is another mammoth hedgehog idea.

SUSTAINABLE DEVELOPMENT

The poverty–conservation nexus

Sustainable development, though defined in myriad ways, retains two core principles: that our actions must take into account effects on the environment, economy, and society, and that what we do today should not compromise the well-being of future generations (WCED, 1987, Strange and Bayley, 2008). Although once hailed as an innovative way of thinking through and managing human impacts on the world, there is little evidence that it has worked for conservation. Not only has it failed to engender a reduction in rates of environmental destruction and biodiversity loss, but I join others (e.g., Robinson, 1993; Willers, 1994; Collar, 2003) in arguing that, in terms of conservation, it is a fundamentally flawed concept that cannot lead to human beings choosing to conserve over destroying the environment when push comes to shove.

Conservationists often assert that efforts to preserve biodiversity can (or do) also benefit the people who rely on natural resources for food and income (Sanderson, 2005; Gilbert, 2010), assuming almost a causal link between biodiversity loss and poverty alleviation (Sachs *et al.*, 2009). Yet, except for indications of poverty–biodiversity overlaps (Balmford *et al.*, 2001; Sunderlin *et al.*, 2005) and a scatter of examples showing biodiversity projects improving living standards, it is at best a mixed picture, at worst a picture of failure: failure in the sense that biodiversity conservation does not actually lead to an improvement in people's lives, economically or culturally. Keller (2008) for instance, argued that biodiversity conservation in Madagascar left local people with a sense of having been defeated in the purpose of life as they understood it. Although conservation and poverty alleviation are said not be fundamentally incompatible, most ways of accomplishing poverty alleviation are not good for the environment (Gilbert, 2010).

My take? Embracing the sustainable development agenda nicely played into the economists' utilitarian hands, rendering conservationists curiously diffident to defend the core underpinning of the conservation movement (Collar, 2003). Nowadays, we are even warned that failing to view ecosystems as an economic asset is undermining our efforts towards sustainable development (Miththapala, 2008). Now, despite its inherent problems (Spangenberg and Settelea, 2010), the concept of ecosystem services is a useful one in thinking about nature; paying for them is even better! And there is nothing wrong with generating income through some ecotourism venture. But these are not the deep-seated reasons why many of us strive to conserve nature: they are useful and ought to be pursued, but as secondary reasons. Making conservation calls based wholly on them is not only innately dishonest (Collar, 2003), but also provides fertile ground for skeptics and for blundering. For instance, it was cynically noted that CBD's implicit equation that what's good for the planet is good for the poor is suspiciously convenient: "next we'll hear that biodiversity can help bring peace to the Middle East" (Tuhus-Dubrow, 2010). Own goals by conservation advocates, such as the blatantly incorrect statement in the Fourth Assessment Report of the IPCC that the Himalayan glaciers could completely disappear by 2035 (Black, 2010), and conniving acts such as "Climategate" (www.climategate.com) do not help. They render conservationists contemporary Cassandras (Redford and Sanjayan, 2003); no one seems to grasp the gravity of our calls anymore, however true they may be. These incidences of shooting our own feet are symptomatic of this inability to express and stand up for why, really, we care. Let us now evaluate a commonly flaunted basis for sustainable development-type rationale with a particular relevance for Africa, and see a real example of how it plays out, before ending with where I think possible solutions may lie.

The Environmental Kuznets Curve

One frequently invoked justification for focusing on economic development to achieve environmental conservation goals is based on the Environmental Kuznets Curve (EKC) hypothesis (Box 9.1). EKC proposes that indicators of environmental degradation first rise and then fall with increasing income per capita – that is, economic growth (Figure 9.1). This has been variously disputed (e.g., Roca, 2003; Stern, 2004; Sunderlin *et al.*, 2005), the core argument being that "the world simply does not have the resources, renewable or otherwise, to sustain Western lifestyles across the globe" (O'Hara, 2007). Yet, calls for increased consumption to underpin environmental conservation keep growing louder, even from conservationists (Arrow *et al.*, 2004; Christensen, 2005). This is despite economic growth being historically correlated closely with increased consumption especially of energy (Ayres, 1995), and besides the EKC often failing to provide a good basis for predicting consumer behavior due to externalities and displacement of environmental costs (Roca, 2003). In any case, this argument only explains a reduction in environmental pressures *per unit of GDP* as income increases, and not a straightforward reduction in the pressures in absolute terms (Roca, 2003).

Box 9.1 The Environmental Kuznets Curve

The EKC hypothesizes that the relationship between per capita income and the use of natural resources and/or the emission of wastes has an inverted U-shape. Therefore, at relatively low levels of income the use of natural resources and/or the emission of wastes increase with income. Beyond some point of inflection, the use of the natural resources and/or the emission of wastes decline with income. Reasons for this inverted U-shaped relationship are thought to include income-driven changes in: (1) the composition of production and/or consumption; (2) the preference for environmental quality; (3) institutions that are needed to internalize externalities; and/or (4) increasing returns to scale associated with pollution abatement.

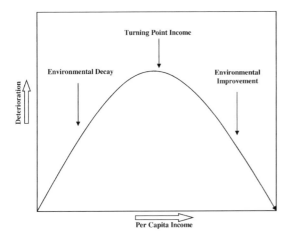

Figure 9.1 Environmental Kuznets curve schematic.

Now, little of this is really new. However, it is particularly relevant for Africa because most of the continent lies on the left side of the EKC graph, suggesting that we damage the environment because of our low per capita incomes. This has fuelled calls for stepping up economic development, the endemic claim going along

these lines: "How do you expect poor people to care about conservation? They first have to place food on the table." Calls for Africa to choose a different developmental trajectory are muffled by a lethal combination of personal and political greed, genuine quest for "a better life," plus token, often barren, technology transfer that would facilitate that. A case in point is currently playing out in Tanzania.

An EKC upshot: The Serengeti highway

The government of Tanzania is planning a major commercial highway across the Serengeti National Park, linking the Lake Victoria area in the northwest with eastern Tanzania (Figure 9.2). The proposed road will directly traverse the wildebeest migration route in Serengeti. Understandably, conservationists are up in arms asserting that increased traffic poses a great threat to the wildebeest migration and the integrity of Serengeti as the number one natural wonder of the world, *upon which Tanzania's tourism depends* (e.g., Dobson, Borner and Sinclair, 2010). In perhaps all too familiar doublespeak, the country's president promised the dissenters: "I am also a conservation ally and I assure you I'm not going to allow something that will ruin the ecosystem to be built." To this avowal, however, he also added that people residing in areas through which the road will pass through deserve better social services like improved infrastructure. Now, without a doubt, public policy is a mixed bag of competing demands; ultimately, leaning more towards the EKC's sustainable development agenda, the president gave the proposed road a nod.

To me, one thing stands out: even the dissenters base their argument ultimately on economic grounds – mainly from tourism – and not on the gnus themselves or the ecosystem. This is crucial, in my view, because for the four main, *human* stakeholders the road heralds different losses and gains: politicians (interested in votes from these communities), local communities (promise of socio-cultural transformation for the better), conservationists (research and career interests, sentimental attachment to wildlife), and businessmen and the Tanzanian economy (promised growth). Besides possible negative impacts of socio-cultural transformation, it is easy to see that only the conservationists' sentiments are a near-guaranteed negative outcome. Strictly speaking, researchers could always now set up before-and-after experiments, akin to the

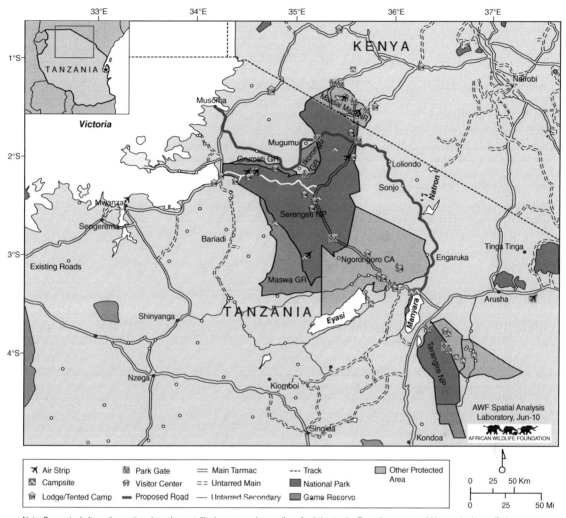

Note: Conceptual alternative routings based on satellite imagery and upgrading of existing tracks. Formal surveys would be required to verify these routes.
Sources: AWF, EAC, USGS, Tracks4Africa, SRTM, WDPA 2009.

Figure 9.2 The proposed highway (in red) cutting through Serengeti National Park in northern Tanzania.
Source: African Wildlife Foundation (AWF)

Biological Dynamics of Forest Fragment Project (BDFFP) in Brazil (e.g., Laurance *et al.*, 2004) so need not lose out, career-wise at least. The only real losers in this circumstance are the gnus and the ecosystem itself. I deliberately left them out of the stakeholders earlier because their interests are typically smothered by human interests. Yet, it is for them that conservationists should base the core of their opposition. Without their interests being strongly in the frame of negotiations and *distinguished* from their links to tourism and foreign earnings, most other arguments are likely to be ably and justifiably countered by human needs and associated economic arguments.

It is fairly easy to understand the basis of choice made by the Tanzanian president. Governments face the complex challenge of finding the right balance between the competing demands of natural and social resources, *without sacrificing economic progress* (OECD, 2001). This is often also entrenched in policy: the Environment section of Kenya's development blueprint, for example, states that "Specific strategies will involve promoting environmental conservation *in order to provide better support to the economic pillar flagship projects*" (my emphasis added) (Republic of Kenya, 2008). Thus, it is imperative to pursue economic progress as long as associated environmental costs are the minimum necessary or acceptable. But there are no thresholds set, even mentioned; no limits to growth are considered, explicitly or otherwise. As such, it is clear what environmental conservation is for, and what, therefore, should yield in case of a clash between conservation and economic progress. From the conservation perspective, this is the fundamental flaw.

A CASE FOR SUSTAINABLE CONSERVATION

A Leap of faith

John Humphrys, in his book *In God We Doubt*, tried hard to pick a proper argument with a theology tutor about theology; the tutor freely acknowledged that theology is not some sort of intellectual platform on which faith can be built, asserting:

> Theology is "faith seeking understanding" – *which means you get your faith first and then try to make sense of it.* Faith is not grounded in rational argument and neither is there any good line of reasoning that can persuade one to believe. Belief just isn't like that, and theology is not the foundation of faith. *People believe because they believe.* This is not about intellect or learning: it's more basic than that. It is both more profound and simpler. Strip from Christianity the notion of proof, evidence and historical events (or non-events) and what drives belief has little to do with the head and a great deal to do with the heart (my emphasis added). (p. 368)

Admittedly, I am taking a leap of faith here, but I trust that, likewise, people conserve simply because they conserve. At least for me, fascination with nature and the accompanying sense for protecting or preserving it came well before my understanding of how it works or ways in which we can harness it for our improved well-

being. Even if some form of biodiversity exploitation especially for threatened species is shown to be more profitable than conserving them, my bet, or hope, is that many conservationists will still opt for conservation. It is a conscious – even moral – choice that need not always be beneficial. If this is indeed the basis, then we must be bold to say so and defend it like the theology tutor, however absurd it might appear – at first.

The heart for conservation

In Business Process Reengineering, while the goal is to improve processes for service delivery, the primal question an organization asks is whether it needs a particular process in the first place; only if it deems it necessary does it bother with asking whether the process needs reengineering. Likewise, perhaps the primary question to ask in conservation ought to be whether or not we need to conserve the species or ecosystem in question. A related question was recently posed by Andrew Dobson, an ecologist at Princeton University: "Do you want your children to grow up in a world without elephants?" (Gettleman, 2012). It reminded me of a question that a friend once asked me in the middle of a pub chat: "By the way, why do we even need (to conserve) the rhino? What would happen or change for me if they went extinct?"

Many thoughts went through my mind, including Nigel Collar's paper asserting that the mere existence of species – not just the direct experience of them, but the knowledge that they are there – confers on us all an essential human liberty (Collar, 1988). In the end, I just told him that ultimately it was not, neither should it be, a human decision. I trust that rhinos, like humans, simply strive to survive and in fact, because we are their greatest threat, we should help bring them back from the brink, then let them be. Schmaltzy! Too sappy, he retorted. Needless to say, he was not much impressed with my response and I realized just how far-flung human sentiments on conservation were, given I was speaking to a Harvard-educated economist!

While it is easy to base the need to conserve on tangible uses directly linked with human well-being, it is not easy to find consensus as to whether biodiversity has value over and above what humanity confers to it, and infinitely harder to use it as a basis for conservation. Simple respect or love of nature as a conservation rallying call has been considered a hopelessly fragile stick to beat any conscience with (Collar, 1988). Yet, to me, arriving at such a consensus promises the only universal and unifying principle for truly endorsing species conservation. We should argue about how to go about undertaking conservation, but not whether or not we need to. Saying this sounds incredulously naive even to me. But I have tried to look, and this seems to me the only enduring solution. Though, philosophically, this is still a decision made by humans, it is made *despite* of them, and deliberately made in admission of adverse human impacts thus far.

This frees conservationists of the burden of proof; instead of asking "why do we need rhino?" we ask why not. Except for a few "pest" species, it is clear that, through this, conservation would easily become mainstreamed: as widely understood as hygiene, and voluntarily practiced as bathing (Western, 2000). Only transcending to such agreement level would guarantee safeguarding of non-human interests. Like faith, it is both more profound and simpler. It is an altruistic, benevolent act that speaks to our humanity and humanness, identity and social organization. As Hardin (1968) observed – more than 40 years ago! – some environmental problems just do not have a technical solution; they demand more in the way of change in human values or ideas of morality. I do not purport to provide answers here; I only hope to contribute towards clarifying this basic conservation issue, and painting a picture of a desirable destination that is crucial if we are to figure out the best way to get there.

This would be truly *prioritizing* conservation; it becomes sustainable when conservation science and technological advancements help us find the best ways of conserving and protecting that which we have decided to conserve. They also provide solutions when we have to make the inevitable sacrifices this calls for, which shows that this need not stifle innovativeness or technological advancements, as it is often claimed or feared. Nonetheless, it is apparent to me that, going forward, we must be prepared to accept lower living standards than we currently enjoy or we would otherwise attain (especially for Africa), curtailed or steady economic growth, and possibly simpler lives. The upside is that this does not necessarily imply wretchedness; in fact it might turn out that people will be happier, with more leisure time (and great places to spend it), and will rest easy and fulfilled, knowing it is all by choice. Because if we do not attain this by deliberate choice, the tragedy of the commons seems set to remorselessly play out (Hardin, 1968), leaving us, anyway, in an impoverished world of lower incomes,

few places to enjoy it (due to degradation and conflicts), and the distress of yearning for what will be by then lost forever.

Yet, perhaps like most people reading this, I am privileged enough to be thinking – and worrying – about conservation of biodiversity from such a seemingly sentimental basis. My world is different from that of the rural farmer grappling with the elephant menace; it is also different from the politician who is seeking votes, the economist interested in dollars, and quite different from the average European teenager who cares more about Pokémon than rhinos (Balmford *et al.*, 2002). The challenge I see is how to reconcile the worlds of these people, and rally them around this single intention; how to connect our diverse perceptions, needs, and fears. Indeed, one of the most important unanswered questions in evolutionary biology, and social sciences, is how cooperative behavior evolved and can be maintained in human groups and societies (May, 2005). We certainly need some good, pragmatic ideas that the majority can actually live with, before figuring out how to rally public support around them (Vasi, 2012).

CONCLUSION: WHICH WAY FOR AFRICA?

A few years ago while on the Archbishop Tutu Africa Leadership Programme, Graça Machel engaged us in a conversation where she stressed the importance of celebrating our "Africaness" in our leadership. She had just opened a can of worms that led to long, heated discussions! On the one hand, as pertains to the oft-mentioned "African leadership" and "leadership in Africa" – debate centered around if there really is such a thing as African leadership. On the other hand, albeit sub-Saharan Africa is too big and diverse for a single African leadership style to be defined, we questioned whether there are more similarities between the way a Ghanaian and a Zimbabwean lead their subordinates than there may be between a Ghanaian and a German or an Italian. *Just what is this Africaness?*

While the peoples and cultures across Africa are diverse indeed, I also find them peculiarly similar in several basic foundations including spirituality, adherence to societal norms, and submission to elders. Usually, these engender almost-complete loyalty; subservience to some form of authority (whether spiritual, political, eldership, or other) seems to me ubiquitous across Africa (see e.g., Bernard and Kumalo, 2004). For instance, in the run-up to the last Rwandese elections, one diplomat noted that "it is in Rwandese culture to listen to authority." Similarly, in recognition of religion's potency, some people have also recently questioned whether religion could indeed save Africa's wildlife (Straziuso, 2012). Typically, the external locus of control is much stronger than the internal one in Africa, which can be attributed to these elements of spirituality, subservience, and societal dominance over the individual. For me, therein lies an opportunity to harness in spreading a conservation agenda and ensuring it is ingrained throughout society and freely practiced as a way of life as suggested by Western (2000).

In many traditional African systems, the link between leadership, land, and life was undisputed, implicit, and strong. Indeed, Africa is said to have had complex ecologically-based politico-religious structures governed by the "African philosophy of the earth" (Bernard and Kumalo, 2004). Scientifically-informed and Western notions of landscape are more concerned with its ecological, functional, or aesthetic values, rather than its more complex relational (spiritual, political, and social) representations within such a historical framework. Today, in the murky waters of development and conservation, Africa seems lost midstream. We do not know how to get back to shore (these traditional systems) and the destination (whether West or, as is increasingly the case, East) either appears further than we imagined, or does not seem all that appealing the closer we get. Do we have to keep moving? To me the best option is to stop! Figure out how to best remain where we are using what we have and know – or used to know – then decide how and where we need to go.

It is difficult to conjure up a happy ending. The challenge is stark, and, besides my pub conversation alluded to earlier, had previously been driven home to me following an article arguing that unsustainable lifestyles were our Achilles' heel (O'Hara, 2007). The responses (and there were 50+ fast and often furious ones), though seemingly knee-jerk and at times not very well informed, contained some very germane points: they embraced the fears of the willing, the mirth of the skeptics, and the crushing despair of the concerned. But this is the reality of what those of us who consider ourselves as conservationists must contend with, the hearts we must win, the solutions or answers we must provide in order to ever clinch lasting conservation. As the late Kenyan Nobel Laureate Wangari Maathai

(1940–2011) once said, those of us who understand the complex concept of the environment have the *burden to act*. We must not tire, we must not give up, we must persist; like the fabled hummingbird, we each must do what we can.

REFERENCES

Arrow, K., Dasgupta, P., Goulder, L., Daily, G., Ehrlich, P., Heal, G., Levin, S., Mäler K.-G., Schneider, S., Starrett, D. and Walker, B. (2004) Are we consuming too much? *Journal of Economic Perspectives*, **18**, 147–172.

Ayres, R. U. (1995) Economic growth: politically necessary but not environmentally friendly. *Ecological Economics*, **15**, 97–99.

Balmford, A., Clegg, L., Coulson, T. and Taylor, J. (2002) Why conservationists should heed Pokémon. *Science*, **295**, 2367.

Balmford, A., Moore, J. L., Brooks, T., Burgess, N. D., Hansen, L. A., Williams, P. and Rahbek, C. (2001) Conservation conflicts across Africa. *Science*, **291**, 2616–2619.

Balmford, A., Bennun, L., ten Brink, B., Cooper, D., Côté, I. M., Crane, P., Dobson, A., Dudley, N., Dutton, I., Green, R. E., Gregory, R. D., Harrison, J., Kennedy, E. T., Kremen, C., Leader-Williams, N., Lovejoy, T. E., Mace, G., May, T., Mayaux, P., Morling, P., Phillips, J., Redford, K., Ricketts, T. H., Rodríguez, J. P., Sanjayan, M., Schei, P. J., van Jaarsveld, A. S. and Walther, B. A. (2005) The Convention on Biological Diversity's 2010 target. *Science*, **307**, 212–213.

Bernard, P. and Kumalo, S. (2004) Community-based natural resource management, traditional governance and spiritual ecology in southern Africa: the case of chiefs, diviners and spirit mediums, in *Rights, Resources and Rural Development: Community-based Natural Resource Management in Southern Africa* (eds C. Fabricius and E. Koch), Earthscan, London, pp. 115–126.

Black, R. (2010) UN climate body admits "mistake" on Himalayan glaciers. *BBC News*. http://news.bbc.co.uk/2/hi/8468358.stm (accessed March 19, 2013).

Butchart, S. H. M., Walpole, M., Collen, B., van Strien, A., Scharlemann, J. P. W., Almond, R. E. A., Baillie, J. E. M., Bomhard, B., Brown, C., Bruno, J., Carpenter, K. E., Carr, G. M., Chanson, J., Chenery, A. M., Csirke, J., Davidson, N. C., Dentener, F., Foster, M., Galli, A., Galloway, J. N., Genovesi, P., Gregory, R. D., Hockings, M., Kapos, V., Lamarque, J.-F., Leverington, F., Loh, J., McGeoch, M. A., McRae, L., Minasyan, A., Morcillo, M. H., Oldfield, T. E. E., Pauly, D., Quader, S., Revenga, C., Sauer. J. R., Skolnik, B., Spear, D., Stanwell-Smith, D., Stuart, S. N., Symes, A., Tierney, M., Tyrrell, T. D., Vié, J.-C. and Watson, R. (2010) Global biodiversity: indicators of recent declines. *Science*, **328**, 1164–1168.

Christensen, J. (2005) Are we consuming too much? *Conservation in Practice*. **6**, 15–19.

Collar, N. J. (1988) Life, liberty and the pursuit of happiness. *American Birds.* **42**, 19–22.

Collar, N. J. (2003) Beyond value: biodiversity and the freedom of the mind. *Global Ecology and Biogeography,* **12**, 265–269.

Dobson, A., Borner, M. and Sinclair, A.R.E. (2010) Road will ruin Serengeti. *Nature,* **467**, 272–274.

Gettleman, J. (2012) Elephants dying in epic frenzy as ivory fuels wars and profits. *The New York Times.* http://www.nytimes.com/2012/09/04/world/africa/africas-elephants-are-being-slaughtered-in-poaching-frenzy.html (accessed March 19, 2013).

Gilbert, N. (2010) Can conservation cut poverty? *Nature,* **467**, 264–265.

Hardin, G. (1968) The tragedy of the commons. *Science,* **162**, 1243–1248.

Hulme, M. (2009) What message, and whose, from Copenhagen? *BBC News.* http://news.bbc.co.uk/2/hi/science/nature/7946476.stm (accessed March 19, 2013).

Humphrys, J. (2008) *In God We Doubt: Confessions of a Failed Atheist,* Hodder & Stoughton, London.

Keller, E. (2008) The banana plant and the moon: conservation and the Malagasy ethos of life in Masoala, Madagascar. *American Ethnologist,* **35**, 650–664.

Laurance, W., Mesquita, R., Luizao, R. and Pinto, F. (2004) The biological dynamics of forest fragments project: 25 years of research in the Brazilian Amazon. *Biotropica,* **15**, 1–3.

Maathai, W. (2011) http://www.ciel.org/Staff_Bios/Maathai.html (accessed March 19, 2013).

May, R. M. (2005) Threats to tomorrow's world. The Royal Society Anniversary Address 2005. http://royalsociety.org/uploadedFiles/Royal_Society_Content/about-us/history/Anniversary_Address_2005.pdf (accessed March 19, 2013).

Miththapala, S. (2008) Natural capital. *World Conservation Magazine,* **38**, 21–22.

Mooney, H. and Mace, G. (2009) Biodiversity Policy Challenges. *Science,* **325**, 1474.

O'Hara, E. (2007) Focus on carbon "missing the point". *BBC News.* http://news.bbc.co.uk/2/hi/science/nature/6922065.stm (accessed March 19, 2013).

OECD (2001) Sustainable development: critical issues. OECD. http://www.oecd.org/science/inno/1890501.pdf (accessed 19 March 2013).

Orr, D. W. (2003) Walking North on a Southbound train. *Conservation Biology,* **17**, 348–351.

Perrings, C., Naeem, S., Ahrestani, F., Bunker, D. E., Burkill, P., Canziani, G., Elmqvist, T., Ferrati, R., Fuhrman, J., Jaksic, F., Kawabata, Z., Kinzig, A., Mace, G. M., Milano, F., Mooney, H., Prieur-Richard, A.-H., Tschirhart, J. and Weisser, W. (2010) Ecosystem services for 2020. *Science,* **330**, 323–324.

Pimm, S. L., Russell, G. J., Gittleman, J. L. and Brooks, T. M. (1995) The future of biodiversity. *Science,* **265**, 347–350.

Redford, K. and Sanjayan, M. A. (2003) Retiring Cassandra. *Conservation Biology,* **17**, 1473–1474.

Republic of Kenya (2008) *First Medium Term Plan (2008–2012), Kenya Vision 2030.* Government Printer, Nairobi, Kenya.

Robinson, J. G. (1993) The limits to caring: sustainable living and the loss of biodiversity. *Conservation Biology,* **7**, 20–28.

Roca, J. (2003) Do individual preferences explain the Environmental Kuznets curve? *Ecological Economics,* **45**, 3–10.

Sachs, J. D., Baillie, J. E. M., Sutherland, W. J., Armsworth, P. R., Ash, N., Beddington, J., Blackburn, T. M., Collen, B., Gardiner, B., Gaston, K. J., Godfray, H. C. J., Green, R. E., Harvey, P. H., House, B., Knapp, S., Kümpel, N. F., Macdonald, D. W., Mace, G. M., Mallet, J., Matthews, A., May, R. M., Petchey, O., Purvis, A., Roe1, D., Safi, K., Turner, K., Walpole, M., Watson, R. and Jones, K. E. (2009) Biodiversity conservation and the Millennium Development Goals. *Science,* **325**, 1502–1503.

Sanderson, S. (2005) Poverty and conservation: the new century's "peasant question?" *World Development,* **33**, 323–332.

Spangenberg, J. H. and Settelea, J. (2010) Precisely incorrect? Monetising the value of ecosystem services. *Ecological Complexity,* **7**, 327–337.

Stern, D. I. (2004) The rise and fall of the environmental Kuznets curve. *World Development,* **32**, 1419–1439.

Stokstad, E. (2010) Despite progress, biodiversity declines. *Science,* **329**, 1272–1273.

Strange, T. and Bayley, A. (2008) *Sustainable Development: Linking Economy, Society, Environment.* OECD Publishing, Paris, France.

Straziuso, J. (2012) Can religion save Africa's elephants and rhinos? *Seattle Times.* http://seattletimes.com/html/nationworld/2019220918_apafricareligionvspoachers.html (accessed March 19, 2013).

Sunderlin, W. D., Angelsen, A., Belcher, B., Burgers, P., Nasi, R., Santosa, L. and Wunder, S. (2005) Livelihoods, forests, and conservation in developing countries: an overview. *World Development,* **33**, 1383–1402.

Sunter, C. and Illbury, C. (2001) *The Mind of a Fox: Scenario Planning in Action.* Human & Rousseau, Cape Town, SA.

Tuhus-Dubrow, R. (2010) Seeing double-green: can biodiversity conservation reduce poverty? *Slate.* http://www.slate.com/articles/health_and_science/green_room/2010/10/seeing_doublegreen.html (accessed March 19, 2013).

Vasi, I. B. (2012) Public support for sustainable sevelopment: a mile wide, but how deep? *Consilience: The Journal of Sustainable Development,* **8**, 153–170.

Walpole, M., Almond, R. E. A., Besançon, C., Butchart, S. H. M., Campbell-Lendrum, D., Carr, G. M., Collen, B., Collette, L., Davidson, N. C., Dulloo, E., Fazel, A. M., Galloway, J. N., Gill, M., Goverse, T., Hockings, M., Leaman, D. J., Morgan, D. H. W., Revenga, C., Rickwood, C. J., Schutyser, F., Simons,

S., Stattersfield, A. J., Tyrrell, T. D., Vié, J. C. and Zimsky, M. (2009) Tracking progress toward the 2010 biodiversity target and beyond. *Science*, **325**, 1503–1504.

WCED (1987) *Our Common Future: Report of the World Commission on Environment and Development*. UN World Commission on Environment and Development, WCED, Switzerland.

Western, D. (2000) Conservation in a human-dominated world. *Issues in Science and Technology*, **Spring**, 1–14.

Whitten, T., Holmes, D. and MacKinnon, K. (2001) Conservation biology: a displacement behaviour for academia? *Conservation Biology*, **15**, 1–3.

Willers, B. (1994) Sustainable development: A new world deception. *Conservation Biology*, **8**, 1146–1148.

Section 2: Americas

CHAPTER 10

CHALLENGES AND OPPORTUNITIES FOR BRIDGING THE RESEARCH-IMPLEMENTATION GAP IN ECOLOGICAL SCIENCE AND MANAGEMENT IN BRAZIL

Renata Pardini[1], Pedro L.B. da Rocha[2], Charbel El-Hani[2] and Flavia Pardini[3]

[1]Departamento de Zoologia, Instituto de Biociências, Universidade de São Paulo, São Paulo, SP, Brazil
[2]Programa de Pós-Graduação em Ecologia e Biomonitoramento, Universidade Federal da Bahia, Salvador, BA, Brazil
[3]Página 22, São Paulo, SP, Brazil

SUMMARY

In this chapter, we argue that Brazil faces the challenge, but also has the opportunity, of producing creative and effective solutions to bridge the research–implementation gap in the effort to solve environmental problems and achieve sustainable development. We discuss this proposition from the perspectives of the uncertainties of ecological knowledge and the general failure in translating such knowledge into powerful management tools. We claim that awareness on both the limits of knowledge and the central role of questioning to learning, science, and conscientious decision making, in conjunction with the creation of collaborative teams gathering students, researchers, and practitioners, represents a fruitful strategy to bridge the research–implementation gap. To exemplify such endeavor, we describe a set of activities and results achieved at the Federal University of Bahia through the Graduate Studies in Ecology and Biomonitoring.

INTRODUCTION

Despite the respect and trust that society as a whole holds in the promise that science can help solve environmental problems, societal and political inaction remains a great obstacle to the resolution of the complex and drastic environmental problems we face today (Groffman et al., 2010). Part of the explanation to this apparent paradox lies in what has been called the research–implementation, research–practice, or knowing–doing gap; that is, the fact that scientific knowledge is usually neither driven by an effort to solve particular real-world problems nor effectively communicated and transferred to society in general and to decision makers in particular (Knight *et al.*, 2008; Shackleton, Cundill and Knight, 2009; and references therein). This is not restricted to ecological and environmental sciences; it is indeed a widely acknowledged problem hampering the application of scientific knowledge to the needs of society in general (Knight *et al.*, 2008; Shackleton, Cundill and Knight, 2009).

The research-implementation gap might be even more pressing in developing countries such as Brazil than elsewhere. Brazil not only harbors the largest area of tropical forest in the world (Amazonia), but it also holds great biological diversity in this biome and six others in terrestrial and marine environments. Awareness of how valuable such assets are for present and future generations is growing, and environmental issues, as well as their connections to society and the economy, have been increasingly highlighted in the Brazilian media. Presidential elections in 2010 put environmental matters at the center of the political agenda: Marina Silva, a former environment minister, won 20% of the votes in the first round with an electoral platform based on sustainability issues. Following decades of investments in higher education and research, research institutes, government bodies, and non-governmental organizations are better structured and increasingly well qualified to deal with environmental challenges. There is, however, still much to do. Poverty dominates in several regions, democracy is still considered "flawed" (Economist Intelligence Unit, 2010), and powerful lobbies continue to challenge environmental legislation in Congress (e.g., Metzger *et al.*, 2010). Because of its natural wealth, its young and growing science, and the deep social challenges it faces, Brazil is well positioned to reap the opportunities that can arise from bridging the research-implementation gap in the pursuit of sustainable development.

Here we explore the idea that some of the greatest challenges and opportunities for nature conservation and sustainable development in Brazil lie in the development of an applied ecological science of high quality and effectiveness. We envision this development taking place in close association with graduate education. By high quality, we mean a science that, by recognizing the limits of our knowledge concerning our complex ecological systems, stimulates questioning and innovation and develops protocols to deal with the problems faced by decision makers: urgency, limited resources, and uncertainty. A science that can provide a sound base for decision making in the face of limited knowledge and uncertainty. By high effectiveness, we mean

a science that flourishes by taking into account and nourishing the horizontal interaction and mutual learning between academic and practical knowledge, as well as between the scientific community and decision-making institutions. Although other factors also contribute to the societal and political inaction regarding environmental issues, we believe that the ways scientific knowledge is built and disseminated is a major concern, and one that scientists are in position to confront.

CHALLENGES AND OPPORTUNITIES

Given the ongoing expansion of its agricultural and urban frontiers, it is urgent that Brazil develops a high-quality and effective ecological science. The challenge is to do so in the face of the current paucity of knowledge on the structure, function, and resilience of tropical ecosystems, and of the dissociation of such knowledge from management practices. Nevertheless, the combination of rich biodiversity with young and growing science and scientific institutions can also be viewed as an opportunity.

The quality of ecological science: Limits of knowledge and scientific uncertainty

The application of ecological knowledge to environmental management practices in Brazil is indeed a particularly good example of the importance of recognizing the limits of knowledge and thus acting with prudence (Sousa Santos, 2007; Vitek and Jackson, 2008), as well as of dealing with uncertainty and being aware of surprises (e.g., Peterson, Cumming and Carpenter, 2003). This is because our tropical ecosystems are not only complex – systems for which predictions and forecasts are unusually difficult given the large number of drivers, non-linear relationships and interactions among them, contingency on the particular context, and interactions at different temporal and spatial scales (Carpenter, 2002) – but also very poorly known. Basic information such as which or how many species are present is often limited and is generally biased toward particular groups of organisms (Lewinsohn and Prado, 2005). There are few models of the dynamics and functioning of these ecosystems, and the relationship between diversity and function or resilience is poorly known (Scarano, 2007). Obviously the paucity of knowledge greatly increases uncertainty about the responses of our complex ecological systems to human disturbances or management.

However, this knowledge is clearly increasing. Despite the relatively short history of Brazilian ecological science, the country is experiencing exponential growth in the quantity and scientific impact of research (Scarano, 2008), and increasing numbers of scientists are dedicating themselves to studying the impacts of human activities on ecological systems (Grelle et al., 2009). The number of all graduate courses in ecology and environment in Brazil grew from three in 1976 (Martins et al., 2007) to 37 in 2010 (CAPES, 2011), doubling with each decade (Scarano, 2008). The latest data show that 343 students completed their doctorates in the discipline between 2007 and 2009, while 1081 students obtained a Master's degree (CAPES, 2011). The impact of research also increased – 6.5% of 4900 papers published by Brazilian scientists in those three years were accepted in high-impact international journals. As a whole, publications by Brazilian scientists in ecology and environment rank twentieth in the world in terms of number of citations (CAPES, 2011), and Brazil's ambition is to become first in the world in producing ecological information and knowledge (Scarano, 2008).

Given the research-implementation gap, such increase in quantity and impact of research does not necessarily result in effective solving of practical environmental problems. However, the numbers suggest that adjustments to such young and growing science institutions should have an impact on building sound environmental policy (see "Recommendations").

The effectiveness of ecological science: The need for horizontal interaction with decision makers

Although ecological knowledge is surely relevant to decision making (Carpenter and Folke, 2006) and academia is the best equipped societal sector to gather ecological information (Whitmer et al., 2010), the paucity of such knowledge is not the main problem preventing action. Far more influential is the way in which knowledge is usually constructed, decoupled from practice as well as from stakeholders and decision makers (i.e., knowing is not sufficient for doing, Knight et al., 2008; Shackleton, Cundill and Knight, 2009). This is at the base of the research-implementation gap,

limiting not only the access to and the utility of the ecological knowledge, but more importantly the potential for learning and adaptation as ways of dealing with uncertainties inherently linked to the management of complex socio-ecological systems (Knight et al., 2008; Shackleton, Cundill and Knight, 2009). The vertical, one-way transference of knowledge or skills from scientists to practitioners is insufficient to bridge this gap (Knight et al., 2008), since decision making depends fundamentally on the practical knowledge, skills, and behavior of the individuals who are directly involved (Shackleton, Cundill and Knight, 2009). To do a better job in this area, we need to reevaluate the approaches currently used to disseminate the results of scientific research.

Although higher education in ecology in Brazil is still mostly academically oriented, and does not train students or scholars specifically for engagement and communication with society, this picture could change in a few years in view of recent governmental stimuli for the establishment of partnerships between researchers and applied sectors. One of the first initiatives of this kind was the creation of a new modality of graduate education in the National System of Graduate Studies – the Professional Master courses. They are suited to help bridge the gap between academic research and decision making because they allow this connection to be built during the training of people who already are (or wish to be) engaged in decision making (Box 10.1; Scarano and Oliveira, 2005; Scarano, 2007). The first two such Brazilian courses in ecology are only a few years old. Only one of them is offered by a public university in close association with academic courses (Graduate Studies in Ecology and Biomonitoring, Federal University of Bahia, www.ecologia.ufba.br, Box 10.1). However, similar courses are recently being established in other universities.

Nonetheless, the limited federal funding to professional courses compared with academic courses brings serious difficulties to their full implementation. In public universities, where tuition must be free, partnerships with other private or public institutions interested in training their employees would offer one way of funding these professional courses. Since the criteria for admission in public universities must be strictly merit-based, however, it is not possible to guarantee positions to the employees of the sponsoring institutions, making the establishment of such partnerships difficult. Consequently, although professional courses might represent one of the instruments most suited to

Box 10.1 An example of a strategy to bridge the research-implementation gap inside universities

A group of researchers from the Federal University of Bahia realized that about a quarter of its graduates were or had been working for state environmental agencies. They often returned to the university, expressing difficulty in applying acquired scientific knowledge to the real problems they faced at work. It was clear that the help they required would demand an integrated and institutionalized strategy to change teaching and training practices. The strategy included: (a) short-term courses and outreach projects in which practitioners together with students from the academic graduate course in Ecology and Biomonitoring developed scientifically informed solutions to concrete problems; (b) the creation of a professional master course (Ecology applied to Environmental Management) for practitioners; (c) the adoption of problem-based learning methods as the main tool in all these activities, since it increases curiosity, helps perceiving the curriculum as pertinent to professional activities, and integrates learning from different components (Barrett and Moore, 2010); and (d) the creation of an online free journal to publish papers focused on the use of scientific knowledge to solve practical problems faced by environmental practitioners (Revista Caititu, http://www.portalseer.ufba.br/index.php/revcaititu).

This experience has been fruitful at many levels: (a) several products derived from these actions were incorporated into everyday work at environmental agencies, making the decision-making process better informed scientifically; (b) practitioners felt capable of better formulating questions and procedures to address the environmental issues they deal with at work; (c) networks of social learning (among practitioners, students, and researchers) were built, outlasting the particular problems around which they were created to solve; (d) the students perceived their efforts in understanding the theories, concepts, and methods in ecology as valuable to address and solve environmental issues; (e) the university is now seen as an effective partner by environmental agencies; and (f)

the researchers devote more of their time and thinking to research and teaching activities associated with local environmental problems.

This teaching/training process leading to creative, applicable solutions to practical problems can be exemplified by the results of one of the short-term, problem-based learning courses, which focused on how to evaluate requests for suppression of native vegetation. According to practitioners, a lack of strong theoretical grounds and clear protocols had led to the quasi-mechanical approval of requests at the maximum legal limits of suppression. The goal of the short-term course was to produce a scientifically informed protocol, congruent with the law, which could be applied easily in the context of urgency and lack of information and resources typically faced by practitioners. This was achieved by sharing previous relevant knowledge (and lack of knowledge) and searching for relevant information, interspersed with lectures and consultation with researchers. At the end, the group produced a text (Rigueira et al., 2013) that: (a) reviews all pertinent legislation, and the scientific literature that relates habitat amount to biodiversity maintenance, ecosystem services, and human welfare; (b) establishes a protocol for decisions based on three spatial scales (Figure 10.1), models of landscape thresholds (Pardini et al., 2010), and precautionary criteria; (c) discusses the practical implications of using this system in the environmental agencies.

The protocol has been used to evaluate suppression requests and some of its criteria are being considered to be incorporated into the guidelines for developing director plans for drainage basins in the state of Bahia. According to practitioners, the ideas and rationale in the text produced further reverberated within environmental agencies, initiating discussions on biodiversity monitoring programs, payment for ecological services schemes, integration of previously disconnected evaluation processes (e.g., fauna and vegetation), and ecological–economic

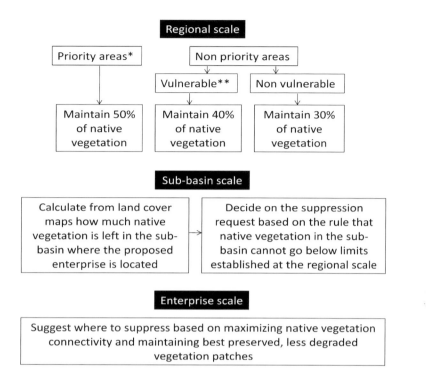

Figure 10.1 Schematic representation of the protocol for evaluating requests for suppression of native vegetation developed by graduate students and practitioners.
*Based on available maps of priority areas for biodiversity conservation
**Based on available maps of vulnerability to desertification
Modified from Rigueira *et al*. Reproduced with permission from Rigueira et al., (2013) Perda de hábitat, leis ambientais e conhecimento científico: proposta de critérios para a avaliação dos pedidos de supressão de vegetação. Revista Caititu.

zoning plans. The effective implementation of these policies, as well as the quality of their results, still needs to be evaluated. However, the experience of this short-term course and other activities developed by researchers from the Federal University of Bahia indicates a solid way to bridge the research-implementation gap.

build creative bridges between research and implementation (Box 10.1), there is still need to establish proper conditions for them to flourish. The inclusion of professional courses in the traditional quality-based federal funding applied to academic courses would solve this problem.

Procedures and standards used by Brazilian universities and funding agencies to evaluate and reward scientists also contribute to the research-implementation gap, as is generally the case in other countries (Knight et al., 2008; Shackleton, Cundill and Knight, 2009; Whitmer et al., 2010). These standards focus on scientific communication within academia, and do not stimulate – and in some instances might even prevent – the planning and implementation of strategies for societal engagement. The evaluation system's standards for both researchers (in the form of a productivity fellowship to scholars) and graduate courses, which directly influence their ability to obtain funding, are focused mainly on scientific publications in peer-reviewed journals. For example, indicators used to evaluate researchers in ecology, botany, and zoology include only the number of published scientific papers, the impact of the scientific journals where they are published, the number of supervised Master's and PhD theses, the coherence of the research program, and the merit of the submitted scientific project (http://www.cnpq.br/web/guest/criterios-de-julgamento). Criteria to evaluate graduate courses in these disciplines follow the same general logic (http://trienal.capes.gov.br/?page_id=568); although social impact is nowadays considered, it contributes only 10% of the evaluation. It is noteworthy, however, that the scientific community itself has a central role in defining and applying those criteria – 877 experienced researchers acted as evaluators in the latest round of evaluation of graduate courses in 2010. Recent changes in the evaluation and funding systems, with the creation of fellowships for technological and industrial development, and transference of technology, are welcome but still timid to overcome the inertia of years fomenting intra-walls communication.

The paralysis in action and decision making

Although there have been some important exceptions in Brazil (e.g., Joly *et al.*, 2010), the limits and uncertainty of scientific knowledge and its disconnection from management practices may frequently result in a reinforcing cycle of paralysis in decision making regarding environmental issues (Peterson, Cumming and Carpenter, 2003). A pressing example concerns how environmental impacts are measured, monitored, and evaluated during the process of licensing enterprises such as hydroelectric dams and expansions of the highway system. The ecological component of such evaluations is mainly based on species inventories and lists (CONAMA – Conselho Nacional do Meio Ambiente – Resolutions number 001/86 and 237/97; IBAMA – Instituto Brasileiro do Meio Ambiente e dos Recursos Naturais Renováveis – Normative Instruction number 146 2007). Such lists are expensive to produce and often in themselves insufficient to estimate the ecological impact of proposed enterprises and to define mitigating actions (Gardner, 2010). The reliance on species lists has probably several roots, including the legal importance of the National Red List in regulatory policy in Brazil (for criticism on this usage of Red Lists, see Possingham *et al.*, 2002). However, the paucity of scientific knowledge on our biodiversity and ecological systems, as well as the poor communication between

scientists and practitioners, also contributes to the inertia or resistance to use scientifically driven ways of evaluating environmental impacts. This resistance to rely on ecological hypothesis and theory increases the disconnection of the ecological, the social, and the economic components of environmental impact assessments, reinforcing the political discredit of ecological information and knowledge.

Breaking this reinforcing cycle requires the involvement of multiple institutions and actors (Shackleton, Cundill and Knight, 2009), but it crucially depends on changing the ways in which ecological knowledge is generated and transferred, and on breaking the prevalent endogenous conversation within both academic and decision-making sectors.

RECOMMENDATIONS

The quality of ecological science: Limits of knowledge and scientific uncertainty

A fundamental step towards developing a high-quality ecological science concerns education and scientific training, which should embrace failure, ignorance, and uncertainty as opportunities to learn and adapt, and to push forward both theoretical and applied knowledge (Peterson, Cumming and Carpenter, 2003; Root-Bernstein, 2008; Knight, 2009). This certainly requires reformulation of curricula and changing the focus of teaching from addressing only what we know and how we can solve known problems to considering also what we do not know and how we can detect unknown problems. Such approach highlights the fundamental importance of questioning to learning and science (Root-Bernstein, 2008; Witte *et al.*, 2008), as well as to conscientious action and decision making (Vitek and Jackson, 2008).

In the process of pushing forward ecological knowledge, innovation is particularly important for confronting the pressing environmental problems in tropical, developing countries such as Brazil. The use of existing theoretical models and practical protocols to deal with urgency, limited resources, and uncertainty may be inadequate or impractical in our context. We should avoid the apparent easiness of applying already developed protocols or frameworks that rely on detailed but unavailable information, and instead foster the development of viable tools. For example, despite the legal importance of the Brazilian National Red List, the listing process that is based on the IUCN criteria and protocol (Mace *et al.*, 2008) is inadequate, given that basic information needed for accurate assessment is unavailable for the great majority of species in Brazil (e.g., Scarano and Martinelli, 2010). As a result, the National Red List is dominated by naturally rare, restricted-range species, and often does not consider species that, although common, are strongly affected by human activities (Bueno, 2008), decreasing the value of this list to evaluate environmental impacts. In this context, looking for examples and tools in other applied sciences, such as medicine and business, is a promising option (e.g., medical triage, see Bottrill *et al.*, 2008; scenarios of decision making in business, see Bennett *et al.*, 2003; Peterson, Cumming and Carpenter, 2003).

Besides the development of viable tools and protocols, general principles such as the precautionary principle are paramount to decision making concerning environmental issues in the face of limited knowledge and uncertainty. It holds that we should take precautionary measures regarding potential threats that could be irreversible and dangerous (United Nations, 1992). Although the principle has been criticized as excessively risk-averse and unscientific (Brombacher, 1999), there are proper guidelines for its rational use, including epistemic and practical criteria for evaluating if the threat is plausible and if the proposed response is reasonable (Resnik, 2003).

We should also increase efficiency by fostering and rewarding the sharing and synthesis of ecological data as well as interdisciplinary and collaborative work, aiming at developing new analyses and approaches (Kinzig, 2001; Carpenter *et al.*, 2009), which are urgently needed to mitigate environmental problems. Such analyses and approaches are usually not valued highly by traditional academic evaluation systems (Fox *et al.*, 2006).

The effectiveness of ecological science: The need for horizontal interaction with decision makers

To confront complex environmental issues, we should develop "user-inspired" and "user-useful" management approaches that consider both local (practical) and scientific knowledge (Raymond *et al.*, 2010). By breaking with the vertical, hierarchical relationship between researchers and practitioners, in which the latter are limited to the passive role of consumers of scientific products, and building a horizontal approach, in which researchers and practitioners can act as peers in the construction of knowledge, we can confront the problems of relevance and accessibility of knowledge at the base of the research-implementation gap (Kennedy, 1997). Because researchers and practitioners think about environmental issues from different perspectives, dialogue is essential for an integrated picture. Moreover, scientific knowledge is often difficult to access by practitioners because of the complex framework within which it is embedded, both theoretically and empirically. Building avenues for practitioners to get hold – at least partly – of ecological knowledge without passing through the whole trajectory of scientific training traveled by ecologists themselves is thus very important.

The formation of collaborative teams including researchers and practitioners seems to be an effective way to advance in the resolution of both problems. This requires striving to overcome the difficulty of integrating two different kinds of knowledge: academic, which is abstract, generalized, and theoretically grounded, and practical, which is more concrete, situated, and grounded in everyday decision making. There are lessons from the field of education that can help ecologists and other scientists to engage more effectively with decision makers (Shackleton, Cundill and Knight, 2009). Lave and Wenger's theory of situated learning (Lave and Wenger, 1991; Wenger, 1998) is one of many theoretical frameworks that treats learning as a necessarily social process. Social theories of learning take learning not only as a cognitive process, but, above all, as a social practice that shapes what we do, who we are, and how we interpret what we do (Wenger, 1998). From this perspective, learning is seen as an integral and indivisible part of social practice, and, thus, is taken to be situated in "communities of practice" – groups of individuals with distinct knowledge, abilities, and experiences, who actively participate in collaborative processes, sharing knowledge, interests, resources, perspectives, activities, and, above all, practices, thereby building both collective and personal knowledge. The construction of communities of practice integrating researchers and practitioners can provide a way of bridging the research-implementation gap in ecological science and management. If we are successful in doing so, instead of two distinct fields of practice, we will be working in a single field, in which stakeholders and researchers will be engaged in generating relevant research questions, building the knowledge to address them, and implementing the practical protocols to deal with environmental issues and their inherent uncertainty with the necessary prudence (Box 10.1). If students participate in this joint endeavor, they can become full participants in both fields of practice, academic research, and environmental decision making (Box 10.1). Fostering such endeavors depends on reformulating traditional evaluation systems so that researchers are rewarded not only for communicating within the scientific community but also for communicating with other societal sectors that will use the knowledge they acquire for making sound practical decisions.

CONCLUSION

The limited knowledge available concerning tropical ecosystems and the frequent dissociation of such knowledge from management practices are the main challenges, while the rich biodiversity and the young and growing science offer the principal opportunities for bridging the research-implementation gap in Brazil. A fundamental step towards this goal concerns education and scientific training, which should embrace questioning, foster the development of viable tools to deal with urgency, limited resources, and uncertainty, and stimulate the synthesis of ecological data and interdisciplinary work. Far more important though is breaking with the vertical, hierarchical relationship between researchers and practitioners, and consider ways to build communities of practices integrating students, researchers, and practitioners in the construction of our understanding of environmental problems and of the consequences of intervening on them.

ACKNOWLEDGMENTS

We thank Navjot Sodhi, Luke Gibson and Peter Raven for the invitation and opportunity to write this opinion piece, Samanta Levita Coutinho (IBAMA-BA) for providing detailed information on the repercussion, within environmental agencies, of the activities among students, researchers, and practitioners developed at UFBA – Federal University of Bahia, and CNPq – Conselho Nacional de Desenvolvimento Científico e Tecnológico /FAPESB – Fundação de Amparo à Pesquisa do Estado da Bahia for funding several of these activities (PNX0016/2009). The first three authors (RP, PLBR, and CE) were supported by CNPq research fellowships during the production of this text.

REFERENCES

Barrett, T. and Moore, S. (2010) *New Approaches to Problem-Based Learning: Revitalizing your Practice in Higher Education.* Routledge, New York, NY.

Bennett, E. M., Carpenter, S. R., Peterson, G. D., Cumming, G. S., Zurek, M. and Pingali, P. (2003) Why global scenarios need ecology. *Frontiers in Ecology and the Environment*, **1**, 322–329.

Bottrill, M. C., Joseph, L. N., Carwardine, J., Bode, M., Cook, C., Game, E. T., Grantham, H., Kark, S., Linke, S., McDonald-Madden, E., Pressey, R. L., Walker, S., Wilson, K. A., and Possingham, H. P. (2008) Is conservation triage just smart decision making? *Trends in Ecology and Evolution*, **23**, 649–654.

Brombacher, M. (1999) The precautionary principle threatens to replace science. *Pollution Engineering International*, Summer, 32–34.

Bueno, A. A. (2008). Pequenos mamíferos da mata atlântica do Planalto Atlântico Paulista: uma avaliação da ameaça de extinção e da resposta a alterações no contexto e tamanho dos remanescentes. PhD thesis, University of São Paulo, São Paulo, Brazil.

CAPES (2011) Relatório de avaliação 2007–2009 Trienal 2010. http://trienal.capes.gov.br/wp-content/uploads/2011/02/Ecologia-Relatório-de-Avaliação.pdf (accessed March 19, 2013).

Carpenter, S. R. (2002) Ecological futures: building an ecology of the long now. *Ecology*, **83**, 2069–2083.

Carpenter, S. R. and Folke, C. (2006) Ecology for transformation. *Trends in Ecology and Evolution*, **21**, 309–315.

Carpenter, S. R., Armbrust, E. V., Arzberger, P. W., Chapin III, F. S., Elser, J. J., Hackett, E. J., Ives, A. R., Kareiva, P. M., Leibold, M. A., Lundberg, P., Mangel, M., Merchant, N., Murdoch, W. W., Palmer, M. A., Peters, D. P. C., Pickett, S. T. A., Smith, K. K., Wall, D. H. and Zimmerman, A. S. (2009) Accelerate synthesis in ecology and environmental sciences. *BioScience*, **59**, 699–701.

Economist Intelligence Unit (2010). *Democracy Index 2010.* http://graphics.eiu.com/PDF/Democracy_Index_2010_web.pdf (accessed March 19, 2013).

Fox, H. E., Christian, C. J., Nordby, C., Pergams, O. R. W., Peterson, G. D. and Pyke, C. R. (2006) Perceived barriers to integrating social science and conservation. *Conservation Biology*, **20**, 1817–1820.

Gardner, T. A. (2010) *Monitoring Forest Biodiversity: Improving Conservation through Ecologically-Responsible Management.* Earthscan, London, UK.

Grelle, C. E. V., Pinto, M. P., Monteiro, J. and Figueiredo, M. S. L. (2009) Uma década de Biologia da Conservação no Brasil. *Oecologia Brasiliensis*, **13**, 420–433.

Groffman, P. M., Stylinski, C., Nisbet, M. C., Duarte, C. M., Jordan, R., Burgin, A., Previtali, M. A. and Coloso, J. (2010) Restarting the conversation: challenges at the interface between ecology and society. *Frontiers in Ecology and the Environment*, **8**, 284–29.

Joly, C. A., Rodrigues, R. R., Metzger, J. P., Haddad, C. F. B., Verdade, L. M., Oliveira, M. C. and Bolzani, V. S. (2010) Biodiversity conservation research, training, and policy in São Paulo. *Science*, **328**, 1358–1359.

Kennedy, M. M. (1997) The connection between research and practice. *Educational Researcher*, **26**, 4–12.

Kinzig, A. P. (2001) Bridging disciplinary divides to address environmental and intellectual challenges. *Ecosystems*, **4**, 709–715.

Knight, A. T. (2009) Is conservation biology ready to fail? *Conservation Biology*, **23**, 517.

Knight, A. T., Cowling, R. M., Rouget, M., Balmford, A., Lombard, A. T. and Campbell, B. M. (2008) Knowing but not doing: selecting priority conservation areas and the research–implementation gap. *Conservation Biology*, **22**, 610–617.

Lave, J. and Wenger, E. (1991) *Situated Learning: Legitimate Peripheral Practice*. Cambridge University Press, Cambridge, UK.

Lewinsohn, T. M. and Prado, P. I. (2005) How many species are there in Brazil? *Conservation Biology*, **19**, 619–624.

Mace, G. M., Collar, N. J., Gaston, K. J., Hilton-Taylor, C., Akçakaya, H. R., Leader-Williams, N., Milner-Gulland, E. J. and Stuart, S. N. 2008. Quantification of extinction risk: IUCN's system for classifying threatened species. *Conservation Biology*, **22**, 1424–1442.

Martins, R. P., Lewinsohn, T. M., Diniz-Filho, J. A. F., Coutinho, F. A., Fonseca, G. A. B. and Drumond, M. A. (2007) Rumos para a formação de ecólogos no Brasil. *Revista Brasileira de Pós-Graduação*, **4**, 25–41.

Metzger, J. P., Lewinsohn, T. M., Joly, C. A., Verdade, L. M., Martinelli, L. A. and Rodrigues, R. R. (2010) Brazilian law: full speed in reverse? *Science*, **329**, 276–277.

Pardini, R., Bueno, A. A., Gardner, T. A., Prado, P. I. and Metzger, J. P. (2010) Beyond the fragmentation threshold hypothesis: regime shifts in biodiversity across fragmented landscapes. *PLoS ONE*, **5**, e13666.

Peterson, G. D., Cumming, G. S. and Carpenter, S. R. (2003) Scenario planning: a tool for conservation in an uncertain world. *Conservation Biology*, **17**, 358–366.

Possingham H. P., Andelman, S. J., Burgman, M. A., Medellín, R. A., Master, L. L. and Keith D. A. (2002) Limits to the use of threatened species lists. *Trends in Ecology and Evolution*, **17**, 503–507.

Raymond, C. M., Fazey, I., Reed, M. S., Stringer, L. C., Robinson, G. M. and Evely A. C. (2010) Integrating local and scientific knowledge for environmental management. *Journal of Environmental Management*, **91**, 1766–1777.

Resnik, D. B. (2003) Is the precautionary principle unscientific? *Studies in the History and Philosophy of Biological and Biomedical Sciences*, **34**, 329–344.

Rigueira, D. M. G, Coutinho, S. L., Pinto-Leite, C. M., Sarno, V. L. C., Estavillo, C., Campos, S., Dias, V. S., de Barros, C. and Chastinet, A. (2013) Perda de hábitat, leis ambientais e conhecimento científico: proposta de critérios para a avaliação dos pedidos de supressão de vegetação. *Revista Caititu*, **1**, 21–42. doi: 10.7724/caititu.2013.v1.n1.d03

Root-Bernstein, R. (2008) I don't know! In *The Virtues of Ignorance* (eds B. Vitek and W. Jackson), University Press of Kentucky, Lexington, KY, pp. 233–250.

Scarano, F. R. (2007) Perspectives on biodiversity science in Brazil. *Scientia Agricola*, **64**, 439–447.

Scarano, F. R. (2008) A expansão e as perspectivas da pós-graduação em Ecologia no Brasil. *Revista Brasileira de Pós-Graduação* **5**, 89–102.

Scarano, F. R. and Martinelli, G. (2010) Brazilian list of threatened plant species: reconciling scientific uncertainty and political decision-making. *Natureza & Conservação*, **8**, 13–18.

Scarano, F. R. and Oliveira, P. E. A. M. (2005) Sobre a importância da criação de mestrados profissionais na área de ecologia e meio ambiente. *Revista Brasileira de Pós-Graduação*, **2**, 90–96.

Shackleton, C. M., Cundill, G. and Knight, A. T. (2009) Beyond just research: experiences from southern Africa in developing social learning partnerships for resource conservation initiatives. *Biotropica*, **41**, 563–570.

Sousa Santos, B. de (2007) *Cognitive Justice in a Global World: Prudent Knowledge for a Decent Life*. Lexington Books, Lanham, MD.

United Nations (1992) *Agenda 21: The UN Programme of Action from Rio*. United Nations, New York, NY.

Vitek, B. and Jackson, W. (2008) *The Virtues of Ignorance*. University Press of Kentucky, Lexington, KY.

Wenger, E. (1998) *Communities of Practice: Learning, Meaning, and Identity*. Cambridge University Press, Cambridge.

Whitmer, A., Ogden, L., Lawton, J., Sturner, P., Groffman, P. M., Schneider, L., Hart, D., Halpern, B., Schlesinger, W., Raciti, S., Bettez, N., Ortega, S., Rustad, L., Pickett, S. T. A. and Killilea, M. (2010) The engaged university: providing a platform for research that transforms society. *Frontiers in Ecology and the Environment*, **8**, 314–321.

Witte, M. H., Crown, P., Bernas, M. and Witte C. L. (2008) Lessons learned from ignorance: the curriculum on medical (and other) ignorance, in *The Virtues of Ignorance* (eds B. Vitek and W. Jackson), University Press of Kentucky, Lexington, KY, pp. 251–272.

CHAPTER 11

CONSERVING BIODIVERSITY IN A COMPLEX BIOLOGICAL AND SOCIAL SETTING: THE CASE OF COLOMBIA

Carolina Murcia[1], Gustavo H. Kattan[2], and Germán Ignacio Andrade-Pérez[3]

[1]Department of Biology, University of Florida, Florida, USA
[2]Departamento de Ciencias Naturales y Matemáticas, Pontificia Universidad Javeriana-Cali, Cali, Colombia
[3]School of Management, Universidad de Los Andes, Bogotá, Colombia

SUMMARY

The multi-temporal and spatial scale interaction of history, geography, and evolution are responsible for Colombia's extreme diversity. In spite of a partially comprehensive protected area system, much of the territory remains exposed to the negative impacts of development and ecosystem transformation. Colombian biodiversity in the twenty-first century, in addition to consolidating conventional conservation strategies, must face up to emerging new challenges deriving from: (a) the vulnerability of montane biodiversity to fragmentation and climate change; (b) the transformation and contamination of natural and seminatural ecosystems by large-scale mining activities; (c) the displacement of its large cattle industry from the more appropriate savannas in the eastern plains to the Caribbean plains, replacing dry forest with savannas, while turning the original savannas into high-intensity agricultural monocultures; and (d) the consolidation of a conservation regime in the extensive communal lands in the Chocó and Amazon regions, and incorporating the extensive marine territory into its conservation plans.

INTRODUCTION

Two words best describe Colombia's biological heritage: diversity and complexity. Its standing as a megadiverse country is clear: with only 0.22% of the planet's land surface, Colombia harbors between 10 and 20% of the world's species, depending on taxonomic group (Mittermeier, Robles and Goettsch, 1997). This diversity is the result of geographical, historical, ecological, and evolutionary factors that generate a complex mosaic with high beta-diversity at several geographical scales. Superimposed on this biological complexity is a socioeconomic mosaic resulting from cultural diversity, social unbalance, and political unrest. Thus, the challenges for conserving this country's biological diversity are quite complex. Here, we first describe Colombia's biological complexity and its conservation context. Then, we present some of the main challenges for conservation professionals, managers, and decision makers in their aim to preserve Colombia's rich biodiversity for generations to come.

Colombia's biological richness is the product of three main factors: location, history, and geomorphology. Located at the crossroads of the Americas, and abut-

Conservation Biology: Voices from the Tropics, First Edition. Navjot S. Sodhi, Luke Gibson, and Peter H. Raven.
© 2013 John Wiley & Sons, Ltd. Published 2013 by John Wiley & Sons, Ltd.

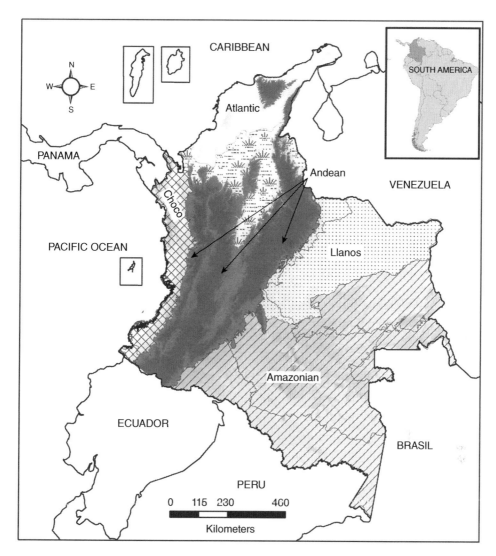

Figure 11.1 Map of Colombia showing five terrestrial regions: Chocó, Andean, Amazonian, Llanos and Atlantic. Stippled area in Atlantic illustrates the wetland portion of this region. Insert shows Colombia's general location in South America.

ting two oceans, it encompasses five terrestrial and two marine ecoregions (Figure 11.1): (1) the Chocó region, renowned for its highly diverse tropical wet forest with high endemism; (2) the xeric Caribbean plains, comprising a mosaic of savannas, dry forests, and extensive wetland complexes; (3) the lowland rainforests of the northwestern corner of the Amazon basin; (4) the eastern plains and western portion of the Guiana shield, which include savannas, speckled with table mountains (tepuis) and inselbergs, with high endemism levels; and (5) the three highly dissected ranges of the northern Andes, some rising to an elevation of more than 5000 m, which creates montane forest mosaics with wet and dry climates. Above tree line is the *páramo*, a geologically recent mountaintop archipelago of grassland and shrubland ecosystems that harbor the most diverse tropical highland flora of the world (Cleef, 1981), one filled with endemic species. The Caribbean and Pacific Oceans, separated since the formation of the Panama Isthmus, encompass different biotas.

Furthermore, history and location interplayed to define several diverse species pools for Colombia. First, the original biota inhabiting present-day northwest South America evolved in isolation as South America drifted away from Africa (Gentry, 1995). Later, a complex sequence of northern Andean uplifting generated three separate ranges and several peripheral mountain systems (Graham, 2009) that produced intricate, small-scale mosaics of habitats in which high levels of species diversity and endemism evolved (Gentry, 1995, Kattan *et al.*, 2004). More recently, the formation of the Isthmus of Panama linked South America with North America, resulting in the Great American Biotic Interchange, which enriched the Neotropical biota with Neartic elements that mainly diversified in the northern Andes (Gentry, 1982, Hooghiemstra and Cleef, 1995). In addition, some of the lineages that evolved in the southern Andes migrated north and again diversified (Gentry, 1995, Hooghiemstra and Cleef, 1995).

The result of Colombia's location and geological history is a complex biological landscape with high beta diversity observed at all spatial scales. At the country scale, biotic differentiation among the five ecoregions described earlier is very high. There is also internal heterogeneity within these ecoregions, caused by soil, humidity, and temperature gradients that generate biotic mosaics. This is most dramatic in the Andes, where each mountain range and inter-Andean valley contains a largely distinct set of species (Kattan *et al.*, 2004). At smaller scales, species turnover may be 40% or more among elevational belts and adjacent river drainages (Kattan and Franco, 2004; Kattan *et al.*, 2006). Conservation planning in this scenario is complicated, because many scattered conservation areas must be preserved to represent Colombia's high regional diversity adequately. Thus, the conservation challenges presented by the evolutionary legacy, ecological diversity, and complexity of lineages converging in this country are staggering.

CURRENT CONSERVATION CONTEXT

The conservation of Colombian biodiversity faces two types of challenges. The first is the actual protection of the biodiversity from external threats and the second involves the effective planning of the conservation efforts as well as management of the biodiversity inside and outside protected areas (PAs). Protecting biodiversity requires significant political will and economic investments; for example, to declare new areas and control major external threats to these areas, such as illegal encroaching or mining and impacts from major infrastructure development, the national park system must have the willing participation and support of other government agencies such as the Ministries of Development, Energy, and Government. Likewise, to protect biodiversity from poaching or illegal logging requires a reasonable budget and the cooperation of local authorities. In contrast, territorial planning that seeks a balanced use of the land to ensure both development and conservation ideally requires significant knowledge about the spatial needs of species and ecosystems. At the local scale, managers need to achieve minimum management standards to address the biological and social issues within and around their jurisdictions. Both activities are knowledge intensive (Hockings and Dudley, 2011), and without such knowledge the managers will remain powerless to act or make misinformed and erroneous decisions with potentially negative consequences.

Like other modern urbanized countries (McAlpine *et al.*, 2009), Colombia's 48 million inhabitants are quickly changing their consumption patterns and demanding more ecosystem goods and services. To meet these new demands, the government is responding with an economic development model that focuses on foreign trade, large-scale commercial agriculture, mining, and other extractive industries with the concomitant development of transportation infrastructure networks (Government of Colombia, 2011). Producers are abandoning relatively environmentally benign traditional agricultural and cattle ranching practices in favor of technologically intensive systems. The result is an increasing pressure on natural ecosystems (Etter *et al.*, 2006). At the same time, the government has also made clear its commitment to biodiversity conservation, under the premise that biodiversity is an essential component of the country's natural assets (Government of Colombia, 2011). Biodiversity management in a rapidly changing socio-ecological scenario of globalization emerges as a major and complex endeavor that needs to take place both inside and outside of PAs.

Colombia is relatively advanced in terms of PAs. In the last decade, the PA system has grown to 56 areas, its budget increasing three-fold between 2007 and 2010 (Hockings and Dudley, 2011). Currently, 11% of the terrestrial surface area (12,580,000 ha) is under protection as national parks (Figure 11.2), and an additional 3% is protected or has special management status under provincial, municipal, or private jurisdictions. However, like many developing countries, actual protection is often weak or variable because of the insufficient human and economic resources available for it. Therefore, in spite of these achievements, social complexity and limited governance still makes isolation from regions of rapid development a park's

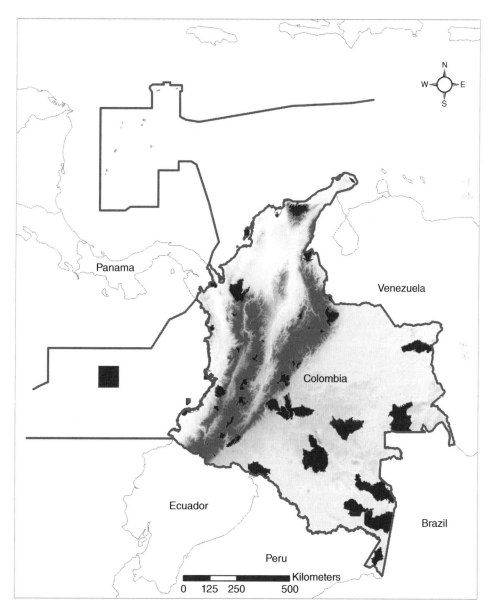

Figure 11.2 Map illustrating Colombia's terrestrial and marine territory (dark line), and location of current national parks (dark grey).

best defense – at least for the time being. In addition to PAs, around 37% of Colombia's land area, mostly covered by natural ecosystems such as savannas, rainforests, and wetlands, is under jurisdiction of indigenous and Afro-descendant communities, and thus excluded from agricultural development (Vélez, 2011). Although these communal lands do not receive the same levels of protection as PAs, they nonetheless represent an important conservation opportunity involving local community governance and management (Andrade, 2003).

Half the land and almost all the oceans in Colombia are unprotected (Figure 11.2), yet they harbor a significant biodiversity that is threatened by development and population growth. Thus, a change in paradigm in conservation management needs to take place, to include the productive matrix in a conservation framework. To this end, we need good knowledge of ecological processes in these varied and complex ecosystems, how they couple with human systems, and how both can be managed to maintain ecosystem services. Many challenges lie ahead, exacerbated by a boom of economic growth and globalization, as well as global climate change. The challenge relates not only to the quantity of relevant information, but also to the barriers for its application (Murcia and Kattan, 2009). We fear that the gap between the slow pace of knowledge production and the vertiginous rhythm of ecosystem transformation will continue to increase. To overcome these challenges, there is need for an integrated effort among scientists and research agencies, policy makers and management institutions, and decision makers at all levels. A first step in this direction was taken by the National Parks Unit in 2000, when scientists and park staff jointly identified the information needs in conservation biology that limited their capacity to effectively manage the areas under their care and to mitigate negative impacts from outside (Kattan and Murcia, 2000). However, the initative addressed only the needs of PAs, leaving those that applied to the rest of the nation's territory unaddressed. Furthermore, many of these information needs are still unfulfilled and, therefore, we present them, along with new ones, as parallel challenges in the next section of this chapter.

CONSERVATION AND MANAGEMENT CHALLENGES IN THE TERRITORY

We divide the challenges for biodiversity conservation in Colombia into two categories: (1) territorial issues

and current development models; and (2) threats specific to the different ecoregions. In addition, we discuss challenges associated with conservation strategies, specifically the creation and management of PAs and ecological restoration.

TERRITORIAL ISSUES AND SOCIOECONOMIC MODELS

Communal lands: Biodiversity implications of the payment of a social debt

In an attempt to do justice after centuries of neglect, Colombia granted to minority ethnic groups autonomous management of 42 million hectares (ha), of which approximately 3 million ha lie within the boundaries of previously established national parks. This overlap represents a major conservation challenge. Cultural changes and the aspiration to eradicate poverty impose management models that might be incompatible with conservation aims (Vélez, 2011). The challenge is to assist these communities in economic and cultural transitions that incorporate conservation as a basic principle.

The use of communal lands and collective resources represents a major challenge for sustainable management. Particularly, better management techniques are required for hardwood forests. Also required are alternative sources of income from forest products as well as alternative production systems, and the development of culturally appropriate sustainable-management techniques. Current research about the evolving mechanisms of resource allocation and governance of community-owned resources (Vélez 2011) could be complemented with studies about the relationship between social change and biodiversity conservation.

Hydroelectric power – A clean energy alternative?

As the human population and its technological demands increase, so does the need to generate electricity. Because of its topography and rich river network, Colombia holds great potential for hydroelectricity generation, particularly in the Andes. Hydroelectric power generates substantially lower greenhouse gas emissions compared with coal and other energy sources, but hydroelectric plants are not free of negative impacts. Dam construction causes extensive habitat loss and also alters the dynamics of aquatic communities, particularly migratory fish. Changes in hydrological cycles also affect flooding regimes downstream and have a heavy impact on overall rivers and floodplain ecology (Barthem and Goulding, 1997).

Hydroelectric projects, however, may be turned into opportunities for habitat conservation. The need to keep dams free of sediments and extend their useful lives is an incentive for watershed protection. Conservationists could work with hydroelectric enterprises to include protection schemes in the projects, funded by the proceeds from electricity sales. These areas also are excellent scenarios for long-term scientific research programs, particularly on topics such as assessments of ecosystem services and monitoring the effects of climate change on tropical montane forests. At the same time, the rapid melting of glaciers and permanent snowbanks in the Andes that is accompanying global climate change is likely soon to limit the availability of hydropower and even the provision of water to the larger cities along the Andes.

Oil and mineral exploration and extraction

Technological advances and economic growth demand ever more oil and minerals. A Colombian economic leap forward is foreseen based on the expansion of oil and mining enterprises. Although *páramos* and PAs are legally excluded from mining (Ponce de León, 2005), these laws are not sufficiently enforced. Mining poses a great threat to biodiversity and human communities, with a wealth of environmental, health, and social impacts.

Almost 2 million ha have been allocated for mining prospecting (Rudas, 2001), placing a serious threat to Colombia's often very local biodiversity. The mining concessions are set up without concern for biodiversity conservation, sector policies are uninformed, there are no geographically explicit models of the distribution of biodiversity in the country and there is no mechanism to link these two with the process of license granting. Greatly improved regional planning is needed to exclude mining projects from sensitive areas, particularly in montane ecosystems. Of particular concern in this regard are *páramo*-restricted species, which face the additional stress of climate change (Forero, Joppa and Pimm, 2010). With an interdisciplinary orientation, conservation biology needs to develop methods

and protocols for the integration of biodiversity considerations into strategic planning for the energy and mining sectors. Requirements for restoration and rehabilitation of lands, and mechanisms for compensation with very high offset ratios, should be defined and implemented for the mining industry.

ECOSYSTEM AND ECOREGION-SPECIFIC THREATS

Ocean and coastal ecosystems

Marine territorial waters represent 50% of Colombia's area (Figure 11.2). However, marine ecosystems have received little attention despite ongoing and unrestricted degradation caused by fisheries and other forms of marine resource exploitation. Symptoms of the collapse of some fisheries and other marine resources are already evident. The biggest challenges occur in the continental shelf and shorelines, where the future is uncertain for important habitats including mangroves, coral reefs, and beaches. A 1-m rise in sea level during the remainder of this century, now regarded as a minimum estimate, would permanently flood 4900 km^2 of coastal areas, and alter the hydrological cycles of an additional 5100 km^2 (IDEAM, 2010). Thus, of particular importance is the need to model the effect of sea level rise on the configuration of ecosystems at the extensive river deltas (e.g., the Magdalena and Atrato Rivers), to predict future spatial arrangements, and to reassess the conservation status of species under these new scenarios. New PAs and corridors will be needed to cope with the effects of rising sea levels. The need to address the future of human infrastructure near the rising sea is obvious.

In the seas themselves, research is needed on population management to help establish proper and sustainable fishing quotas and protected zones. Knowledge gained by studies throughout the world needs to be brought to Colombia and used to protect its resources for the future. At smaller scales, there is need for information on the impacts of climate and overfishing on species-rich, complex systems such as coral reefs and mangroves. It is also important to evaluate the effects of the often irreversible integration of alien species, creating new species assemblages and food web configurations. A primary challenge in oceanic islands is the prevention and control of invasive species, but sea level rise will assume key importance as the years go by.

Caribbean plains: The emergent savanna ecosystem

Great expanses of the Caribbean lowlands (Figure 11.1), originally covered with tropical dry forest, continue to be transformed into cattle pastures. These pastures are often dominated by non-native African grasses, which, combined with altered disturbance regimes, block the regeneration of native forests (Cavelier *et al.*, 1998). The conversion of the Caribbean lowlands is likely to be accelerated as cattle ranging in the eastern plains (see later) decreases due to changes in land use.

Biodiversity assessments have yet to be conducted in the Caribbean lowlands to identify the most sensitive areas. The limited inventories that are available indicate a high level of diversity and underline the urgent need to determine the viability of isolated populations of dry forest plants and animals. Few PAs exist in the region, so the native biota is likely to be relegated to small fragments where it will decay unless plans to reduce fragmentation and restore connectivity are put in place. The occurrence of similar development patterns in similar vegetation types around the world would seem to make international cooperation highly desirable in confronting the problems presented in such areas.

Andean ecosystems: Forest fragmentation and the rural landscape

The Andes contain the most productive soils and benign climates of the country (Figure 11.1). Over the past five centuries, these regions have been transformed to accommodate the increasing demand for European-style agriculture and the growth of urban settlements. Over the past few decades, new agricultural models are emerging to keep up with the demands for productivity as well as to plan regional adaptation to global climate change. In the intensive agroecosystems in these regions, new crops are being developed to satisfy new markets and adapt the agricultural production to a constantly changing economy. Consideration is also being given to the likely effects of climate change on agricultural productivity overall, particularly in view of the probable future water shortages that will be associated with it. For these reasons, the productive matrix that surrounds forest remnants and PAs is in a permanent state of flux. Agricultural innovations ought to be evaluated carefully in terms of their negative impacts on remnant ecosystems, as well as possible positive effects (e.g., shade coffee improving connectivity among forest patches). In addition to incorporating new productive trends, farmers will need to adapt to higher temperatures and drier conditions to maintain their income level (Mendelsohn and Dinar, 1999). These adaptations might represent a threat to biodiversity if farmers clear new areas as the climate changes. This is very likely to be the case with montane crops such as tea or coffee that require mild temperatures. To maintain production for them, the path of least resistance would seem to be to move the areas of cultivation to higher elevations, where most of the remnant forests are. All of these factors must be taken into account in planning a sustainable rural landscape for the future and to meet regional conservation goals. In the future, changing conditions will be the rule, not the exception.

Research on the effects of forest fragmentation on Andean biodiversity (Kattan and Álvarez-López, 1996), as well as the effect of the agricultural lands that surround the fragments (Renjifo, 2001; Muriel and Kattan, 2009), has led to the definition of "landscape management tools" (Lozano, 2009), which are currently being applied in some rural productive settings. Additional research is needed to develop a comprehensive view of the long-term effects of rural landscape management on biodiversity, especially in coping with the effects of climate change along altitudinal gradients.

The eastern plains: Large-scale agro-industries at the frontier

The eastern plains (*llanos orientales*) (Figure 11.1), originally covered with a mosaic of dry and wet savannas and gallery forests, have been occupied since the 1700s with extensive cattle ranches that had a relatively low impact on the original terrestrial vertebrates (Hoogesteijn and Chapman, 1997; Scognamillo *et al.*, 2003). However, the promise of quick cash from biofuels and other new crops during recent years lured the government to promote the transformation of this region. In less than a decade, large expanses of natural savannas have been converted to oil palm, sorghum, and other forms of large-scale agriculture, displacing cattle ranching to other regions like the Caribbean plains that are less appropriate for this activity. The *llanos* have experienced conversion rates of 200,000 ha/year, rates that rank among the world's highest for topical savannas (Romero *et al.*, 2010). The challenge posed by the expansion of these land-use changes is to define an ecological–economic equilibrium in the transformation process at the landscape level. This calls for an expansion of the PA system in this region, and for requiring entrepreneurs to develop both ecologically and environmentally sound management protocols for these emerging agricultural landscapes. The homogeneous, large-scale crops produced over wide areas by industrial agriculture are not likely to be suitable for the survival of native plants and animals unless specific steps are taken to modify the landscape-scale effects of such practices.

Biological knowledge on this extensive region of the country is scarce, so that there is limited capacity to design conservation efforts that would forestall the impacts of large agro-ecosystems replacing large savannas and fragile riparian forests (Lasso *et al.*, 2010). For example, this shared Colombia–Venezuela *llanos* ecoregion depends on particular fire and flooding regimes to subsist (Romero *et al.*, 2010), yet with the establishment of commercial crops both regimes are being altered, with unknown consequences to the remnant ecosystems and their biota. Lack of information on the spatial patterns of biodiversity along habitat gradients at the regional scale are necessary to direct land planning such that habitat fragmentation is minimized and areas of strategic importance are left untouched by the massive land conversion. The effects of global climate change on this region also need to be considered carefully in constructing any plans for its future sustainability.

Wetlands: An often-neglected biodiversity component

Along the lower Cauca River in the Caribbean plains (Figure 11.1), an extensive matrix of permanent wetlands and seasonally flooded plains sprawls. This region, rich in wildlife, is also occupied by extremely poor and vulnerable human communities. Dams, overfishing, pollution, and the accumulation of toxic substances collected by the rivers that feed them are some of the threats of this region. Floods associated with La-Niña of 2010–2011 left 2.2 million people (5% of the country's population) afflicted, raising national awareness of the need for better territorial and ecological management for this region.

As rivers are dammed and modified with canals and hydrological cycles altered, ecosystems dependent on river pulses become threatened. A large-scale research agenda for this extensive region (circa 20 million ha) should be developed to understand the patch dynamics of these complex systems and how are they being affected. Population viability analyses for threatened species restricted to these wetlands, coupled with models of ecosystem change under human drivers, are required for conservation planning.

A major threat for freshwater systems in Colombia, as throughout the world, is the spread of introduced fish species (Gutiérrez, 2006), which has caused the collapse of commercial and artisanal fisheries (Mojica, 2002). The challenge is to integrate species-level information with ecosystem parameters, in order to gain an understanding of threats and the management implications of countering them. Although our knowledge of fish migrations and their management in tropical American rivers has increased, especially for the Amazon basin (Barthem and Goulding, 1997), specific knowledge for many Colombian river systems is scant. Because most wetlands are dynamic open systems, their management requires a whole-catchment approach, especially for rivers regulated at the headwaters such as the Cauca River system.

Wetlands also occur in the Andean highlands and valleys, where they play an important role in hydrological cycles and contain remarkable endemics that evolved in these distinctive local ecosystems. Particularly threatened wetlands occur in the Bogotá Plateau and the Cauca Valley, where rapidly expanding metropolitan areas and agriculture threaten them with extinction.

CONSERVATION STRATEGIES

Protected areas: The challenges in a heterogeneous context

The National System of Protected Areas (SINAP) includes the national park system as well as regional, state, municipal, and private PAs (Figure 11.2). Altogether, they include over 22 million ha of land (Vásquez and Serrano, 2009). In spite of this impressive expanse and representation of major biomes (Forero and Joppa, 2010), some ecosystems such as mid-elevation Andean forests, wetlands, dry forests, deserts and dry scrub, and *páramo*, are underrepresented (Armenteras, Gast and Villarreal, 2003). This is particularly challenging because a PA system should incorporate not just all ecosystem types, but also the ecological and evolutionary factors that generate and maintain biodiversity at different spatial scales (Kattan *et al.*, 2006), in order to ensure species flows and ecological resilience.

All parks in the national park system and some regional PAs have management plans that include a research agenda. Most of this research is oriented towards establishing the long-term viability of conservation targets (species, habitats, and ecosystems). Research is urgently needed to understand the effects of global change on ecological processes that support biodiversity in PAs. Region-wide planning to adapt the current SINAP to changing threats, and the functional integration of national, regional, and municipal conservation areas is also needed. As isolation of natural ecosystems increases, it is important to conduct research and develop techniques aimed at integrating the management of PAs with the productive matrix.

Ecological restoration: A complement to habitat protection

Ecological restoration is increasingly resonating as a strategy in landscape management and conservation. Fragmentation and habitat loss are threatening ecosystems such as Andean montane forests, which are on the verge of losing ecological integrity because their extent and connectivity are insufficient to sustain populations of key species (top predators, regional migrants, and rare or large-bodied animals and trees) (Kattan, *et al.*, 1994) and to sustain dynamic processes such as succession that counteract the effects of natural and human disturbance (Kattan and Murcia, 2003). In addition, decades of deforestation and land exploitation have left many areas in a fragile state and unable to recover on their own. Therefore, a significant restoration effort will be required for recovering large areas from the effects of uncontrolled conversion and poor land management.

A FINAL THOUGHT

Superimposed on all the problems directly caused by human actions are the pervasive effects of climate change. Biodiversity conservation planning in Colombia has to be revised to address this new challenge. Although general scenarios of potential climate change have been developed, there are few analyses of the vulnerability of biodiversity to this phenomenon (e.g., Aguirre *et al.*, 2011). Some trends are evident, such as increasing extinction risks with elevation (Buytaert, Cuesta and Tobón, 2011), which could be stronger for mid-elevation species that might get into thermal traps due to the irregular topography (Forero, Joppa and Pimm, 2010). As species move at different rates in search of adequate conditions, communities will likely decouple and become more susceptible to invasions. It is anticipated that many species will be lost, and that in many areas synergies between climate stresses and human impacts will result in the emergence of novel simplified ecosystems (Hobbs *et al.*, 2006). Of particular concern in this context are PAs, especially those created to conserve fragile and unique ecosystems and endangered species, when they are surrounded by extensive human-dominated landscapes, and the communities they have been set up to preserve will often, in the face of climate change, literally have nowhere to go.

CONCLUSIONS

Colombia's extreme biodiversity is the product of complex historical, geographical, and evolutionary processes that have resulted in a very heterogeneous and finely partitioned mosaic of ecosystems laden with endemic species. The country is also host to complex social and political processes that superimpose both constraints and opportunities for conservation. Aside from the overarching threat of global climate change, the country faces the pressure of responding to global economic models that require significant infrastructure development and large-scale productive systems, as well as demands from a growing educated and upward-scaling population that demands more resources.

To address these challenges, the country has a partially comprehensive PA system, with some impetus and strength, but insufficient for its needs; reasonably strong environmental legislation, but insufficient political will and capacity for it to be enforced; and a growing research community that is still not sufficiently strong to generate the vast information necessary for management decisions. At the current rate of change, it is likely the country will lose a significant amount of biodiversity before it has all the necessary tools to protect it. A significant paradigm change is required to ensure that all productive systems are incorporated into a conservation framework that allows development while minimizing environmental impact.

REFERENCES

Aguirre, L. F., Anderson, E. P., Brehm, G., Herzog, S. K., Jørgensen, P. M., Kattan, G. H., Maldonado, M., Martínez, R., Mena, J. L., Pabón, J. D., Seimon, A. and Toledo, C. (2011) Phenology and interspecific ecological interactions of Andean biota in the face of climate change, in *Climate Change Effects on the Biodiversity of the Tropical Andes: An Assessment of the Status of Scientific Knowledge* (eds S. K. Herzog, R. Martínez, P. M. Jørgensen and H. Tiessen), Inter-American Institute of Global Change Research (IAI) and Scientific Committee on Problems of the Environment (SCOPE), São José dos Campos and Paris, pp. 68–92.

Andrade, G. I. (2003) National parks versus protected landscapes? Legitimacy, values, and the management of the Colombian tropical wildlands, in *The Full Value of Parks: From Economics to the Intangible* (eds A. Putney and D. Harmon), Rowman & Littlefield Publishers Inc., Lanham, MD, 169–184.

Armenteras, D., Gast, F. and Villarreal, H. (2003) Andean forest fragmentation and the representativeness of protected natural areas. *Biological Conservation*, **113**, 245–256.

Barthem, R. and Goulding, M. (1997) *The Catfish Connection: Ecology, Migration and Conservation of Amazon Predators.* Columbia University Press, New York, NY.

Buytaert, W., Cuesta, F. and Tobón, C. (2011) Potential impacts of climate change on the environmental services of humid tropical alpine regions. *Global Ecology and Biogeography*, **20**, 19–33.

Cavelier, J., Aide, T. M., Santos, C., Eusse, A. M. and Dupuy, J. M. (1998) The savannization of moist forests in the Sierra Nevada de Santa Marta, Colombia. *Journal of Biogeography*, **25**, 901–912.

Cleef, A. M. (1981) The vegetation of the Colombian Cordillera Oriental. Rijksuniversiteit Utrecht. The Quaternary of Colombia no. 9. Dissertationes Botanicae no. 61. Vaduz, Cramer.

Etter, A., McAlpine, C., Phinn, S., Pullar, D. and Possingham, H. (2006) Unplanned land clearing of Colombian rainforests: spreading like disease? *Landscape and Urban Planning*, **77**, 240–254.

Forero, G. and Joppa, L. (2010) Representation of global and national conservation priorities by Colombia's protected area network. *PLoS ONE*, **5**, e13210.

Forero, G., Joppa, L. and Pimm, S. D. (2010) Constraints to species' elevational range shifts as climate changes. *Conservation Biology*, **25**, 163–171.

Gentry, A. H. (1982) Neotropical floristic diversity: phytogeographical connections between Central and South America, Pleistocene climatic fluctuations, or an accident of the Andean orogeny? *Annals of the Missouri Botanical Garden*, **69**, 557–593.

Gentry, A. H. (1995) Patterns of diversity and floristic composition in neotropical montane forests, in *Biodiversity and Conservation of Neotropical Montane Forests* (eds S. P. Churchill, H. Balslev, E. Forero, and J. L. Luteyn), The New York Botanical Garden, Bronx, NY, pp. 103–126.

Government of Colombia (2011) National Development Plan (Plan Nacional de Desarrollo 2011–22015). Bogotá, Colombia.

Graham, A. (2009). The Andes: a geological overview from a biological perspective. *Annals of the Missouri Botanical Garden*, **96**, 371–385.

Gutiérrez, F. (2006) *Estado de conocimiento de las especies invasoras. Propuesta de lineamientos para el control de impactos.* Instituto de Investigación de Recursos Biológicos Alexander von Humboldt, Bogotá, Colombia.

Hobbs, R. J., Arico, S., Aronson, J., Baron, J. S., Bridgewater, P., Cramer, V. A., Epstein, P. R., Ewel, J., Klink, C. A., Lugo, A. E., Norton, D., Ojima, D., Richardson, D. M., Sanderson, E. W., Valladares, F., Vila, M., Zamora, R. and Zobel, M. (2006) Novel ecosystems: theoretical and management aspects of the new ecological world order. *Global Ecology and Biogeography*, **15**, 1–7.

Hockings, M. and Dudley, N. (2011) *Efectividad de manejo del Sistema de Parques Nacionales Naturales de Colombia (Management effectiveness of the Colombian National Park System).* Internal report. Unidad Administrativa Especial Sistema de Parques Nacionales Naturales. Bogotá, Colombia.

Hoogesteijn, R. and Chapman, C. A. (1997) Large ranches as conservation tools in the Venezuelan llanos. *Oryx*, **31**, 274–284.

Hooghiemstra, H. and Cleef, A. M. 1995. Pleistocene climatic change and environmental and generic dynamics in the North Andean montane forest and páramo, in *Biodiversity and Conservation of Neotropical Montane Forests* (eds S. P. Churchill, H. Balslev, E. Forero, and J. L.Luteyn), The New York Botanical Garden, Bronx, New York, NY, pp. 35–49.

IDEAM (2010) *Segunda comunicación nacional ante la Convención Marco de las Naciones Unidas sobre Cambio Climático.* Instituto de Estudios Ambientales, Bogotá, Colombia.

Kattan, G. H., Álvarez-López, H. and Giraldo, M. (1994) Forest fragmentation and bird extinctions: San Antonio eighty years later. *Conservation Biology*, **8**, 138–146.

Kattan, G. H. and Álvarez-López, H. (1996) Preservation and management of biodiversity in fragmented landscapes in the Colombian Andes, in *Forest Remnants in the Tropical Landscape* (eds J. Schelhas and R. Greenberg), Island Press, Washington, D.C. pp. 3–18.

Kattan, G. H. and Murcia, C. (2000) *Desarrollo de una estrategia de investigación en biología de la conservación en el sistema de Parques Nacionales.* Internal report. Unidad Administrativa Especial de Parques Nacionales, Bogotá, Colombia.

Kattan, G. H. and Murcia, C. (2003) A review and synthesis of conceptual frameworks for the study of forest fragmentation, in *How Landscapes Change: Human Disturbance and Ecosystem Disruptions in the Americas* (eds G. A. Bradshaw and P. A. Marquet), Ecological Studies 162, Springer-Verlag, Berlin, pp. 183–200.

Kattan, G. H. and Franco, P. (2004) Bird diversity along elevational gradients in the Andes of Colombia: area and mass effects. *Global Ecology and Biogeography*, **13**, 451–458.

Kattan, G. H., Franco, P., Rojas, V. and Morales, G. (2004) Biological diversification in a complex region: a spatial analysis of faunistic diversity and biogeography of the Andes of Colombia. *Journal of Biogeography*, **31**, 1829–1839.

Kattan, G., Franco, P., Saavedra, C. A., Valderrama, C., Rojas, V., Osorio, D. and Martínez, J. (2006) Spatial components of bird diversity in the Andes of Colombia: implications for designing a regional reserve system. *Conservation Biology*, **20**, 1203–1211.

Lasso, C. A., Usma, J. S., Trujillo, F. and Rial, A. (2010) *Biodiversidad de la cuenca del Orinoco. Bases científicas para la identificación de áreas prioritarias para la conservación y uso sostenible de la biodiversidad.* Instituto Humboldt, WWF Colombia, Fundación Omacha, Fundación La Salle e Instituto de Estudios de la Orinoquia, Bogotá, Colombia.

Lozano, F. (ed.) (2009) *Herramientas de manejo para la conservación de la biodiversidad en paisajes rurales.* Instituto de Investigación de Recursos Biológicos Alexander von Humboldt y Corporación Autónoma Regional de Cundinamarca (CAR), Bogotá, Colombia.

McAlpine, C. A., Etter, A., Fearnside, P. M., Seabrook, L. and Laurance, W. L. (2009) Increasing world consumption of beef as a driver of regional and global change: a call for policy actions based on evidence from Queensland (Australia), Colombia, and Brazil. *Global Environmental Change*, **19**, 21–33.

Mendelsohn, R. and Dinar. A. (1999) Climate change, agriculture, and developing countries: does adaptation matter? *World Bank Research Observer*, **14**, 277–93.

Mittermeier, R. A., Robles, P. and Goettsch, C. (1997) *Megadiversidad. Los países biológicamente más ricos del mundo.* Cemex, México D.F.

Mojica, J. I. (2002) Las pesquerías del río Magdalena: ejemplo a no repetir, in *Libro rojo de peces dulceacuícolas de Colombia* (eds J. I. Mojica, C. Castellanos, J. S. Usma and R. Álvarez), Serie Libros Rojos de Especies Amenazadas de Colombia. Instituto de Ciencias Naturales, Universidad Nacional de Colombia, Ministerio del Medio Ambiente, Bogotá, Colombia, pp. 35–43.

Murcia, C. and Kattan, G. H. (2009) Application of science to protected area management: overcoming the barriers. *Annals of the Missouri Botanical Garden*, **96**, 508–520.

Muriel, S. B. and Kattan, G. H. (2009) Effects of patch size and type of coffee matrix on Ithomiine butterfly diversity and dispersal in cloud-forest fragments. *Conservation Biology*, **23**, 948–956.

Ponce de León, E. (2005) *Estudio jurídico sobre categorías regionales de áreas protegidas.* Instituto de Investigación de Recursos Biológicos Alexander von Humboldt, Bogotá, Colombia.

Renjifo, L. M. (2001) Effect of natural and anthropogenic landscape matrices on the abundance of subandean bird species. *Ecological Applications*, **11**, 14–31.

Romero, M., Etter, A., Sarmiento, A. and Tansey, K. (2010) Spatial and temporal variability of fires in relation to ecosystems, land tenure and rainfall in savannas of northern South America. *Global Change Biology*, **16**, 2013–2023.

Rudas, G. (2001) *Minería y cambio climático, una señal de alarma.* La Silla Vacía. Economía y Sociedad, Bogotá, Colombia.

Scognamillo, D., Maxit, I. E., Sunquist, M. and Polisar, J. (2003) Coexistence of jaguar (*Panthera onca*) and puma (*Puma concolor*) in a mosaic landscape in the Venezuelan llanos. *Journal of Zoology*, **259**, 269–279.

Vásquez, V. and Serrano, M. A. (2009) *Las áreas naturales protegidas de Colombia.* Conservación Internacional y Biocolombia, Bogotá, Colombia.

Vélez, M. A. (2011) Collective titling and the process of institution building: the new common property regime in the Colombian Pacific. *Human Ecology*, **39**, 117–129.

CHAPTER 12

INDIGENOUS RIGHTS, CONSERVATION, AND CLIMATE CHANGE STRATEGIES IN GUYANA

Michelle Kalamandeen

Department of Biology, University of Guyana, Georgetown, Guyana

SUMMARY

Historically, conservation practices in the tropics have paid little attention to the rights of indigenous peoples and their special relationship with the land. As the colonial period wound down, however, indigenous rights have often and appropriately assumed a greater prominence than they had held previously. Land titles are often still confused, however, and the rights of indigenous people to participate in the global discussion of assigning values to relatively intact forests should receive much more emphasis than they have to date. This paper examines how Guyana, a small South American country with relatively large intact rainforests, is coping with the matter of indigenous rights in protected areas generally, and in the implementation of the nation's recently initiated Low Carbon Development Strategy.

INTRODUCTION

Conservation science recognizes that the well-being of society is closely linked with the well-being of natural ecosystems. There are strong ethical and scientific reasons for conserving and enhancing biodiversity, which is now properly understood to be essential for sustainable development and economic prosperity. Traditionally, conservation has focused primarily on priority species, by identifying their ecological requirements and implementing the necessary government management and/or supported interventions via different categories of protected areas (PAs). The template for PAs was first developed in the US, starting in 1872 and based on the "wilderness" ideology (see Kalamandeen and Gillson, 2007). PAs were seen as the fastest growing land change movement in history, with the establishment of over 120,000 PAs covering approximately 21 million km^2 of land and sea worldwide by 2008 (see Figure 12.1; UNEP, 2010).

In the course of their establishment, the formation of PAs often paid little attention to the rights of indigenous peoples[1] and their special relationship with their

[1]For the purpose of this paper, the International Labour Organisation (ILO) Convention No. 169 (1989) is the foremost document used for identifying the rights of Indigenous People, which is legally binding for states that ratify it. Where applicable, the 2007 UN Declaration on the Rights of Indigenous People (UNDRIP) (United Nations, 2013) is the supporting document. UNDRIP provides a fresh take, reaffirming the importance of the principles of ILO 169. UNDRIP is not subjected to ratification and therefore not legally binding. Guyana is not a signatory of ILO 169.

Conservation Biology: Voices from the Tropics, First Edition. Navjot S. Sodhi, Luke Gibson, and Peter H. Raven.
© 2013 John Wiley & Sons, Ltd. Published 2013 by John Wiley & Sons, Ltd.

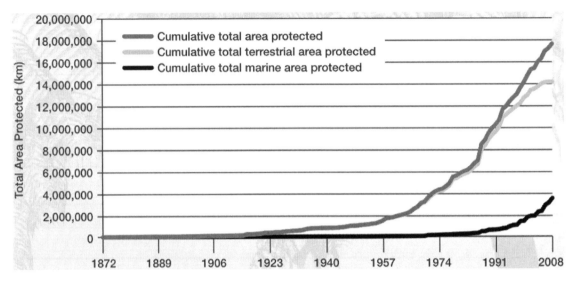

Figure 12.1 Growth in nationally designated protected areas from 1872 to 2008. Reproduced with permission from UNEP-WCMC and BIP (2009).

lands.[2] In that sense, the establishment and management of the PAs tended to prolong the colonial attitude that wildlife came first and people second. In doing so, problems of conflict between human beings and wildlife often became more severe than they had been previously. Management authorities were often drawn into these conflicts with undesirable consequences and limited conservation success. Pursued in the traditional way, the establishment of PAs often forced the displacement of local communities, denied their access to resources, limited their human rights generally[3] and undermined their livelihoods (McShane, 2003; Colchester, 2004). Despite their taking such a Draconian approach, PAs often failed to achieve the conservation goals (Haynes and Ostrom, 2005).

Over the last two decades, a paradigm shift occurred that increasingly acknowledged the important role

indigenous peoples play in the conservation of natural resources and their inherent place in natural communities. Today, although attitudes vary widely from country to country, relatively few people would argue against the essential role of communities in conservation actions, and even fewer would defend conservation as a realm where human rights are less important than species protection. Humans and nature together form a whole, and must operate as a system in protecting the functioning of the whole system properly and sustainably.

Over the past 10 years or so, there has been a global tendency to consider tropical forests not only as important repositories of biodiversity and places where people live, but also as vitally important storage "sinks" for carbon that might otherwise enter the atmosphere and exacerbate the effects of global climate change on these and other ecosystems. Now that it is understood and accepted that indigenous people are, in some sense, part of the forest, their role in making decisions to set aside large areas of forest for carbon retention needs much more careful consideration than was the case in initial discussions of such storage mechanisms. Many species will face extinction as a result of the projected changes in temperature, and the forest itself may col-

[2] As recognized in ILO 169: 13(1).
[3] Human Rights is defined as "*rights inherent to all human beings, whatever our nationality, place of residence, sex, national or ethnic origin, colour, religion, language, or any other status. We are all equally entitled to our human rights without discrimination. These rights are all interrelated, interdependent and indivisible*" (OHCHR, 2011).

lapse in the face of climate change. In that case, there would be massive extinction coupled with the release of huge amounts of carbon into the atmosphere. There are many reasons to want to maintain the forest, but the role of local human communities, which will also suffer from the effects of the climate changes, in helping to form comprehensive solutions in which their own rights will be preserved should certainly be considered fully.

There are many ways in which indigenous communities can suffer consequences associated with climate change itself. For example, the indigenous Wai-Wais of Guyana, a semi-nomadic people living in remote forests, were forced to relocate their village of Akotopono in 2000 after unprecedented flooding. Whether this instance of flooding was caused by global climate change or not, it illustrates the kind of situation that will occur with increasing frequency in the future. The situation of the Wai-Wai is by no means unique. In May 2011, several communities within Region 9, Guyana, were forced to evacuate as a result of continuous overflowing of the Takutu River coupled with incessant rainfall. Water rose in communities such as Lethem to approximately 20 inches (*Kaieteur News*, 2011). In changing the environmental conditions on which their livelihoods and culture depend, climate change can have devastating effects on indigenous communities. This relationship is especially true for agriculture and the harvesting of particular natural resources for the livelihood of the people.

The danger we face is that if the agreements about forest maintenance themselves (in lieu of payments, usually from abroad) or the direct effects of climate change lead to displaced and disenfranchised local communities, violations of customary land and territorial rights, and corruption (Griffiths, 2008; Cotula and Mayers, 2009), the enforcement of indigenous rights might become increasingly difficult to achieve. In this sense, taking an indigenous rights approach to the conservation of natural resources and the development of strategies to mitigate climate changes needs to involve addressing current and future impacts on human rights. Indigenous rights belong to the "third generation" of human rights, characterized through their collective nature, and recognized by the right to self-determination and development,[4] and ultimately the

right to self-government (see Gutmann, 1994).[5] There may be reluctance on the part of governments to recognize these collective rights fully in their national laws,[6] particularly if they lead to territorial independence. However, such resistance is not consistent with reality. No community in Guyana has ever declared or embarked on declarations of "independence" from the Guyanese State. The more realistic goal, which is not entertained by the Government of Guyana (GoG), would be the "autonomy" of the communities within the country. The important point for discussion then would center on the interpretation of autonomy, including what such a definition would mean in practice. Furthermore, in attempting to understand the rights of indigenous people in conservation and climate change mechanisms promoted by a state, it is necessary to acknowledge the wider social and political context under which those rights were sought.

Essentially, human rights are state related, in that states have an obligation to establish minimal standards for dealing with vulnerable groups such as indigenous people. But what does this entail? Conservation actions have helped communities protect their watershed, sustain healthy populations of hunted species, and protect traditional sources of medicine. Likewise, conservation might exclude certain segments from their livelihood or even result in their displacement and resettlement. In both cases, there will be impacts on social pointers such as access to safe drinking water, improvement in health, and access to natural resources for livelihoods, all of which can indicate the level of human rights achieved.[7]

ADDRESSING CONSERVATION OF NATURAL RESOURCES AND CLIMATE CHANGE IN GUYANA

Guyana, located between Suriname and Venezuela, gained independence in 1966.[8] As a condition of independence, the British Government and the Government

---

[4] As per ILO 169: 7(1).

[5] As per UNDRIP: Article 4.
[6] Collective rights are recognized in the Amerindian Act 2006 of Guyana via the collective titling to land and collective identity as a people.
[7] As noted in Article 25(1) of the Universal Declaration of Human Rights.
[8] A detailed analysis of the social and political context within which the wider conservation movement emerged in Guyana is not within the scope of this paper.

of then British Guiana agreed that indigenous people of Guyana (termed "Amerindians") should be granted "legal ownership or rights of occupancy over areas and reservations . . . where they now by tradition or custom de facto enjoy freedoms and permissions corresponding to rights of that nature" (British Guiana Independence Conference Report, 1965).

As part of this requirement, the newly created Amerindian Land Commissions (ALC) undertook visits to Amerindian communities in 1967. The ALC recommended that 24,000 out of the Amerindian-requested 43,000 square miles be granted titles. To date, only 17,000 square miles have been titled (Office of the President, 2008), the outline and amount of which "bears little relationships to Amerindian subsistence practices, the extent of ancestral lands, or indigenous land rights as defined by international human rights law" (APA, 2000). The process of legal recognition of Amerindian lands was essentially one of each subsequent step shrinking the recommendations made at the previous stage: the ALC report systematically recommended less land than every village requested; the law that approved land titles in 1976 approved (in every case) less land than had been recommended in the ALC report.

In the 1970s, the socialist Burnham administration commenced on a period of "nation building" in an attempt to assimilate Amerindians into the wider Guyanese society as a consequence of their call for titling over ancestral lands (Spinner, 1984; Ifill, 2009). The land title documents distributed with much ceremony by President Burnham contained descriptions that frequently bore no relation to the lands the communities had traditionally occupied. In some cases, the description bore no relation even to the physical territory. Without looking at individual community titles, it is difficult to say how widespread the practice of erroneous description of land titles was. Such discrepancies gave impetus to the call for demarcation of lands, and led some communities such as those in the Upper Mazaruni to seek assistance in 1995 from local Amerindian groups such as the Amerindian Peoples Association (APA) and foreign non-governmental organizations such as Forest Peoples Programme and Local Earth Observation in the process of demarcating their boundaries (see Griffiths, 2002). Griffiths (2002) writes:

> Over nine months the whole territory was mapped to show boundaries, past and present-day settlements, natural resources and cultural sites using names and categories defined by the communities themselves in accordance with their language and traditions. The final community map showed the whole Upper Mazaruni basin to be covered in an impressive blanket of indigenous place names, extensive and multiple indigenous land uses, burial grounds and special traditional areas such as bodawa: "hunting and fishing reserves". Since the map was published in 1998 it has been praised by many individuals and organizations including the Organization of American States and the World Bank. Sadly, however, the GoG still refuses to acknowledge the map as a legitimate claim to indigenous land ownership. (Paragraph 5)

From 1998 onwards, four more communities underwent a similar mapping process under the helm of the APA, covering 14,000 square miles.

From independence onwards, Guyana continued to maintain a political system divided along racial lines, particularly between other ethnicities such as Indians and Africans. This division complicated the adoption of neoliberal policies in the 1980s from a financial perspective. This saw the privatization of state holdings by numerous foreign companies in the mining, timber, and agriculture sectors in Guyana's interior, where the majority of Amerindians reside (Browne, 1997). While the politics from socialism to democracy changed, the overall national strategy of assimilation of Amerindians continued (Ifill, 2009).

Similar to the political systems within Guyana, conservation can place indigenous people on a historic trend of exclusion from ownership and control of resources (Stevens, 1997). Historically, two PAs were established under separate Acts in Guyana (the Kaieteur National Park Amendment Act and the Iwokrama International Centre for Rain Forest Conservation and Development Act), with an additional five sites earmarked for PA status. As of September 2011, Parliament passed the Protected Area Act, which gives precedence for the creation of a PA system. Additionally, two new protected areas were also named in October, 2011 – Kanuku Mountains and Shell Beach.

The first two PAs, Kaieteur National Park, established in 1929, and Iwokrama Rainforest, created in 1996, adopted a "colonial conservation" approach that for the most part excluded indigenous people from planning considerations (see Colchester, 1993; La Rose, 2004). However, since then, conservation efforts have largely shifted to incorporate indigenous people in the planning, management, and decision-making processes for conservation activities. The conceptualization and formation of Konashen Community-Owned Conservation Area (COCA), managed solely by the indigenous Wai-Wais, is an example. Inclusion of indigenous rights into PAs in Guyana was largely facilitated by the passage of the Amerindian Act of 2006, which prevented the state from establishing PAs without the consent of local communities.[9] The Amerindian Act of 2006 also gave recognition to the continuation of traditional subsistence activities within PAs.[10] However, it is unclear what traditional subsistence activities entail in PAs. Would shifting agricultural practices and historical poisoning of ponds for fish still continue?

Simultaneous to the promotion of PAs, the GoG in 2009 launched its Low Carbon Development Strategy (LCDS), which provided the broad framework of Guyana's response to climate change, with the secondary goal of conserving biodiversity. With 85% of its forested land (15 million ha) still relatively intact, the LCDS would encourage developed countries to pay for the continued conservation of Guyana's forests by putting finances into carbon credit schemes under the UN's Reduced Emissions from Deforestation and Forest Degradation (REDD+) programme. The Government estimates that the value of Guyana's state-owned rainforest, if harvested for timber, rice, and palm oil, is approximately US$5.8 billion, pending fluctuation in pricing for these commodities (Office of the President – Republic of Guyana, 2008). If left standing, conservative valuations of the Economic Value to the World (EVW) as estimated by the McKinsey Report in 2007 based on carbon sequestration, suggests that Guyana's forests may contribute US$40 billion to the global economy annually (Office of the President – Republic of Guyana, 2008)

Subsequently, the Norwegian Government contributed US $30 million in 2010 and pledged up to US $250 million in total through 2015 in support for Guyana's efforts to limit forest-based greenhouse gas emissions (*Stabroek News*, 2009). The Norwegian funds were channeled to a new fund called the Guyana REDD+ Investment Fund (GRIF) and managed exclusively by GoG. Indigenous groups, which constitute approximately 9.1% of Guyana's population within 169 communities and which own approximately 14% of the land, can either "opt in" or "opt out" of the LCDS.

Presently, there are no clear indications of what are the requirements and role of communities as they "opt in or out" of the LCDS. Little has been said of their participation in the decision-making process involved in forming the LCDS and its associated policies. Initially, communities were unaware of the then President Jagdeo's "Take over our forests" plan to offer Guyana's forests to the world as a means of climate change mitigation (see Howden, 2007, Jagdeo, 2008). While the GoG has clarified that only state lands will be included in the LCDS, there is little recognition that numerous communities are still without title, and this could ultimately affect resource use and livelihood activities. Amerindian-titled lands have been estimated to cover over 3 million hectares, with forested lands of over 2.5 million hectares (see GFC and Indufor, 2012). No doubt the Norwegians will want to see community forests "opt in" so as to give further legitimacy to the scheme.

In the Amerindian Act of 2006, indigenous peoples were given the right to traditional subsistence use on state lands.[11] This raises questions of their right to free, prior, and informed consent regarding any decision affecting their traditional lands, territories, and way of life.[12] Furthermore, numerous communities have also raised concerns about the rapid "sensitization" consultations that ensued, and "the 'independent' monitors present at the LCDS meetings who were actively engaged in discussions and even encouraging 'support' of government proposals, instead of observing in a neutral manner," as reported by the Forest Peoples Programme (2009).

It must be highlighted that the principles of the LCDS revolve around clear land ownership and management. This may encourage the state to retain ownership of traditional lands in order to obtain revenue from forests. Even where ownership of indigenous lands is apparent, the state might insist that revenues obtained from REDD+ initiatives be paid directly to them rather than given to local communities, as seen in the case of GRIF and the GoG. In turn, the GoG will disburse associated funds to indigenous communities for development and other projects. This has the potential to lead to nepotism and other forms of misuse of the funds, and the whole system would need to be monitored carefully to contribute fully to its objectives.

[9] Article 58(2) of the Amerindian Act 2006 of Guyana.
[10] Article 58(3) of the Amerindian Act 2006 of Guyana.

[11] Amerindian Act 2006 of Guyana: 57
[12] As per ILO 169: 6 (or UNDRIP: 19 which relates to FPIC).

Eventually, this strategy fails to recognize that indigenous communities are the absolute owners of forests as per their collective land title under the Amerindian Act of 2006. It further undermines indigenous rights to determine and develop priorities associated with their way of life. Under the Amerindian Act of 2006, there is some indirect recognition of community ownership of sub-soil minerals by incremental or limited control over mining on community lands, albeit control over sub-soil minerals. Ultimately, vesting residual power in the hands of the Minister of Amerindian Affairs undermines the political autonomy of the communities (Section 50 of the Amerindian Act of 2006). While the land titling process should not be completely written off, it is limited. In developing countries such as Guyana where governance and management capacities for forests are weak (see Davis, 2010), involving local communities in the management of forest resources can be an effective tool against increasing deforestation.

ROLE OF TRADITIONAL PRACTICES IN CLIMATE CHANGE AND CONSERVATION IN GUYANA

As Guyana develops strategies to mitigate and adapt to climate change, it is important that weak and unscientific analysis of drivers of deforestation does not disregard historical, cultural, and existing community efforts to protect forests, as well as their inherent rights. In its REDD+ plan, Guyana identified traditional farming or shifting agriculture as a major cause of forest and biodiversity loss (Griffiths, 2008). Application of faulty analyses of forest loss and degradation risks depriving indigenous people of their legitimate means of livelihood, security, and the right to their way of life.

Shifting cultivation is a traditional farming practice used by many of the indigenous peoples in Guyana, supporting the production of much-needed food to relatively financially poor but resource-rich communities. Because small plots are used, coupled with long fallow periods and rotation of the lands used, this form of traditional land use represents a sustainable strategy (see Fox et al., 2000; Toledo et al., 2003). An analysis of bird species diversity in farmed, forested, and transition area in the indigenous Makushi village of Moco-Moco indicated that a higher number of species were present in farmed areas (566) using shifting cultivation method, followed by intermediate (444) and forested

(421) areas (Kalamandeen and Gomes, in review). Some 47% of farmers utilize a particular area for 4 to 6 years, while 28% utilize their farms between 7 and 10+ years, and 25% utilize their farms between 1 and 3 years. Farms of about two acres or fewer, like 92% of farms in Moco-Moco, and with a 4–7 year planting (or cropping) period, are termed as sustainable (Kalamandeen and Gomes, in review; Fox et al., 2000).

According to Griffiths (2008), most REDD concepts fail to distinguish between permanent and temporary forest loss and none acknowledges that these practices are often carbon neutral or carbon positive, while at the same time sustaining important biodiversity and cultural values. Primary forests were rarely felled for agricultural purposes in Moco-Moco, with different sections of farms being rotated for a 4 to 6 year crop phase and an average fallow period of 10 years.

The techniques used in farming various crops play a key role in farming success and in the sustainability of the area; such techniques are often derived from decades of traditional knowledge based on practice. Traditional knowledge provides an established way of managing natural resources, taking into consideration ecosystem dynamics as observed over generations. For this reason, it often can assist in monitoring long-term ecological changes and mitigating adverse effects.

The usefulness and value of traditional knowledge at this junction in our history is clear, as economies worldwide are in crisis, which results in less funding available for traditional species-only conservation in developing countries.[13] Coupled with this, there has been a dramatic shift of conservation funding and focus away from South America and a corresponding increase in conservation funding for projects in Africa and Southeast Asia (Venter et al., 2009).

The decline in available funding for preventing biodiversity loss can be provided by carbon funding initiatives if they are appropriately executed and respectful of indigenous rights. These areas of high value could be protected through carbon-based conservation, while others could benefit from complementary funding arising from their carbon content (Strassburg et al., 2009). Efforts to tackle climate change are thus becoming increasingly entwined with efforts to address biodiversity loss, particularly those that shrink carbon emissions via forest conservation. What implications

---

[13]Less than 12% of $6 billion is spent yearly on managing protected area is spent in less developed countries where most biodiversity occurs (Balmford et al. 2003).

this means for local communities living in and around areas earmarked for carbon projects are questions still to be researched.

CONCLUSION

As mentioned previously, indigenous rights as part of the wider human rights framework is relevant because efforts to achieve conservation and to mitigate climate change may at the same time cause human rights violations. Human rights can be helpful in approaching, adapting, and managing climate change through the inherent right for different needs being met: for example, rights to health, food, access to safe drinking water, rights associated with livelihood and culture, and rights associated with migration and resettlement (ICHR, 2008).

While climate change affects everyone, it will no doubt hit the poorest and most marginalized groups the hardest. It is key that existing REDD+ schemes such as Guyana's LCDS do not exacerbate these problems and crush the rights sought after by indigenous peoples for centuries. Few REDD models represent the views or needs of forest-dependent communities or consider adaptive responses, and mitigation strategies, and the traditional practices of indigenous people (Salick and Byg, 2007). Respect for indigenous knowledge and its integration into policies that relate to REDD suggests that new ways of looking at natural resources and the world may be developed along these lines.

In order to be truly effective, efforts to conserve natural resources and adapt to climate change, as in Guyana, can incorporate strategies learned from the traditional practices of indigenous people and their associated institutional mechanisms, including cooperation and collective action; intergenerational transmission of knowledge, skills and strategies; concern for well-being of present and future generations; reliance on local resources, which in turn enables restraint in resource exploitation and respect for nature; and consideration of traditional adaptation to change climates. Utilizing a rights-based approach[14] to conservation and climate change ensures that the strategies employed do not come at the expense of indigenous rights.

---

[14] A rights-based approach to development is a framework that integrates the norms, principles, standards and goals of the international human rights system into the plans and processes of development (Boesen and Martin, 2007).

REFERENCES

APA (Amerindian Peoples Association) (2000) Upper Mazaruni Amerindian District Council and the Forest Peoples Programme (2000) *Indigenous Peoples, Land Rights, and Mining in the Upper Mazaruni*. Global Law Association, Netherlands.

Balmford, A., Gaston, K. J., Blyth, S., James, A. and Kapos, V. (2003) Global variation in terrestrial conservation costs, conservation benefits, and unmet conservation needs. *Proceedings of the National Academy of Sciences of the United States of America*, **100**, 1046–1050.

Boesen, J. K. and Martin, T. (2007) *Applying a Rights-based Approach: An Inspirational Guide for Civil Society*. The Danish Institute for Human Rights. http://www.humanrights.dk/files/pdf/Publikationer/applying%20a%20rights%20based%20approach.pdf (accessed March 19, 2013).

British Guiana Independence Conference Report (1965) Quoted in: *Indigenous Peoples, Land Rights, and Mining in the Upper Mazaruni* (eds Amerindian Peoples Association, Upper Mazaruni Amerindian District Council and Forest Peoples Programme) (2001) Global Law Association. Netherlands.

Browne, S. (1997) *The Rise and Fall of Development Aid*. UNU World, Helsinki. Institute for Development Economics Research, UNDP.

Colchester, M. (1993) *Who's Who in Guyana's Forests?* Report prepared for the Amerindian Peoples Association. World Rainforest Movement, Penang, Malaysia.

Colchester, M. (2004) Conservation policy and indigenous people. *Cultural Survival Quarterly*, **28**, 17–22.

Cotula, L. and Mayers, J. (2009) Tenure in REDD – Startpoint or afterthought? Natural Resource Issues no. 15. International Institute for Environment and Development. London.

Davis, C. (2010) *Governance in REDD+: Taking Stock of Governance Issues Raised In Readiness Proposals*. Monitoring Governance Safeguards in REDD+ Expert Workshop, Chatham House, London. http://www.fao.org/climatechange/21145-091981d43d2eb7409b8a710e700c6571.pdf (accessed March 19, 2013).

Forest Peoples Programme (2009) *Guyana: Indigenous Peoples, Forests and Climate Initiatives*. Rights, forests and climate briefing series – November 2009. http://www.forestpeoples.org/sites/fpp/files/publication/2010/02/guyanabriefingnov09eng.pdf (accessed March 19, 2013).

Fox, J., Truong, D. M., Rambo, A. T., Tuyen, N.P. and Cuc, L. T. (2000) Shifting cultivation: a new old paradigm for managing tropical forests. *Bioscience*, **50**, 521–528.

GFC and Indufor (2012) *Guyana REDD+ Monitoring Reporting and Verification System* (MRVS) Interim Measures Report 01 October 2010 – 31 December 2011 Version 1. June 15, 2012. Guyana Forestry Commission publication.

Griffiths, T. (2002) *Guyana: Empowerment of Indigenous Peoples through Participatory Mapping*. WRM's bulletin no. 62,

September. http://www.wrm.org.uy/bulletin/62/Guyana.html (accessed March 19, 2013).

Griffiths, T. (2008) *Seeing REDD? Forests, Climate Change Mitigation and the Rights of Indigenous Peoples and Local Communities*. Forest Peoples Programme, UK.

Gutmann, A. (ed.) (1994) *Multiculturalism: Examining the Politics of Recognition*. Princeton University Press, Princeton, NJ.

Haynes, T. and Ostrom, E. (2005) Conserving the world's forests: are protected areas the only way? *Indiana Law Review*, **38**, 595–617.

Howden, D. (2007) *Take over our Rainforest*. 24 November. http://www.independent.co.uk/environment/climate-change/take-over-our-rainforest-760211.html (accessed March 19, 2013).

Ifill, M. (2009) *The Indigenous Struggle: Challenging and Undermining Capitalism and Liberal Democracy*. http://www.devstud.org.uk/aqadmin/media/uploads/4ab8f1299ecbe_SA5-ifill-dsa09.pdf (accessed March 19, 2013).

ICHR (International Council on Human Rights) (2008) *Climate Change and Human Rights: A Rough Guide*. International Council on Human Rights Policy, Versoix, Switzerland.

ILO (International Labour Organisation) (1993–2013) *Convention No. 169 – Indigenous and Tribal Peoples Convention, 1989 (169)*. http://www.ilo.org/dyn/normlex/en/f?p=1000:12100:0::NO::P12100_ILO_CODE:C169 (accessed March 19, 2013).

Jagdeo, B. (2008) Why the West should put money in the trees. *BBC News*. http://news.bbc.co.uk/1/hi/sci/tech/7603695.stm (accessed March 19, 2013).

Kaieteur News (2011) Flood-hit Lethem out of electricity and potable water. http://www.kaieteurnewsonline.com/2011/06/06/flood-hit-lethem-out-of-electricity-and-potable-water (accessed March 19, 2013). June 6.

Kalamandeen, M. and Gillson, L. (2007) Demything "wilderness": implications for protected area designation and management. *Biodiversity and Conservation*, **16**, 165–182.

Kalamandeen, M. and Gomes, F. (in review) Impact of shifting agriculture on avifaunal diversity in Moco-Moco Village, Guyana. BSc thesis, University of Guyana.

LaRose, J. (2004) Indigenous Lands or National Park. *Cultural Survival Quarterly*. http://www.culturalsurvival.org/publications/cultural-survival-quarterly/guyana/guyana-indigenous-peoples-fight-join-conservation-ef (accessed March 19, 2013).

McShane, T. O. (2003) Protected areas and poverty: the linkages and how to address them. *Policy Matters*, **12**, 52–53.

Office of the President – Republic of Guyana (2008) *Saving the World's Rainforests Today. Creating Incentives to Avoid Deforestation*. Office of the President, Georgetown, Guyana.

OHCHR (2011) *What are Human Rights?* Office of the High Commissioner for Human Rights http://www.ohchr.org/en/issues/Pages/WhatareHumanRights.aspx (accessed March 19, 2013).

Salick, J. and Byg, A. (eds) (2007) *Indigenous Peoples and Climate Change*. Tyndall Centre for Climate Change Research, Oxford. http://tyndall.ac.uk/sites/default/files/Indigenous%20Peoples%20and%20Climate%20Change_0.pdf (accessed March 19, 2013).

Spinner, T. J., Jr. (1984) *A Political and Social History of Guyana, 1945–1983*. Westview Press, Boulder, CO.

Stabroek News (2009) *Jagdeo Hails Norway Forest Deal as "Our Copenhagen."* http://www.stabroeknews.com/2009/news/stories/11/11/jagdeo-hails-norway-forest-deal-as-'our-copenhagen (accessed March 19, 2013).

Stevens, S. (ed.) (1997) *Conservation Through Cultural Survival. Indigenous Peoples and Protected Areas*. Island Press, Washington, D.C.

Strassburg, B. B. N., Kelly, A., Balmford, A., Davies, R. G., Gibbs, H. K, Lovett, A., Miles, L., Orme, C. D. L., Price, J., Turner, R. K. and Rodrigues, A. S. L. (2009) Global congruence of carbon storage and biodiversity in terrestrial ecosystems. *Conservation Letters* **3**, 98–105.

Toledo, V. M., Ortiz-Espejel, B., Cortés, L., Moguel, P. and Ordoñez, M. D. J. (2003) The multiple use of tropical forests by indigenous peoples in Mexico: a case of adaptive management. *Conservation Ecology*, **7**, 9.

UNEP (2010) *Coverage of Protected Areas*. http://www.twentyten.net/pacoverage (accessed March 19, 2013).

United Nations (2013) *Declaration on the Rights of Indigenous Peoples*. http://social.un.org/index/IndigenousPeoples/DeclarationontheRightsofIndigenousPeoples.aspx (accessed March 19, 2013).

Venter, O., Laurance, W. F., Iwamura, T., Wilson, K. A., Fuller, R. A. and Possingham, H. P. (2009) Harnessing carbon payments to protect biodiversity. *Science*, **326**, 1368.

CHAPTER 13

CHALLENGES AND OPPORTUNITIES FOR CONSERVATION OF MEXICAN BIODIVERSITY

Gerardo Ceballos[1] and Andrés García[2]

[1]Instituto de Ecología, Universidad Nacional Autónoma de México, México, D.F. México
[2]Estación de Biología Chamela, Instituto de Biología, Universidad Nacional Autónoma de México, Melaque, Jalisco, México

SUMMARY

The world is clearly at a major turning point. Now, for the first time in human history, scientists are concerned about the possible collapse of global civilization. That is because the local components of the global ecosystem that support our lives and economy are tightly interconnected. In this context, the future of biodiversity largely depends on the future of human population growth, levels of individual consumption, the choice of particular technologies, and the conservation actions we take at national and regional scales. Mexico is one of the top five most biodiverse countries in the world, hosting up to 10% of all living species, many of them endemic to the country. However, the country faces severe environmental problems derived from its rapid human population growth, high poverty levels, and the war on drugs, among other factors. Climate change, habitat fragmentation, illegal hunting, overexploitation of wild populations, pollution, and invasive species are some of the major drivers of biodiversity loss in Mexico. For example, at least 70 vertebrate species in the country have become extinct or extirpated in the last 100 years and deforestation rates remain at about 260,000 hectares (ha) per year. The main conservation challenge is to reduce environmental degradation to maintain biodiversity, ecosystem services, and human well-being. In this context, Mexico has made a heroic effort in the last two decades to preserve its biodiversity despite its economic, social, and political problems. Conservation policies have been focused at both species and ecosystems, and much progress has been achieved in the form of suitable conservation policies, the development of a protected area network, and the identification of endangered and priority species.

BIODIVERSITY IN MEXICO

Mexico covers 2 million square kilometers, ranking it the thirteenth largest country in the world. It is considered a megadiverse country because it maintains up to10% of all extant species in the world in less than 1% of the global land mass (Ceballos and Brown, 1995; Mittermeier et al., 1999). It ranks first in species richness of reptiles and amphibians combined, third for mammals, fourth for vascular plants, and eighth for birds (Ceballos and Oliva, 2005; Flores-Villela and

Canseco-Márquez 2004; Navarro-Sigüenza and Gordillo 2006; Villaseñor, 2004). Mexican biodiversity is unique in that about 20% of its vertebrate species are endemic to the country. Endemism is higher than 40% for amphibians, reptiles and vascular plants, and also remarkable in particular groups such as cacti (84%), orchids (48%) and pines (43%; Sarukhán *et al.*, 2009). Its biodiversity is composed of both tropical and temperate species, being the only country on Earth in which two major biogeographic realms (i.e., the Nearctic and the Neotropical) completely intergrade. There is a gradient of humidity from north to south, with arid lands and deserts dominating the northern landscapes, temperate forests covering the mountains, and tropical forests dominating the south and coastal regions. There are 50 vegetation types representing four main biomes: arid shrubland, temperate forests, tropical forests, and grasslands (Sarukhán *et al.*, 2009). Mexico is an important center of origin and domestication for crop plants; about 10% of the 128 most important plants for human use worldwide were domesticated in Mexico including corn (*Zea mays*), beans (*Phaseolus vulgaris*), squashes (e.g., *Cuburbita pepo*), chillis (e.g., *Capsicum annum*), cotton (*Gossypium hirsutum*), and cacao (*Theobroma cacao*), among others (Perales and Aguirre, 2008).

CONSERVATION PROBLEMS

Mexico has a very large human population of 112 million people, ranking it eleventh in the world. Its economy is also large, ranked fourteenth in the world, with a gross domestic product of US$1.2 trillion. But roughly 50% of its people live in poverty and there are huge inequities in income distribution (INEGI, 2010; Hanson, 2007). The national population is projected to increase to about 144 million people by 2050. Mexico faces profound environmental problems leading to the loss of biological diversity. Deforestation rates have been decreasing but still approximately 400,000 ha of natural vegetation are lost annually. Tropical ecosystems have been greatly damaged; for example, rain forests have been devastated, with only 1 million hectares surviving, covering just 5% of the area they covered in 1940. Tropical dry forests are now suffering even greater annual losses (Ceballos and Garcia, 1995; Challenger, 1998; Maass *et al.*, 2005).

Erosion is severe throughout the country, with an estimated 54 million ha (25% of the country) having severe erosion problems (Challenger, 1998). Grasslands and scrublands in central and northern Mexico are becoming desertified because of overgrazing and cropland expansion (Ceballos *et al.*, 2010). Practically all major aquifers in the central and northern regions are unsustainably used, pumping more water than the amount that naturally enters them. Not surprisingly, a large percentage of the human population lives in those regions; southern Mexico has more available water, but is less densely populated. Most hydrological basins and rivers are polluted (Challenger, 1998). Illegal hunting and trade of species, invasive species, and emerging and remerging diseases are also major environmental problems, for both natural populations and people. Climate disruption will soon become even more evident as a major force negatively affecting the fate of biodiversity (e.g., Sarukhán, 2006).

Biodiversity loss is severe in tropical and developing countries. Mexico is not an exception, and environmental problems have caused the known extirpation or extinction of more than 70 vertebrate species (Ceballos and Navarro, 1991; Ceballos and Oliva, 2005). Freshwater fishes have suffered the greatest species losses, followed by birds and mammals, with no recorded extinctions of reptiles and amphibians. There is no solid information on extinctions in plants, but hundreds of species have not been collected for decades. More than 2600 species of plants and animals are officially considered at risk of extinction (SEMARNAT, 2010), but there is no doubt that many more, especially plants and invertebrates, are also threatened. Some groups such as freshwater fishes (200 species) and cacti (255 species), many endemic to Mexico, are among the most threatened because of pollution of lakes and rivers, and illegal trade, respectively. There are no estimates of population extinctions, but very likely thousands of vertebrate species populations, such as those of black bears (*Ursus americanus*) and scarlet macaws (*Ara macao*) have already become extinct (see Ceballos and Ehrlich, 2006).

Recent studies have evaluated the loss or degradation of environmental services and the economic impacts associated with the loss of ecosystems and species. For example, a single hectare of deforested tropical dry forest in western Mexico loses 800 kilograms of soil annually (Maass *et al.*, 2005); similarly, the extirpation of prairie dogs leads to desertification, loss of soil fertility, water infiltration, and other services in northwestern Mexico (L. Martinez, personal communication). Losing most natural vegetation such as tropical rain forests, on one hand, and species such as the bighorn sheep (*Ovis canadensis*), on the other, can cause negative impacts at local, regional, and national scales, such as flooding and loss of environmental services and income, and thus pose severe economic, social, and political problems.

OVERVIEW OF CONSERVATION LINES

Even though the history of conservation in Mexico goes back more than one hundred years with the establishment of a national park system and the creation of environment-related governmental institutions, the pace of such activities has increased over the past two decades. Although much remains to be done, Mexico has developed a number of strong and far-reaching conservation programs that are based on solid scientific knowledge (e.g., Sarukhán, 2006). Scientists and conservationists do understand that a major challenge is to couple biological conservation and human development. Our study group has been greatly involved in shaping a comprehensive conservation strategy that incorporates mechanisms to protect both priority areas and species for conservation, and to enhance the value of human-dominated landscapes for the maintenance of biodiversity (Daily *et al.*, 2003; García 2006; Ceballos, 2007; Ceballos *et al.*, 2009; Ceballos *et al.*, 2010). Major conservation strategic axes are focused to evaluate the state of biodiversity, develop legislation and conservation policies (e.g., impact assessments and ecological zoning laws), protect endemic and endangered species through the Mexican Norm on Endangered Species, create the National Protected Areas System (NAPAS), and incentivize a better use of human-dominated landscapes through a variety of schemes, such as the creation of a system of wildlife use in conservation units (UMAs) and the establishment of a national environmental services payment program.

Evaluating the state of biodiversity

A first step to developing a national conservation strategy is to have a solid understanding of the state of nature and biodiversity. Mexico is one of a handful of countries that has made a major effort to achieve that goal. The formation of the National Commission for the Knowledge and Use of Biodiversity (CONABIO, www.conabio.gob.mx), created in 1992, was a turning

point in history of conservation in Mexico. It was established with the aims of generating and compiling information on biodiversity and developing the capacity of biodiversity informatics to make this knowledge available to the general public and decision makers (Sarukhán and Dirzo, 1992). Among the many notable achievements of CONABIO, the most remarkable have been the creation of a National Information Network on Biodiversity and the development of a National Biodiversity Strategy. These activities are complementary to those of other federal government agencies such as the Ministry of Environment and Natural Resources (SEMARNAT, www.semarnat.gob.mx) and the National Institute of Ecology (INE, www.ine.gob.mx), which share the goal of gaining information about Mexico's ecosystems and biodiversity in order to conserve them effectively.

Legislation and conservation policies

Mexico's environmental legislation includes the constitution, laws, and other legislation such as executive orders and official Mexican standards related to the environment enacted by the Mexican government. The General Law of Ecological Equilibrium and Environmental Protection (LGEEPA) was Mexico's first comprehensive environmental law made in 1988. It provides a legal framework to try to ensure the preservation and restoration of ecological balance, sustainable development, and preservation of biodiversity (SEMARNAT, 2007). The LGEEPA contains provisions for land-use regulations such as environmental impact assessments (EIAs), risk determination, and ecological zoning to encourage sustainable development of certain regions. By 2012, about 23 million ha (12% of Mexico territory) had already been classified in an ecological zoning program. About 126 environmental official Mexican standards have been issued since 1992, and 30 are exclusively related to the protection of Mexican flora and fauna in general and for endangered species, forest exploitation, and control of invasive species in particular (SEMARNAT, 2003).

Endangered and priority species for conservation

Conservation strategies to minimize the loss of species in Mexico are based on the National Endangered Species Act and on protected areas (SEMARNAT,

2010). The Endangered Species Act includes 2606 species, mostly vertebrates (1524 species or 58%) and plants (987 species or >5%). The list is continuously being revised to ensure updated information on the conservation status of each species, to include recently described species, and to apply the criteria to include or exclude species from the list based on a multicriterion analysis. Different analyses have been developed to identify priority areas for the conservation of threatened species. Priority species for conservation include endemic, threatened, and economically important species (Figure 13.1).

Global analyses for entire taxonomic groups have recently used new datasets to identify diversity hotspots and proposed conservation strategies. Some of these have concentrated on groups of vertebrates. A recent assessment of the distributional patterns of non-marine mammals of Mexico found the highest concentration of Mexican endemics in Central Mexico (Ceballos and Ehrlich, 2006). The global analysis of distribution patterns for 129 marine mammals identified 20 key conservation sites, nine of which are located along the coast of Baja California. The marine species with the most restricted range globally in the world is the vaquita (*Phocoena sinus*), a porpoise endemic to just $4000\,km^2$ in the northern Gulf of California (Pompa, Ehrlich and Ceballos, 2011). This kind of evaluation is extremely important in a megadiverse country such as Mexico, which sustains high species richness, endemism and endangerment of mammals (Ceballos, Arroyo-Cabrales and Medellín, 2002), reptiles and amphibians (Garcia 2006; Ochoa-Ochoa and Flores-Villela, 2006), and birds (Navarro-Singüenza and Sánchez-González, 2003). There are higher concentrations of vertebrate species in southern Mexico (in the states of Oaxaca, Veracruz, Chiapas, and Tabasco), whereas endemic species are concentrated in Balsas Basin, the Trans-Mexican Volcanic Belt, and Western Mexico. Recently, the identification of critical and high-risk sites associated with endemic species with restricted ranges as a comprehensive conservation strategy for Mexican vertebrate taxa (i.e., Zero Extinction Sites, Figure 13.2) reported 415 sites protecting 485 species (18% of all Mexican vertebrate species); 64 sites included 115 mammal species, 43 sites included 80 bird species, 119 sites included 169 reptile species, and 92 sites protected 121 amphibian species. This evaluation demonstrates the effectiveness of NAPAS to represent a large number of threatened species (Ceballos *et al.*, 2009).

Nature protected areas

Although most countries have decreed protected areas for conservation, few continue to do so under financial constraints and conflict with development activities. Mexico is an exception, conducting a very aggressive program to improve and expand its NAPAS (Ceballos 2007). NAPAS includes 174 areas under different categories of protection covering a total area of 25 million ha, which accounts for 12% of the land in Mexico (Figure 13.2). NAPAS is increasing the area protected under reserves, with the objective of having 30% of the land and waters of Mexico officially protected within the next 10 years. Until the 1970s, however, most nature reserves were selected for reasons not having to do with biological diversity, but mostly for preserving beautiful, frequently mountainous landscapes, that often have relatively limited value for biodiversity conservation. Consequently, the protected area system did not include all ecoregions and diversity hotspots; it was biased towards temperate and high-altitude ecosystems, with tropical dry forests, pine-oak forests, and wetlands underrepresented. Those deficiencies are now being corrected (CONABIO, 2007a, b). The evaluation of the effectiveness of NAPAS to protect species richness, endemism, and endangerment showed that 82% of the mammal species, 96% of birds, 61% of reptiles, and 38% of amphibians have been recorded in federal protected areas and are effectively protected (Ceballos, 1999; Ceballos and Ehrlich, 2006; Ceballos *et al.*, 2009). The effectiveness of the system in protecting reptiles and amphibians is inadequate. Selecting priority areas for conservation based on the representation of diverse ecosystems and species helps to optimize the use of scarce resources. So, Mexico's reserve network can be improved by designating complementary protected areas and priority sites (Ceballos and Ehrlich, 2006). Currently, such optimization schemes are being used to select new reserves (Garcia 2006; Ceballos, 2007).

Conservation in human dominated landscapes

The future of biodiversity largely depends on the future of food production and conservation actions in human-dominated landscapes (e.g., Daily *et al.*, 2003). In Mexico, there are several national programs that have been implemented to increase the value of human-

Figure 13.1 Examples of different kinds of priority species for conservation in Mexico. Species on the verge of extinction such as the Black-footed ferret (*Mustela nigripes*) and bison (*Bison bison*) are recovering through reintroduction programs. There are 400 species of vertebrates considered to be endangered in Mexico such as the black tailed prairie dog (*Cynomys ludovicianus*), the Scarlet macaw (*Ara macao*), and the Mexican beaded lizard (*Heloderma horridum*). Hundreds of endemic species to Mexico include the Toluca silverside (*Chirostoma jordani*), Tropical anolis (*Anolis taylori*), and the Black jack rabbit (*Lepus insularis*). Economically important species are the jaguar (*Panthera onca*) and mule deer (*Odocoileus hemionus*). Photos courtesy of © Gerardo Ceballos

dominated landscapes for conservation and to provide incentives to maintain natural vegetation in areas not officially protected. The system of wildlife use and conservation units (UMAs) is a strategy to protect both species and their habitats as a part of the National Program for Wildlife Conservation and Productive Diversification of the Rural Sector (SEMARNAT, 1997). UMAs are the legal mechanism for the use, capture, and trade of wildlife species. Hunting, legal wildlife trade, and large-scale commercial use of species are

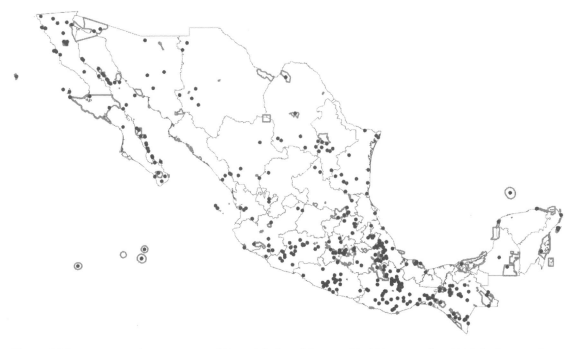

Figure 13.2 Priority areas for conservation of Mexico's biological diversity. NAPAS has more than 170 federal protected areas (gray boundaries) covering almost 12% of the Mexican territory. There are 415 Zero Extinction Sites (black dots) that represent areas where 485 micro-endemic species (i.e., known from one to three localities) are found.

possible only in UMAs, which require a management program. At the end of the year 2011, there were about 10,607 UMAs containing 1130 vertebrate species and subspecies under intensive (605 species and subspecies) or extensive (697) management. Land surface within UMAs is around 35.7 million ha that account for 18% of the Mexican territory (www.semarnat.gob.mx). Together, NAPAS and UMAs cover more than a third (31%) of the Mexican territory.

Another successful program to incentivize the conservation of habitats and species in human-dominated landscapes has been the environmental services easement program. The federal government has a budget of more than 600 million dollars to pay landowners to maintain natural ecosystems on their lands. The program is still in an experimental phase, but it promises to become one of the most important conservation mechanisms in the next decade.

CONCLUDING REMARKS

Mexico is facing extremely difficult social times that put a severe burden on the environment, economics, and political stability of the country. A misguided effort to reduce the influence of drug cartels has resulted in increasing insecurity, corruption, impunity, and lack of law enforcement. Despite these problems, some sectors of civil society, private initiative, and government have made heroic efforts to halt and restore environmental degradation in general and the loss of biodiversity in particular. Although much remains to be done, there has been great progress in protecting biodiversity in Mexico over the last two decades. Major challenges lie ahead such as climate disruption, habitat fragmentation, pollution, emerging and re-emerging diseases, and invasive species – all of them root causes of biodiversity loss. But the Mexican people and government have shown that they possess the will, the science, the technology, and that they recognize the need to protect their ecosystems and species in order to maintain economic growth, ecosystem services, and human well-being. Successful programs such as the establishment of protected areas, the designation of priority species for conservation, the setting up of functional wildlife management units, and the provision of economic incentives for conservation should be carried out to form the backbone for guiding conservation action. Coupled with strategic efforts to halt some of the root causes of biodiversity loss, they may offer hope to Mexico's people and for the survival of its outstanding biological diversity. The next decade will probably define the future of the incredible plants, animals, and ecosystems that occur within the boundaries of Mexico. As it does so, it will also likely define the economic, social, and political viability of the country.

ACKNOWLEDGMENTS

We would like to thank the Programa de Apoyo a Proyectos de Investigación e Innovación Tecnológica Proyecto IN 211811 de la Universidad Nacional Autónoma de México for providing funds to develop this chapter.

REFERENCES

Ceballos, G. (1999) Áreas prioritarias para la conservación de los mamíferos de México. *Biodiversitas*, **27**, 1–8.

Ceballos, G. (2007) Conservation priorities for mammals in megadiverse Mexico: the efficiency of reserve networks. *Ecological Applications*, **17**, 569–578.

Ceballos, G. and Brown, J. H. (1995) Global patterns of mammalian diversity, endemism, and endangerment. *Conservation Biology*, **9**, 559–568.

Ceballos, G. and Ehrlich, P. R. (2006) Global mammal distributions, biodiversity hotspots, and conservation. *Proceedings of the National Academy of Sciences of the U. S. A.*, **103**, 19374–19379.

Ceballos, G. and García, A. (1995) Conserving Neotropical biodiversity: the role of dry forests in Western Mexico. *Conservation Biology*, **9**, 1349–1353.

Ceballos G. and Navarro, D. (1991) Diversity and conservation of Mexican mammals, in *Topics in Latin American Mammalogy: History, Biodiversity, and Education* (eds M. A. Mares and D. J. Schmidly). University of Oklahoma Press, Norman, OK, pp. 167–198.

Ceballos, G. and Oliva, G. (eds) (2005) *Los mamíferos silvestres de México.* CONABIO-Fondo de Cultura Económica, México, D.F.

Ceballos, G., Arroyo-Cabrales, J. and Medellín, R. A. (2002) The mammals of México: composition, distribution, and status. *Occasional Papers, Texas Tech University*, **218**, 1–27.

Ceballos, G., Davidson, A., List, R., Pacheco, J., Manzano-Fischer, P. and Santos, G. (2010) Rapid collapse of a grassland system and its large scale ecological and conservation implications. *PLoS ONE*, **5**, e8562.

Ceballos, G., Díaz Pardo, E., Espinosa, H., Flores Villela, O. F., García, A., Martínez , L., Martínez Meyer, E., Navarro, A., Ochoa, L., Salazar, I. and Santos Barrera, G. (2009) Zonas críticas y de alto riesgo para la conservación de la biodiversidad de México, in *Capital Natural de México*, Vol. II: *Estado de Conservación y Tendencias de Cambio* (ed. J. Sarukhán), CONABIO, México City, México, pp. 575–600.

Challenger, A. (1998) *Utilización y conservación de los ecosistemas terrestres de México: pasado, presente y futuro.* CONABIO, México City, Mexico.

CONABIO (2007a) *Análisis de vacíos y omisiones en conservación de la biodiversidad marina de México: océanos, costas e islas.* CONABIO, México City, Mexico.

CONABIO (2007b) *Análisis de vacíos y omisiones en conservación de la biodiversidad terrestre de México: espacios y especies.* CONABIO, México City, Mexico.

Daily, G., Ceballos, G., Pacheco, J., Suzan, G., and López, A. (2003) Countryside biogeography of Neotropical mammals: conservation opportunities in agricultural landscapes of Costa Rica. *Conservation Biology*, **17**, 1814–1826.

Flores-Villela, O. and Canseco-Márquez, L. (2004) Nuevas Especies y Cambios Taxonómicos para la Herpetofauna de México. *Acta Zoológica Mexicana*, **20**, 115–144.

García, A. (2006) Using ecological niche modeling to identify diversity hotspots of the herpetofauna of Pacific lowlands and adjacent interior valleys of Mexico. *Biological Conservation*, **130**, 25–46.

Hanson, G. H. (2007) Globalization, labor income, and poverty in Mexico, in *Globalization and Poverty* (ed. A. Harrison), University of Chicago Press, Chicago, pp. 417–454.

INEGI (2010) *Principales resultados del censo de población y vivienda.* INEGI, Mexico City, Mexico.

Maass, J. P., Balvanera, P., Castillo, A., Daily, G. C., Mooney, H. A., Ehrlich, P., Quesada, M., Miranda, A., Jaramillo, V. J., García-Oliva, F., Martínez-Yrizar, A., Cotler, H., López-Blanco, J., Pérez-Jiménez, A., Búrquez, A., Tinoco, C., Ceballos, G., Barraza, L., Ayala, R., and Sarukhán, J. (2005) Ecosystem services delivered by tropical dry forests: a case study from the Pacific coast of Mexico. *Ecology and Society*, **10**, 17.

Mittermeier, R. A., Myers, N., Mittermeier, C. G. and Robles Gil, P. (1999) *Hotspots: Earth's biologically richest and most endangered terrestrial ecoregions.* CEMEX – Agrupación Sierra Madre, Mexico City, Mexico.

Navarro-Sigüenza, A. G. and Gordillo, A. (2006) *Catalogo de autoridad taxonómica de la avifauna de México*. Museo de Zoología, Facultad de Ciencias, UNAM, México D.F.

Navarro-Sigüenza, A.G. and Sánchez-González, L. A. (2003) La diversidad de las aves, in *Conservación de Aves: Experiencias en México* (eds H. Gomez de Silva and A. Oliveras de Ita), CONABIO-CIPAMEX, México City, Mexico, pp. 24–86.

Ochoa-Ochoa, L. M. and Flores-Villela, O. (2006) *Áreas de diversidad y endemismo de la herpetofauna mexicana*. UNAM-CONABIO, México.

Perales, H. R. and Aguirre, J. R. (2008) Biodiversidad humanizada, in *Capital Natural de México*, Vol. I: *Conocimiento Actual de la Biodiversidad* (ed. J. Sarukhán), CONABIO, Mexico, pp. 565–603.

Pompa, S., Ehrlich, P. R., Ceballos, G. (2011) Global distribution and conservation of marine mammals. *Proceedings of the National Academy of Sciences of the U. S. A.*, **108**, 13600–13605.

Sarukhán, J. (ed.) (2006) *Capital Natural y Bienestar Social*. CONABIO, México City, México.

Sarukhán, J. and Dirzo, R. (eds) (1992) *México ante los retos de la biodiversidad*. CONABIO, México City, México.

Sarukhán, J., Koleff, P., Carabias, J., Soberón, J., Dirzo, R., Llorente-Bousquets, J., Halffter, G., González, R., March, I., Mohar, A., Anta, S. and de la Maza, J. (2009) *Capital natural de México. Síntesis: conocimiento actual, evaluación y perspectivas de sustentabilidad*. Comisión Nacional para el Conocimiento y Uso de la Biodiversidad, México.

SEMARNAT (1997) *Programa de conservación de la vida silvestre y diversificación productiva del sector rural: 1997–2000*. SEMARNAT, México City, México.

SEMARNAT (2003) *Norma Oficial Mexicana NOM-022-SEMARNAT-2003, que establece las especificaciones para la preservación, conservación, aprovechamiento sustentable y restauración de los humedales costeros*. SEMARNAT, México City, México.

SEMARNAT (2007) *Reformas a la Ley General de Vida Silvestre, relativas a la protección del manglar*. SEMARNAT, México City, México.

SEMARNAT (2010) *Norma Oficial Mexicana NOM-059-SEMARNAT-2010, Protección ambiental-Especies nativa de México de flora y fauna silvestres-Categorías de riesgo y especificaciones para su inclusión, exclusión o cambio-Lista de especies en riesgo*. SEMARNAT. México City, México.

Villaseñor, J.L. 2004. Los géneros de plantas vasculares de la flora de México. *Boletín de la Sociedad Botánica de México*, **75**, 105–135.

CHAPTER 14

PARAGUAY'S CHALLENGE OF CONSERVING NATURAL HABITATS AND BIODIVERSITY WITH GLOBAL MARKETS DEMANDING FOR PRODUCTS

Alberto Yanosky

Guyra Paraguay, Gaetano Martino, Asunción, Paraguay

SUMMARY

Since the 1970s, the immigration of Brazilian commercial farmers producing soybeans for export has led to the massive deforestation of the Atlantic forest in the eastern region of Paraguay. By the start of the new millennium, this change in land use had caused major ecological damage and had also exacerbated the problem of landlessness and rural poverty as small farm holdings were increasingly wiped out by large-scale commercial agribusiness. By contrast, until recently the semi-arid western Chaco region, comprising 60% of the national territory but only 2% of the national population, had been largely protected due to its relative isolation. I have been one of the leading campaigners seeking to protect the rich biodiversity of the Chaco region. This region is now also under serious threat from an escalation in deforestation by Brazilian cattle companies.

INTRODUCTION

Paraguay holds rich natural resource assets but, like many other developing countries, is rapidly expanding its economy in such a way as to negatively impact these natural resources and compromise the ecological stability of much of the nation. These natural resources are threatened and in most cases destroyed to maximize gains in the short term, which might prohibit their future use. The lack of knowledge about the disappearance of natural resources might also affect human welfare. Soils, water, and biodiversity and its many ecological services are jeopardized by the increasing spread of agricultural production. Paraguay epitomizes many developing countries: it is trying to adopt democracy to allow the participation and free speech of its citizens while also supporting sound legal environmental frameworks, but completely lacks effective law enforcement. Paraguay also lacks the national capacity needed to address the environmental matters that are addressed by various international conventions.

UNIQUENESS OF PARAGUAY

Paraguay is ecologically unique, being located at the confluence of six ecoregions: the Atlantic Forest, the Humid Chaco, the Chaco Woodland or Dry Chaco, the Pantanal,

Conservation Biology: Voices from the Tropics, First Edition. Navjot S. Sodhi, Luke Gibson, and Peter H. Raven.

the Southern Grasslands, and the Cerrado. This gives the country a rich biodiversity in a relatively small territory ($406,752 \, km^2$), but also the need to have diverse land-use policies that encompass the production of livestock in the semi-arid Chaco and the humid Atlantic Forest, two extremely different environments that require very different agricultural practices. Applying the practices that work well in areas with rich forest soils to dry forest areas could cause serious damage to both the natural habitats and the species that inhabit them. Clearly, diverse agricultural policies are crucial for the country to develop in an environmentally friendly fashion. The projected growth of the national population from 6.7 to 10.3 million people by the middle of the century (Population Reference Bureau, 2012), coupled with the desire for increased standards of living, will make the development and implementation of carefully considered plans that couple development with conservation obviously necessary.

The Paraguay River, flowing through the middle of the country from north to south, divides the country into two distinct biomes. To the east of the river are the remnants of the Alto Paraná Atlantic Forest, an ecological region containing many endemic subtropical tree species as well as some tropical and some Cerrado and Pampas species. To the west of the river are vast alluvial plains supporting the Chaco Woodland, a habitat for many tree species rarely seen elsewhere in South America and home to many endangered animal species. For instance, the reportedly extinct Chacoan peccary (*Catagonus wagneri*) was discovered in the Dry Chaco of western Paraguay in 1975 (Wetzel *et al*, 1975). In fact, the Chaco is perhaps one of the little known wildernesses on our planet in which one of the last remaining groups of uncontacted human beings lives. In the Chaco, scientific activity in areas of high value for biodiversity and of crucial importance for indigenous communities, such as the Ayoreo, Ishir, Ñandeva, should also balance the needs of these cultures, which need to be consulted about activities in the territories where they live. Undoubtedly, in this region there are "corridors" or areas that are not only important for biodiversity but also for cultural and social aspects of human use of the land. These biocultural-corridors, which have been identified by academics, have rarely been taken into account for land-use planning by the government either at a national or a local level. Clearly, both the biodiversity and the cultural diversity of Paraguay need preserving: indigenous

people should be part of developing conservation plans for the areas where they live.

DEFORESTATION AND PROTECTED AREAS IN PARAGUAY

Rapid native forest loss is a major threat to Paraguay's rich biodiversity. Paraguay has experienced a massive reduction in forest cover over the last several decades. High rates of forest loss have been reported by international organizations such as the UN Food and Agriculture Organization (FAO). In 1973, 73.4% of the Atlantic Forest region was covered by forest, which was reduced to 40.7% by 1989 and further to 24.9% by 2000 (Huang *et al.*, 2007). By 2010, only around 10% of it remained, according to the non-governmental organizations (NGOs) Guyra Paraguay and WWF-Paraguay.

Two concomitant deforestation processes have contributed to the rapid forest loss, the first driven by settlers and the second driven by large private landowners. Between 1989 and 2000, 80% of the areas deforested were cleared by large private landowners and 20% by settlers. Protected areas (PAs) helped to reduce forest loss within their own boundaries. The average percentage of forest loss in the areas within 5 km of the boundary of Paraguay's major forest PAs was 39% during the 1989–2000 period (Huang *et al.*, 2007). This high rate of forest loss in the areas surrounding PAs has isolated the PAs as ecological "islands" and has also spurred deforestation within the PAs themselves. These PAs are critical to the conservation of the many species endemic to the Atlantic Forest region and surrounding areas. However, in Paraguay as with many other developing countries, the boundaries of the PAs are often ill-defined, and enforcement of the rules for preservation within them non-existent. Further, land tenure in PAs is often unclear. Clearly, there is an urgent need to genuinely "protect" the PAs because they represent the only lifeline for a large proportion of the region's biodiversity. PAs in Paraguay are claimed to cover 14% of the land area, but this represents an exaggeration. In fact, PAs only cover less than 2% of the land area, with the surrounding buffer zones receiving no support for protection. The flagship PA in the Chaco is Defensores del Chaco National Park of around 780,000 hectares (ha), which together with other national parks comprise a complex of more than 1,500,000 ha. For this vast region, only a single ranger is assigned, and there is a complete lack of logistical and financial support for that ranger.

ENCOURAGING SIGNS SINCE 2004

Slowing the high rates of deforestation just described is challenging but there has been an encouraging reversal of this trend since 2004 following a series of actions taken by the government with the support of civil society organizations. These actions included the adoption of a National Environmental Policy that prioritized the conservation of natural resources; the passage of a Zero Deforestation Law in 2004; and the strengthening of the Environmental Protection Agency (SEAM). As a result, the rate of deforestation was reduced substantially, from 100,000 ha annually in 2005 to 53,403 ha annually in 2008 (Guyra Paraguay, 2009). These are certainly good results, even though the total area of forest is still rapidly being reduced. The Zero Deforestation Law prohibits any land-use change and/or conversion of those in eastern Paraguay that still have forest cover. Furthermore, the law provides a more conducive environment to continue implementing programs that support biodiversity conservation in the Atlantic Forest. The law was approved to organize the country's national authorities to provide incentives for forest protection "due to the pending modernization of the forest sector"; the law has been twice renewed. According to Guyra Paraguay, the deforestation in the Atlantic Forest between 2006 and 2008 was reduced to 55,000 ha annually, by illegal deforestation. This deforestation rate does not include properties of fewer than 20 ha, which are in the hands of small producers and are excluded from the law. Governmental expropriation of forested areas for the land reform program is partly responsible for the transformation of extensive tracts of forest into small farms, which are often cleared for agricultural purposes. This is a clear example of political willingness due to the scientific evidence, but, at the same time, of a flexibility required for societal needs. Some of the most important tracts of standing forest were subject to expropriation and fragmentation into small holdings (<20 ha) by the government, in response to requests made by landless people, and this subdivision allowed the land to be deforested legally. We as scientists should not forget that in our democracies the votes are given by people, and politicians are in debt to those giving the votes to them. Legislation such as the Zero Deforestation Law or Moratoria should be based on sound information from both natural and social sciences, and not only on political information. All sectors of society and the academia must be part of such an initiative if it is to be successful.

The massive deforestation and accompanying loss of biological diversity in eastern Paraguay are mainly caused by government policies, such as the promotion of land invasions when the properties are not under production and a legal system that provides incentives for deforestation. Measures to counter increased land clearing for logging, livestock production, and large-scale mechanized soybean farming are almost completely lacking. Undoubtedly some people benefit in the short run from the clearing and development, though there is evidence that, once the forest is gone, the land is soon abandoned. These trends have been exacerbated by weak enforcement of existing laws, a lack of coordination in planning at the Paraguayan national and local level, and the impact of inadequate political and economic policies with regard to the country's natural resources. The establishment of environmental impact assessments has somewhat helped to remedy the situation, but, when ignoring them and even paying the fines still leaves deforestation a profitable activity, there is a disconnection with reality and little incentive to adhere to the environmental regulations. In cases of severe impact, mitigation measures do not replace the ecological services lost. Fines do not compensate for the permanent environmental damage caused. After all, deforestation leads to soil erosion, loss of soil fertility, and a decrease in the quantity and quality of water resources, thereby constraining the livelihoods and economic productivity of farmers. The rate of deforestation and land degradation in eastern Paraguay has slowed during the past decade, especially after the introduction of the Zero Deforestation Law. At the same time, an alarming rate of deforestation has begun to take place in the Chaco of western Paraguay.

THE PREDICAMENT OF THE CHACO REGION

Until recently, the Chaco, and in particular the Western Chaco, represented one of the last extensive undisturbed wilderness areas in Latin America, with ecosystem processes intact at a landscape level and relatively modest changes in the original forest cover. Recently, however, the formerly extensive forests near a major population center in the Central Chaco have now been converted to other land uses (Huang et al., 2009). As a result, the situation has changed radically, and the current minimum estimated rate of deforestation is about 200,000–300,000 ha per year (2005–2009). Land clearance for ranching is now intense, at rates often exceeding 1000 ha per day (Guyra Paraguay's monthly reports, www.guyra.org.py). By mid-2009, 19.1% of the whole Chaco region had been converted to pasture, and licenses to clear additional forests had already been issued to many additional landowners. A recent analysis of economic drivers indicates the very strong likelihood that all suitable land (i.e., land not located within the national PA system or reserved for indigenous communities or as private PAs) will have been transformed for cattle production by 2025. This represents a further 550,000 ha in the east-central Dry Chaco area alone. The clearing is a result of national policies intended to promote the expansion of livestock production in the Chaco with the goal of doubling or tripling production in the coming years to help meet the global demand for beef. If appropriate steps are taken, Paraguay has a major opportunity to provide this beef in an ecologically sustainable fashion, with minimal damage to nature. For example, natural and invasive grasslands could be used as areas for developing cattle ranches in place of areas occupied by forests. Not nearly enough is being done by the government, however, in selecting the most appropriate areas for development. Indeed, the scientific evidence on which to base such decisions is often poorly developed.

Such a phenomenal rate and scale of land-use change carries major environmental consequences. The Dry Chaco currently contains rare and specialized biological communities, with many endemic species. For example, it supports a number of Important Bird Areas (IBAs), which lie partly within PAs but are not limited to them. With the assumption that what we know about bird distribution can mirror that of other groups of endemic species, the IBAs are also internationally recognized as key biodiversity areas (Eken *et al.*, 2004). As further recognition of the region's global environmental importance, a significant part of the Chaco has also been designated a Man and Biosphere Reserve, the Biosphere Reserve of the Great Chaco (which also encompasses land in adjacent Argentina and Bolivia), with the privately owned areas forming the buffer and transition zones. Any reduction in landscape connectivity through habitat loss, degradation, and fragmentation fundamentally jeopardizes the entire ecosystem and its biodiversity, a problem that is receiving inadequate attention at present. Ultimately, the loss of sustainable natural resources and biodiversity would have negative impacts on the economic

Figure 14.1 Recent land-use changes in the Chaco Forest for livestock production. Dirt roads in between cleared plots can be detected. Reservoirs are created to capture and store water and they are established in each cleared plot.
Source: Image of recent land-use change taken from the validation of Guyra Paraguay's monitoring system on Chaco Forest.

productivity of the region, an argument for adopting and enforcing ways of protecting these assets for the long run.

The Dry Chaco ecosystem is fragile, with real long-term risks of degradation following clearance of the natural vegetation cover and hydrological modifications. Loss of soil fertility and raised salinity are two examples, leading to the sort of desertification that is already locally evident. The forest landscape of the Chaco is being transformed into square blocks of exotic pastures (Figure 14.1) by forest removal and burning (Figure 14.2). In most cases, this practice is legal as specified by permit-issuing institutions such as the National Forest Institute (INFONA) and the Secretary of the Environment (SEAM), which provide environmental licenses to remove the native forest up to 75% of the property (Figure 14.3). Such practices certainly lead to reduced opportunities for the survival of biodiversity and of functional biological communities.

In addition, the removal of trees and other vegetation in the dry forest and thicket formations has negative effects on the substantial amounts of carbon stored in the region, and thus exacerbates global warming. Thus, it is estimated that further deforestation in the east-central area of the Chaco region will result in emissions of around 60 million metric tons of CO_2, a highly significant amount (Guyra Paraguay, 2011). The other center of deforestation, in the south-western

Figure 14.2 The process of forest clearing is developed by heavy machinery; land transformers accumulate the cleared forests in lines that are later burned. Once the forest has been removed, the area is ready to receive grass exotic seeds to create forage for livestock.
Source: Image of recent land-use change taken from the validation of Guyra Paraguay's monitoring system on Chaco Forest.

Dry Chaco, carries higher carbon stocks in its natural vegetation, with deforestation already much more advanced there than in the east-central Chaco. A further threat to ecosystems and biodiversity, and ultimately to the sustainability of the region, arises from the exploration for gas and oil that is being carried out throughout the Chaco region, with adequate rules to ameliorate its environmental impacts.

These environmental issues translate into economic risks for the future as well as for indigenous communities, who gradually are losing their cultural and natural heritage. Within the Chaco live the Ayoreo, one of the last indigenous groups anywhere, some of whose people have avoided contact with the outside world. The Ayoreo live in regions containing what appears to be the richest biodiversity in the region, although it has remained poorly known, as well as the most ample supplies of water in the Chaco. More information should be accumulated about these regions and used as evidence to protect them from clearing and other forms of disturbance, both on biological and cultural grounds.

CONSERVATION BALANCES

The dramatic rate of forest clearance in the Dry Chaco has prompted widespread concern, both in Paraguay and elsewhere. A series of moratoria on forest clearance has been proposed by the government, but even if implemented effectively these moratoria afford only temporary solutions to important regional problems. The general trend, driven by strong economic forces, is largely inexorable but can be done with a view to assuring the future sustainability of the region. To assure long-term benefits to all, the concerned parties should all work together to promote a land-use pattern in which a sustainable balance is achieved between environmental and social concerns on the one hand and short-term economic return on the other.

To ensure the preservation of natural habitat in Paraguay, a system of monitoring must be established to gauge properly the impact of economic development on forests and their resident flora and fauna. The health of forests should be monitored through both GIS (geographic information system) and field surveys. Field surveys are important to verify the maintenance of biodiversity in standing forests and other habitats. Scientists need to develop methods not only for monitoring forests from satellites (Asner *et al.*, 2010), but also for field sampling of biodiversity.

The most effective measure in Paraguay, especially when backed by an efficient monitoring system, is the legal requirement to set aside and protect 25% of all forested land on a given landholding during conversion to cattle ranching. The selection of the forest to be retained is left to the rancher and results in various configurations – e.g., single contiguous blocks, multiple geometric strips, or a network of patches contributing to stock management (Figure 14.1). Management action will occur at the level of the individual property holder, a system that gives the owner freedom to develop the land according to his own needs; such management misses the opportunities for further coordination at the landscape scale, a strategy that could vastly increase the environmental benefit without any extra cost to the ranch owner. The scientific underpinnings for such coordination need to be developed fully for the long-term conservation of natural resources in a context of the "adaptive management" of landscapes.

Despite the often-confrontational tone of the debate about deforestation, the ranchers are by no means indifferent to environmental considerations. A proposed increase in the proportion of landholdings to be retained as natural forests to 40% was rescinded under industry pressure, perhaps suggesting the limits of regulation, but the legal requirement of 25% of the land for conservation is now followed scrupulously and indeed often imaginatively. Furthermore, individual members of the ranching community are already key cooperators in conserving IBAs on their properties, while others have expressed interest in the potential for selling carbon credits to maintain more extensive natural cover on their holdings. This raises the possibility of greater use of such incentives as a conservation strategy, in combination with regulatory measures. The UN's Reducing Emissions from Deforestation and Forest Degradation (REDD) programme appears to offer a good mechanism to reduce global greenhouse gas emissions, support developing countries to develop sustainably, generate benefits for local people, and benefit biodiversity. Biodiversity information should come from the scientific community for the validation and accreditation institutions, because sound biodiversity conservation should be important among the many benefits potentially generated by REDD.

Meanwhile, it must also be noted that management funding is deficient for PAs, most noticeably at the national level. To be credible, any initiative to decelerate deforestation on private land should be mirrored by improvements in the management of land already conserved. A blanket prohibition on land-use change in the Chaco is not supported by the NGO Guyra Paraguay. Their monitoring has shown that every time an announcement of a particular prohibition, regulation, or moratoria for the Chaco has been made, the rate of deforestation has actually accelerated. This indicates that the 2004 Zero Deforestation Law for the Atlantic Forest of Eastern Paraguay may not be suitable for the Chaco.

Without science-based management control in the Chaco, there is no possibility of developing the area sustainably. The national debate must be undertaken in such a way as to ensure that the rapid deforestation of the Paraguayan Chaco is slowed down to a considerable degree. For regional sustainability, the Paraguayan Chaco must be managed properly as a major component of the South American Great Chaco. In terms of the economic development, there is clearly a great deal of potential in the Chaco for producing food, especially beef, to help feed the growing human population. As the developments that will enable such productivity take place, however, it is of critical importance to support sustainable development by a system that makes production possible but at the same time leads to the preservation of the most biodiversity-rich and fragile natural areas. Several institutions, such as the NGO Guyra Paraguay, have invested significant

resources into the generation of fortnightly information on land-use changes in the South American Great Chaco, sharing this information with state authorities and citizens through the media. This has generated a broad debate on the future development of the Chaco and has enabled the bringing together of a number of organizations and groups interested in the development of the Chaco. The fact that Guyra Paraguay tracks forest clearances and other changes that are taking place does not mean that such clearances are illegal. The current regulations are a reflection of existing public policy and state action in the Chaco, but that they could be improved in such a way as to simultaneously promote economic growth, improved quality of life, job creation, and higher incomes for its entire population, and at the same time the protection of the native forest cover in part of the area for the use of indigenous communities and the preservation of biodiversity.

There is an urgent need to increase a regional vision in which each case of proposed land development is analyzed and, in critical areas, to increase the rate of protection of preserved areas, such as those of biocultural importance. There is also the need to immediately implement attractive incentives for forest conservation. The criteria for incentives have to be science-based with adequate monitoring mechanisms. There is also an urgent need to clearly define procedures to ensure the active participation of forest owners in providing benefits and environmental services for carbon markets and the corresponding payment to them for environmental services rendered. The protection and control of remaining forests, especially in the territories of biological and cultural importance, cannot be underestimated. The rescheduling of the 3001 Act for Payment for Environmental Services (PES) to compensate for the environmental debt of landowners of the Eastern Region could be directed towards the Chaco territory. This could help to conserve the Chaco instead of reforesting the already denuded Atlantic Forest in Eastern Paraguay through plantations of non-native species such as *Eucalyptus*, which because of their rapid growth prevent the regeneration of much of the previously existing biodiversity. Forest conservation should become a government priority, with the relevant state bodies charged with the conservation and sustainable use of forests given the necessary resources to do the job properly. While these changes are taking place and some of these measures are implemented, the existing legislation should be strictly enforced.

Scientists should also become involved in the implications of biodiversity for business: (1) the regulatory framework, because governments should implement measures and barriers; (2) the reputation, because public opinion weighs in consumers, clients, community, and press, which in turn begins to influence land-use patterns; and (3) finally in the financial sector, because more stringent requirements from the financial community are expected in the years to come (conditioned loans to biodiversity conservation). Scientists also need to understand the forces that are operating outside their offices and laboratories. We scientists need to recognize and add value to private and public organizations that develop and implement initiatives for the conservation and preservation of biodiversity, helping our governments in the decision making based on scientific information. As part of the international community, we have to remember E. O. Wilson's words (Powell, 2002, 2010): "The radical reduction of the world's biodiversity is something for which future generations will least forgive us."

ACKNOWLEDGMENTS

This paper would have not been written without the support of a committed scientific team at Guyra Paraguay (www.guyra.org.py). I would like to especially thank Oscar Rodas and José Luis Cartes, also the National Authorities of Paraguay who have cordially invited us to continue monitoring land-use changes in close cooperation with the environmental agencies of the country such as the Secretary of the Environment, the National Forest Institute, and the Public Ministry.

REFERENCES

Asner, G. P., Powell, G. V. N., Mascaro, J., Knapp, D. E., Clark, J. K., Jacobson, J., Kennedy-Bowdoin, T., Balaji, A., Paez-Acosta, G., Victoria, E., Secada, L., Valqui, M. and Flint Hughes, R. (2010) High-resolution forest carbon stocks and emissions in the Amazon. *Proceedings of National Academy of Sciences of the United States of America*, **107**, 16738–16742.

Eken, G. U., Bennun, L., Brooks, T. M., Darwall, W., Fishpool, L. D. C., Foster, M., Knox, D., Langhammer, P., Matiku, P., Radford, E., Salaman, P., Sechrest, W., Smith, M. L., Spector, S. and Tordoff, A. (2004) Key biodiversity areas as site conservation targets. *BioScience*, **54**, 110–118.

Guyra Paraguay (2009) Elaboración propia en base a interpretación de imágenes satelitales del sensor MODIS, Enero 2009 de la deforestación de la Región Oriental 2005–2008.

Guyra Paraguay (2011) The Paraguay Forest Conservation Project. Reduction of GHG emissions from deforestation and forest degradation in the Chaco-Pantanal ecosystem. PDD. Validation under Climate, Community and Biodiversity Alliance (2nd edn) Standards.

Huang, C., Kim, S., Altstatt, A., Townshend, J. R. G., Davis, P., Song, K., Tucker, C. J., Rodas, O., Yanosky, A., Clay, R., Musinsky, J. (2007) Rapid loss of Paraguay's Atlantic Forest and status of protected areas: a Landsat assessment. *Remote Sensing of Environment*, **106**, 460–466.

Huang, C., Kim, S., Song, K., Townshend, J. R. G., Davis, P., Altstatt, A., Rodas, O., Yanosky, A., Clay, R., Tucker, C. J. and Musinsky, J. (2009) Assessment of Paraguay's forest cover change using Landsat observations. *Global and Planetary Change*, **67**, 1–12.

Population Reference Bureau (2012) *World Population Data Sheet 2012*. http://www.prb.org/pdf12/2012-population-data-sheet_eng.pdf (accessed March 25, 2012.

Powell, A. (2002) A voice for the wilderness: E. O. Wilson talks of striking a balance between humans' needs and nature's. *Harvard Gazette*. http://news.harvard.edu/gazette/?s=A+voice+for+the+wilderness (accessed March 19, 2013).

Powell, A. (2010) *Settle down*, warns E.O. Wilson. [Online] http://news.harvard.edu/gazette/story/2010/04/settle-down-warns-e-o-wilson/

Wetzel, R. M., Robert, E. D., Robert, L. M. and Myers, P. (1975) *Catagonus*, an "extinct" peccary, alive in Paraguay. *Science*, **189**, 379–381.

Section 3: Asia

CHAPTER 15

LAND-USE CHANGE AND CONSERVATION CHALLENGES IN THE INDIAN HIMALAYA

Past, Present, and Future

Maharaj K. Pandit[1,2] and Virendra Kumar[2]

[1]*Department of Environmental Studies, University of Delhi, Delhi, India*
[2]*Centre for Interdisciplinary Studies of Mountain and Hill Environment, University of Delhi, Delhi, India*

SUMMARY

The collision between the Indian and the Eurasian plates, which began about 65 million years ago, led to the gradual uplift of the Himalaya, an uplift that continues at present. During various phases of mountain building, the Himalaya offered diverse edaphic and climatic opportunities to the immigrating biota to colonize the evolving ecosystems. The developing mountains and river divides induced geographic isolation and promoted speciation and the divergence of biota, evident now in the high species diversity and endemism observed in this range. Major changes in the profile of Himalayan biodiversity were initiated about 7000 years ago with the introduction of agriculture and the application of burning to clear the areas under cultivation. Deforestation on a commercial scale began in the Himalaya with the British Raj in the early nineteenth century and has continued in post-independence India. Population explosion, urbanization, and unprecedented hydropower development are transforming natural Himalayan landscapes into fragmented habitats, resulting in the disappearance of some ecosystems and the acceleration of biotic extinctions. Recent studies show that the ongoing deforestation in the Himalaya could reduce forest cover from 84.9% in 1970 to 52.8% in 2100, and the rate and extent of the damage could be much greater. This huge loss of forest cover is estimated to lead to the extinction of about a quarter of the endemic Himalayan species. The habitat fragmentation exacerbated by unprecedented hydropower development in the Himalaya will significantly reduce tree species richness, density, and basal cover. This contribution highlights the importance of land-use controls, the need for higher investments in conservation, education and research, the urgency in resolving human–wildlife conflicts and linking biodiversity and livelihoods to meet future conservation challenges in the Himalaya. Certainly the rapid population growth projected for India over the next few decades will exacerbate the problems of attaining the goal of sustainability and the preservation of biodiversity in India.

THE AREA

The Himalayan ranges stretch for about 2500 kilometers between Nanga Parbat (8126 m) in the west and Namcha Barwa (7756 m) in the east, covering a geographic area of over half a million square kilometers. Located between 26°30'–37° N latitude and 72°–97°30' E longitude, the Himalaya is bordered in the west by the Hindu Kush and Karakoram ranges, in the north by the high Tibetan Plateau, in the south by the plains of India and Pakistan and in the east and southeast by the Bay of Bengal and Myanmar. The width of the Himalaya from south to north varies between 100 and 400 km. The Indian Himalaya is spread over the states of Jammu and Kashmir, Himachal Pradesh, Uttarakhand, Sikkim, Arunachal Pradesh, and the northern part of West Bengal. The rest of the Himalaya falls within the political boundaries of Pakistan, Nepal, Bhutan, Tibetan Autonomous Region and China.

FORMATION OF THE HIMALAYA AND THE BUILD-UP OF BIODIVERSITY

The formation of the Himalaya followed the collision between the Indian and the Eurasian plates in early Eocene (soft collision) and Miocene (hard collision) (Gansser, 1964; Rowley, 1996). The age of the Himalaya is known to range from >65 to <40 million years ago. As the Himalaya has been elevated, it has played a changing but always significant role in affecting Asian climates (Zhisheng *et al.*, 2001). This regional history considerably and progressively influenced the evolution of Himalayan ecosystems and their biological diversity. An elaborate account of phytogeography and palaeohistory of the Himalaya is provided by Vishnu-Mittre (1984), followed by another review by Singh and Singh (1987). A fascinating biogeographic analysis of tropical floras and their links to the break-up of Gondwana is provided by Raven and Axelrod (1974).

The geological factors leading to the Himalayan mountain building brought about a staggered latitudinal/altitudinal zonation accompanying the phased uplift of the range. This provided novel opportunities for the floral and faunal elements in the area to colonize the newly evolving ecosystems and diversify. During the mid-Miocene Period, the Himalaya, with an average height of 2200–2400 m, harbored wet tropical forests, wet subtropical forests, and wet temperate forests at lower, middle, and higher elevations, respectively. The tropical wet evergreen forests consisted predominantly of Malesian and Southeast Asian elements (e.g., *Dipterocarpus, Elaeocarpus, Sterculia, Bursera*), while the temperate forests comprised a number of Palearctic genera (e.g., *Pinus, Abies, Picea, Alnus, Betula, Magnolia*) (Vishnu-Mittre, 1984; Singh and Singh, 1987). The orogenic and tectonic forces that elevated the Himalaya also made its slopes increasingly vulnerable to degradation and erosion. As a result, the newly established ecosystems were frequently disrupted, inducing ecological succession. For instance, during the upper Pliocene the vegetation consisted of palm savanna (with grasses), which are reported to be seral stages of immature and unstable habitats (Vishnu-Mittre, 1984). The post-Pliocene period in the Himalaya was marked by the arrival of African floral elements colonizing the lower reaches of the western Himalaya (Singh and Singh, 1987). Clearly, the angiosperms colonized the Himalaya throughout the course of its geological history; the western Himalaya (particularly Kashmir, but in attenuated form to central Nepal) was colonized by the elements from the Euro-Mediterranean, Caucasian, and Central Asian regions, while the eastern Himalaya acquired floral elements from the Sino-Tibetan-Japanese, Malayo-Burman and Austro-Polynesian regions. The western Himalaya vegetation was marked by the arrival of Mediterranean *Cedrus deodara* during the Pliocene (circa 3.5–2.5 million years ago), while *Larix griffithiana (L. griffithii)* dominated the eastern Himalayan landscape. The final episode of the Himalayan uplift resulted in the development of alpine and sub-alpine environments with *Quercus semecarpifolia* and *Betula utilis* as main components in the timberline forests. The large majority of species that occupied the Himalaya during the last glaciation event nearly 0.7 million years ago, followed by the warm period that has prevailed since, essentially shaped the present-day Himalayan vegetation (see Singh and Singh, 1987).

More recent events dominated by climatic changes with alternating cycles of cold and warm temperatures brought about a vegetational flux in the Himalaya. For instance, *Pinus roxburghii*, an early colonizer, was replaced by *Quercus leucotrichophora* about 8000–4500 years ago in the central Himalayan valleys (Vishnu-Mittre, 1984). It is not clear why oaks invaded the pine forests during this warm phase, considering their present-day distribution at cooler altitudes in the Himalaya (Singh and Singh, 1987). Notably, this period also coincided with the end of the last glacial cycle and a significantly weakened monsoon around 4500 years before present (Morrill, Overpeck and Cole, 2003), but whether the warmer and drier climatic conditions aided the invasion of oaks is unclear. The ensuing consolidation of Himalayan ecosystems during the past several thousand years (post-glaciation) paved the way for new eco-biological mixtures to take over.

As the progressive elevation of the range continued over the past 65–40 million years, immigrant species gradually evolved and diversified. The geographic isolation induced by high mountain peaks and wide river divides aided the processes of speciation and evolutionary divergences (Kumar, 1983). The Himalaya, therefore, played the dual role of acting as a bridge promoting influx of species from east and west, and also functioning as an area in which many endemic species evolved. Numerous workers have reported a high degree of floral diversity and endemism (ranging from 32–40% of the Himalayan flora) in the Himalaya (Nayar, 1996; Jain and Sastry, 1980). Clearly, the composition of the newly formed ecosystems was altered greatly during the latest period of glacial expansion, 110,000 to 10,000 years ago, and the modern aggregations of species in the area are mainly of recent formation and the reassembly of older communities and species that had evolved earlier (Hewitt, 2000; Mitsui et al., 2008).

THE FIRST AXE

The Himalayan biodiversity that survived the last glaciation was exposed to intense human activity as early as about 7000 years ago (Miehe, Miehe and Schlütz, 2009). A detailed account of the past 3000 years of agriculture practices and development in the Himalaya is provided by Knörzer (2000). The first part of this later phase, extending from 3000 to 2400 years ago, began with the cultivation of barley (Hordeum vulgare) and buckwheat (Fagopyrum esculentum), followed by crops such as wheat (Triticum aestivum), true millet (Panicum miliaceum), and peas (Pisum sativum) that reached the area around 2400–2100 years ago. Interestingly, these croplands, up to 4000 m and higher, were reportedly covered by coniferous forests of pines (Pinus wallichiana), junipers (Juniperus indica), and broad-leaf birch (Betula utilis) (see Knörzer, 2000; Miehe, Miehe, and Schlütz, 2009). This probably is the first authentic account of the earliest human-

induced land-use change (deforestation) in the Himalaya; these conclusions are based on robust Palaeoecological and accelerator mass spectrometry (AMS)[14]C dating studies (Kaiser et al., 2008). Additionally, a number of seed samples from the excavation sites provide evidence of early exchange of plants between the Indian subcontinent and Europe and central/east Asia as early as AD 800; many of the subfossil and recent ruderal plants and weeds found here are also found in central Europe, many of them even from prehistoric times (see Knörzer, 2000). This marks the first phase of plant introductions in the Himalaya.

Archaeological accounts of domestication of animals from 2400 years ago in the Himalaya indicate the herding of goats, sheep, and horses by the Himalayan inhabitants and point to early biotic pressures on the natural ecosystems. All of these domestic animals were introduced from the west and northwest. Earlier, it appears that humans were evidently responsible for forest fires nearly 6000–5000 years before present to clear lands for cultivation and also for firewood collection (pines being easily burned). Moreover, severe grazing by the livestock could have largely destroyed the forests, which were transformed into pastures and sometimes became overgrown with unpalatable thorny shrubs and herbs (Miehe, Miehe and Schlütz, 2009). Knörzer (2000) also points out that red deer (Cervus elaphus; called "elk" in North America), which disappeared from the region in medieval times, and blue sheep or bharal (Pseudois nayaur), which live today in the higher Himalaya, were regularly hunted by people.

Subsistence agriculture in the Himalaya expanded following increases in human population, which grew rather slowly in the region until the nineteenth century. With the consolidation of the British Raj in India, commercial deforestation began to unfold as an organized activity. It started with the establishment of tea plantations in the eastern Himalaya and culminated in rapid deforestation of the hardwood western Himalayan forests. Sal (Shorea robusta) forests in the lower Himalayan tracts and deodar (Cedrus deodara) in the higher Himalaya were especially targeted (Tucker, 1987). The governments that came into being after India's independence in 1947 continued policies from British rule, in which forests were seen primarily as a source of timber and thus of easy revenue. Several Indian Himalayan states either granted forest-felling contracts to private firms or later, in the 1970s, established government-owned Forest Development Corporations that engaged in wide-scale logging in the Himalaya.

With the regional human population growing more rapidly after Indian independence than it had earlier, the local communities began to resent the exploitation of forest wealth and its export from the areas where they lived. Herein lie the origins of the now internationally famed *Chipko* movement (see Box 15.1).

THE EXISTING SCENARIO

Census of biodiversity

Table 15.1 provides an overview of the biotic diversity of the Himalaya vis-à-vis India. The eastern Himalaya,

---

Box 15.1 *Chipko*: Success and failure of a Himalayan movement
Maharaj Pandit

In the early 1970s, forest-cutting contracts in the upper reaches of Dhauliganga Valley (in Alaknanda/Ganga catchment) were awarded to private contractors by the Uttar Pradesh State Forest Department, provoking resistance by a group of village women against commercial forest felling. Led by Gaura Devi, this group threatened to tie themselves to trees to prevent their felling, thus earning their movement the name "*chipko*," meaning to hug or embrace. No one actually hugged the trees, as erroneously reported by several authors including publication of vague photographs that show innocent children and women embracing the trees (e.g., Davidar, 2010). However, the state was aware of the developing situation in the area and in 1974 appointed a committee to look into the villagers' grievances and suggest solutions. The committee, headed by Virendra Kumar (this chapter's co-author), met with the two opposing sides making divergent claims on May 27–28, 1974. The Forest Department officials maintained that the trees earmarked for felling were only a few trees per hectare, which did not amount to clearing the forest. On the other side, the village representatives argued that such figures were only a ploy and that most of the forest would go under the contractor's axe. "During the interaction with the village community, Gaura Devi brought along some berries (*Berberis* spp.) and rhizomes (*Polygonatum verticillatum* and *Smilacina purpurea*), and argued that the villagers depended on such herbs during food shortages," reminisces Kumar.

Male members of the community mostly argued about their rights to the forest resources and objected to the import of labor from other states and the plains instead of the employment of locals.

On the first day of the meeting, the protestors lost the argument because the Forest officials claimed that only a few trees and not the whole forest would be felled. "The villagers felt discouraged by the afternoon and their frustration was apparent from their drooping faces," recalls Kumar. While sympathetic to the villagers' cause, Kumar found it hard to discredit the official claims because he did not want to appear biased. He suggested to local leaders to not act emotionally, but rationally, and did his part by spending the entire night going through the official Forest Manual (see Brandis, 1897) on scientific timber harvesting practices. The next day's meeting was crucial and Kumar went thoroughly prepared. Armed with the guidelines on timber harvest contained in the Manual, he informed the committee members that commercial felling – or even sanitary felling (removal of diseased trees) – would not be permitted because the watershed was damaged. Kumar knew the watershed was damaged after having served on another committee investigating flash-floods in the same area in 1970, which resulted in enormous loss of human life (200 people died), property, and engineering structures like canals and agriculture land downstream. The Forest officials were taken aback by this revelation, but gave in to Kumar's suggestion to investigate whether the watershed was really damaged. After a few rounds of deliberations and visits to the watershed, it was found that the upper reaches of it were indeed damaged. Based on this argument, Kumar in 1976 recommended a complete ban on commercial forest felling in the entire 1200 km^2 watershed for a period of 10 years. However, he also recommended the protection of villagers' *haq haqooq* (legal forest rights) to meet their household needs. The government accepted the recommendations and the ban on forest felling in this part of Himalaya was promulgated.

The national media was quick to jump onto this unusual environmental story, which started as a local protest. The movement was also in reaction to the rigid attitudes of the Forest officials who had stopped local communities and village cooperatives from benefiting from forests. "The ecological undertones were a later addition, once the first round of meetings was over and I had raised the importance of ecology and soil erosion in the meetings," insists Kumar. The movement was subsequently variously romanticized by urban elite as "Gandhi's ghost saving Himalayan ecology and forests," "peasant resistance," "empowerment of women," and so forth (see Agarwal, 1975; Guha, 2000), but such claims distract from understanding the forces at play properly. In any case, success came instantaneously; some people hitched onto the *Chipko* bandwagon and harvested rich dividends in terms of awards and rewards nationally and internationally. At the same time, Gaura Devi, whose earlier resistance to loggers initiated government action, was quickly forgotten and later publicly blamed for putting a stop to the area's economic development. In the early 1980s, overzealous government officials, keen to apportion the *Chipko* fame, recommended a total ban on tree cutting (including for the subsistence needs of villagers) and ordered the establishment of a national park in the region. This decision became a flash point, and a conflict developed between state-sponsored conservation initiatives and local communities. This conflict in turn infuriated the villagers against Gaura Devi, who was hence-

forth forced to live as a social outcast. One cannot but compare the close similarity between Gaura Devi and Rachel Carson, who was unfairly treated for her groundbreaking work on environmental protection in the US a decade earlier. While many reaped a rich harvest from the *Chipko* movement, Gaura Devi died forlorn, an unsung hero, in 1991. Today her metallic statuette stands witness to massive destruction of the entire valley by an unprecedented number of hydro-projects, which is in direct conflict with the government's decision to establish Valley of Flowers National Park and Nanda Devi Biosphere Reserve in the region. Gaura's house and simple statue memorial are mute witnesses to incessant blasting for a gigantic tunnel aimed to carry water to the power-house of a hydro-project located on one of the slopes whose forests she earlier stood up to save (Figure 15.1). The entire valley, the birthplace of the world-famous *Chipko* movement today, is a hotspot of hydropower generation. In that sense, *Chipko* is sad story of the failure of an important environmental movement.

Figure 15.1 A view of Reni village from where the resistance to commercial logging began as the *Chipko* movement. The picture shows the memorial bust of Gaura Devi (inset upper right) above her ancestral home (encircled). Note the present condition of the forest and far below (inset lower left) the intake tunnel of one of the several hydropower projects under construction in Dhauliganga Valley. The pictures were taken on June 26, 2010.

Table 15.1 An overview of estimates of Indian and Himalayan biotic diversity and endemism across taxonomic groups. The data is collated from a number of sources (see footnotes)

Taxonomic Group	India		Himalaya	
	Total	Endemic	Total	Endemic
Angiosperms	17, 000*	5700*	8000**	4,000**
Gymnosperms	64*	8*	44**	7**
Pteridophytes	1022*	250*	600**	150**
Bryophytes	2700*	783*	1737**	556**
Lichens	2000–2200✓✓	–	1159**	–
Fungi	20080✓	–	6900**	0
Mammals	350‡	44‡	300±	12±
Birds	1224‡	55‡	977±	15±
Reptiles	408‡	187‡	176±	48±
Amphibians	197‡	110‡	105±	42±
Fish (Freshwater)	667‡‡	284‡‡‡	269±	33±

*Nayar, M.P. (1996)
** Dhar, U. (2002)
✓ Manoharachary, et al., (2005)
✓✓ http://www.tnenvis.nic.in/tnenvis_old/Lichens%5CENVIS-MSSRF.html
± Conservation International
‡ http://ces.iisc.ernet.in/hpg/cesmg/indiabio.html
‡‡ http://zsi.gov.in/checklist/Native%20freshwater%20Fishes%20of%20India.pdf
‡‡‡ http://lntreasures.com/indiaff.html

located closer to tropical latitudes, has higher species richness and endemism (6000 flowering plant species with >30% endemics) compared with the western Himalaya (5000 flowering plant species with 24% endemics) (Nayar, 1996). The reasons for this skewed pattern of endemism in the Himalaya are manifold: (1) during the Pleistocene and Quaternary periods, glaciation affected the western part of the range more than it did the eastern part, presumably resulting in the extinction of more species in the west; (2) the moister mountains in southwestern China and their extensions southward provided much richer sources of species to recolonize the eastern Himalaya than was true in the west; and (3) the closer proximity to tropical latitudes of the eastern Himalaya. As the new regional patterns developed, humid tropical *Dipterocarpus-Anisoptera* forests and the *Podocarpus neriifolius*, a conifer with tropical affinities, disappeared from the western Himalaya during the Quaternary period (Singh and Singh, 1987). Recently invasive species, such as an alien invasive diatom, *Didymosphenia geminata*, in Himalayan rivers, where it has become the dominant constituent of the algal community, have become a potential threat to the native aquatic biodiversity in the region (Bhatt, Bhaskar and Pandit, 2008). Recent exploratory studies in the Himalaya have reported more than 350 new species of plants, vertebrates, and invertebrates (WWF, 2009); there are certainly hundreds of thousands of additional species yet to be discovered in this region (Bawa, 2010). A number of new plant taxa have been reported recently from the eastern Himalaya; strikingly, with a number of them related to commercially important endangered medicinal plants (Pandit and Babu, 1993; Sharma and Pandit, 2009) and many orchid species (Anonymous, 2006).

Current status of biodiversity

A comprehensive assessment of the conservation status of Indian biota is lacking, except for herbarium-based estimates in higher plants (Nayar and Sastry, 1987, 1988) and in animal species based on secondary sources (Tikader, 1983; Ghosh, 1994). Nayar and Sastry's *Red Data Book of Indian Plants* lists 623 taxa under various threat categories, of which 13 are reported to be extinct and 37 possibly extinct. Out of the 13 extinct species, 8 are Himalayan. These include orchids (*Pleione langenaria*, *Vanda wightii*), grasses (*Deyeuxia simlensis*, *Erochrysis rangacharii*), and a sedge (*Carex repanda*). Nearly 40% of the Indian plant taxa listed in the *Red Data Book* are threatened by habitat loss, 10% by over-exploitation, a mere 2% by natural processes, with no specific threat recorded for nearly 37% of the taxa. About 55% of the threatened taxa are forest-dwelling species and 31% belong to open, non-forested areas, a relationship suggesting that deforestation is the most important threat to species survival in the region. Official Indian reports (Ghosh, 1994) suggest that only three vertebrate species are extinct in India – the Asiatic cheetah (*Acinonyx jubatus venaticus*), pink-headed duck (*Rhodonessa caryophyllacea*) and Hima-layan quail (*Ophrysia superciliosa*). Of this group, the two bird species occurred in the Himalaya. Recent reports indicate that as many as 22 Himalayan verte-brate (8 bird and 14 mammal) species are at risk; of these, 3 species are critically endangered, 7 are endan-gered, and 12 are vulnerable (Conservation Interna-tional, 2010). However, higher estimates from other sources, such as IUCN, indicate that there are 44 Hima-layan species (6 birds and 38 mammals) at risk with six additional species (4 birds and 2 mammals) critically endangered.

DRIVERS OF BIODIVERSITY LOSS

Human population growth and urbanization

Land use changes in the Himalaya have a history of over 7000 years and they have become more intense as human populations have grown, rapidly recently. The average human population density has gone up nearly 4.5 times in the Himalayan states, from 20 persons/km^2 in 1951 to 92 persons/km^2 in 2001 (Census, 2001). In this area, the rapid transformation of small villages into towns and cities has been charac-teristic of the past three or four decades (Ghosh, 2007). This growth, followed by agricultural expansion, is largely responsible for the degradation and disappear-ance of natural ecosystems in the Himalaya (Tiwari, 2008; Pandit, 2009). Twin factors – the rapid develop-ment of roads and the availability of relatively inexpen-sive motor cars – have greatly boosted tourism, driving the rampant growth of land-use conversion in the Himalaya (Madan and Rawat, 2000). A recent study revealed that the number of tourists in Sikkim went up

nearly seven times between 1980 and 2007, this pres-sure negatively impacting regional habitats and their biodiversity (Joshi and Dhyani, 2009). One particular kind of pressure on these habitats comes from the recent upsurge in religious travels in India, particularly in the Himalaya, where a large number of shrines are located (Singh, 1993; Pandit, 2009).

The possibility of regaining a sustainable level of resource use in India, including the preservation of biodiversity, must be played out against a projected increase in the human population of the country from an estimated 1.25 billion at present to nearly 1.7 billion in 2050. In other words, the number of people added, 450 million, considerably more than the current population of South America, will amount to about 40,000 additional people net each day for India, com-peting for food and other resources.

Deforestation

Although deforestation in the tropics has been increas-ing sharply for many decades, precise estimates of tropical deforestation are difficult to obtain (Downton, 1995). Clear cutting is one thing, but varying degrees of use and disturbance are more difficult to quantify (Fearnside, 1990; Downton, 1995; Pandit *et al.*, 2007). The forest cover data in India are generated mostly by government agencies (DES, 1957–1993; FSI, 2000, 2005), and some of the reports suffer from serious inconsistencies (see Menon and Bawa, 1998; Jha, Dutt and Bawa, 2000). Discrepancies have also been reported in the rates of deforestation in the Himalaya by a number of independent studies (Kawosa, 1988; Joshi *et al.*, 2001; Anonymous, 2006, 2010; Pandit *et al.*, 2007), and there is no doubt that the problem is a real one. In a study based on satellite data, Pandit *et al.* (2007) projected that at current rates of defor-estation, total forest cover in the Indian Himalaya will be reduced from 84.9% (of the value in 1970) in 2000 to no more than 52.8% in 2100. The dense forest areas, on which many forest taxa critically depend, would decline from 75.4% of total forest area in 2000 to just 34% in 2100, which is estimated to result in the extinction of 23.6% of the taxa restricted to dense forests of the Himalaya. The study predicted that, by the year 2100, 14.8% of the endemic taxa across various taxonomic groups (e.g., plants, mammals, birds, fishes, and butterflies) would be consigned to extinction. The massive deforestation could result in the loss of as many as 366 endemic vascular plant taxa and 35 endemic vertebrate taxa. In sharp contrast, the official data, which suggest little loss of forest cover (FSI, 2005), imply that there would be absolutely no extinctions from habitat loss in the Himalaya. This is clearly not the case, especially taking into account the projected effects of land-use changes due to large-scale dam building and climate change in the region. All of the figures just reviewed are relatively uncertain, but the situation is clearly dire.

Hydropower development

Hydropower development is the single largest economic activity currently unfolding in the Himalaya. The Government of India launched a 50,000 MW hydropower initiative in 2003, wherein nearly 350 hydroelectric projects were envisaged nationwide with the majority (~300) located on Himalayan rivers. No efforts have been made to understand the impact of such an unprecedented developmental activity on the Himalayan ecosystems and their biodiversity. The impacts of river regulation on riverine ecology and aquatic biodiversity are well known, but studies on terrestrial ecosystems are largely lacking. Several studies have recognized river regulation as the most dramatic and cataclysmic event in the life of a riverine ecosystem (Dynesius and Nilsson, 1994; Gup, 1994). It sets in motion a complex chain reaction leading to colonization of former flood plains and increase in the spatial spread of vegetation in an earlier sparsely vegetated habitat (Wieringa and Morton, 1996). Dudgeon (2000) reported that habitat degradation due to flow modification in Asian rivers has threatened a number of taxa such as freshwater dolphins, crocodilians reptiles, and amphibians. Regulation of Himalayan rivers is likely to further worsen the catastrophic impacts on migrating fish species like mahseer (*Tor tor*) (Sehgal, 1999) and Hilsa (*Hilsa ilisha*) (Jhingran, 1982), and the endangered Gangetic dolphin (*Platanista gangetica gangetica*) (Wakid, 2009). A number of protected areas located in the Himalayan foothills, including Kaziranga National Park and Manas National Park, both World Heritage sites in the eastern Himalaya, as well as Jim Corbett National Park and Dudhwa National Park in the western/central Himalaya, will be seriously affected as a result of damming rivers in the upper Himalaya. The altered water flow regimes will doubtless pose a serious threat to the survival of endangered species such as one-horned rhino (*Rhinoceros unicornis*), bristly

hare (*Caprolagus hispidus*), and water buffalo (*Bubalus bubalis arnee*), which inhabit the riparian ecosystems of these protected areas.

A recent study found that the proposed hydropower projects would impact all Himalayan habitats, from tropical forests to alpine valleys; over half of these planned hydropower projects are located in and around the dense forests, which for many taxonomic groups are the richest in species (Pandit and Grumbine, 2012; Grumbine and Pandit, 2013). The studies just cited indicate that the habitat fragmentation caused by hydropower projects could reduce tree species richness by 35%, tree density by 42% and tree basal cover by 30% in the remaining undisturbed forests. The projected hydropower projects in the Himalaya would destroy nearly 1700 km^2 of forests by the submergence and disturbance associated with constructing the dams and reservoirs (Pandit and Grumbine, 2012; Grumbine and Pandit, 2013), thus likely to cause extinctions like those reported earlier in the case of Chagres river dam and Lake Guri (Terborgh, 1974; Terborgh *et al.*, 2001).

It must also be remembered that hydropower is a resource that will diminish over the years as the climate warms, with glaciers and permanent snowfields melting away relatively rapidly, especially in the eastern part of the region (National Research Council, 2012). Planning will have to be especially careful to take these changes into account.

MEETING THE CHALLENGES AHEAD

Land-use controls

Consistent and reliable data on forest cover and its depletion rate would help assist best management practices and facilitate informed land-use decisions. The ongoing advocacy for privatization and/or community ownership of forests, as opposed to state control, might not ensure long-term benefits (Rangan, 1997). It is vital to cap further land-use conversion of the Himalayan forests to achieve or hold on to at least 60% forest cover as outlined in the national forest policy of India (Saxena, 2001). The existing protected area network (8.3% of the geographic area) is inadequate to conserve the entire spectrum of ecosystems and biotic diversity in the Himalaya; thus, there is a strong case for bringing more areas under protective regulation and efficiently managing the existing ones.

Investments in conservation, education, and research

The annual budget (2007–2008) of India's Ministry of Environment and Forests for environmental protection and conservation was $138 million, over 200% higher than the $68 million expended during 2002 and 2003. These figures, however, translate into a budgeted expenditure of just $42 per square kilometer for environmental conservation. More finances need to be earmarked for conservation efforts, including payments and rewards to the management staff and the participating local communities. To ensure long-term benefits, the values of conservation practices need to be made part of curricula at primary, secondary, and tertiary level education. The immediate research agenda at tertiary educational/research institutions needs to focus in part on projecting the fate of the Himalayan biota under the influence of global warming and the impact of biological invasions on the native Himalayan biodiversity (see Pandit, 2009). Researches by independent agencies on regular monitoring of forest cover and Himalayan biodiversity need to be prioritized. There are clear indications of upward and northward plant-species range shifts and range shrinkages during the last century in the Himalaya under the impact of climate change (Nautiyal, Sharma and Pandit, 2009; Telwala et al., 2013); more intensive international collaborative efforts in the Himalayan region are needed to understand this imminent problem. It is critical to understand the link between biodiversity and human livelihoods in the Himalaya (Xu et al., 2009) and the value of the Himalayan ecosystem services. A recent study estimated nearly $1 million worth of ecosystem services accruing from alpine meadows in the western Himalaya and $23 billion from Himalayan forests overall (estimates for the year 1994) (Singh, 2007).

Target species and conservation conflicts

Medicinal plants and wildlife are regularly targeted for overexploitation and illegal trade in the Himalaya. A study by Olsen and Larsen (2003) conservatively estimated annual medicinal plant trade in alpine and subalpine Nepal of 480 to 2500 tons, valued at US$0.8–3.3 million. The unsustainable trends of medicinal plant collections from the wild have also been reported from the Chinese Himalaya (Buntaine, Mullen and Lassoie, 2007). Sadly, the large majority of the traded medici-

nal plants in Indian Himalaya are also threatened species (Kala, 2005). Threats to the Himalayan wildlife from poaching across the Himalayan nations are well known (Yi-Ming et al., 2000; Shepherd and Nijman, 2008; Namgail, 2009). Other serious threats arise from human–wildlife conflicts, which occur due to competition for resources between domestic livestock and wild animals. Reducing livestock populations and establishing livestock-free areas in the Himalayan highlands are seen as the only viable option for conflict resolution (Mishra, 1997; Mishra et al., 2004).

Biodiversity and livelihoods

A number of studies have shown that human communities across the Himalayan nations rely heavily on biotic resources to augment their marginal incomes. Any conservation strategy must address this basic issue. For one, the energy requirements of the dependent human populations in and around wildlife reserves need to be highly subsidized in view of the large-scale hydropower generation from the Himalayan rivers in India, China, and Nepal. The idea that ecotourism would sustain new regional economies needs to be executed with utmost care and deliberation. Some success stories in the Himalaya emphasize socially and culturally embedded tourism, which resists promotion of five-star consumptive culture and encourages local communities to host tourists in their traditional homes, built from locally available material. Such experiments need to be replicated to ensure sustainable livelihoods and to maintain ecological integrity of the Himalayan ecosystems. Certainly the local ownership of tourism facilities should be emphasized to the extent possible so that the benefits accrue to the local communities.

REFERENCES

Agarwal, A. (1975) Ghandi's ghost saves the Himalayan trees. *New Scientist*, **67**, 386–387.

Anonymous (2006) *Report of the Task Force on the Mountain Ecosystems (Environment and Forest Sector) for Eleventh Five-Year Plan*. Planning Commission, Government of India, New Delhi, India.

Anonymous (2010) *Report of the Task Force to Look into the Problems of Hill States and Hill Areas*. Planning Commission, Government of India, New Delhi, India.

Bawa, K. S. (2010) Cataloguing life in India: the taxonomic imperative. *Current Science*, **98**, 151–153.

Bhatt, J. P., Bhaskar, A. and Pandit, M. K. (2008) Biology, distribution and ecology of *Didymosphenia geminata* (Lyngbye) Schmidt an abundant diatom from the Indian Himalayan rivers. *Aquatic Ecology*, **42**, 347–353.

Brandis, D. (1897) *Indian Forestry*. Woking: Oriental Institute.

Buntaine, M. T., Mullen, R. B. and Lassoie, J. P. (2007) Human use and conservation planning in alpine areas of Northwestern Yunnan. *Environment, Development, and Sustainability*, **9**, 305–324.

Census (2001) *Census of India*. Registrar General and Census Commissioner, New Delhi, India.

Conservation International (2010) http://www.conservation. org/where/priority_areas/hotspots/Pages/hotspots_main.aspx (accessed March 19, 2013).

Davidar, P. (2010) Empowering women: the Chipko movement in India, in *Conservation Biology for All* (eds N. S. Sodhi and P. R. Ehrlich), Oxford University Press, New York, NY, pp. 276–277.

DES (Directorate of Economics and Statistics) (1957–1993) *Indian Agricultural Statistics Summary Tables* (Volumes from 1957 to 1993). Ministry of Agriculture and Irrigation, New Delhi, India.

Dhar, U. (2002) Conservation implications of plant endemism in high-altitude Himalaya. *Current Science*, **82**, 141–148.

Downton, M. W. (1995) Measuring tropical deforestation: development of the methods. *Environmental Conservation*, **22**, 229–240.

Dudgeon, D. (2000) The ecology of tropical Asian rivers and streams in relation to biodiversity conservation. *Annual Review of Ecology and Systematics*, **31**, 239–263.

Dynesius, M. and Nilsson, C. (1994) Fragmentation and flow regulation of river systems in the northern third of the world. *Science*, **266**, 753–761.

Fearnside, P. M. (1990) The rate and extent of deforestation in Brazilian Amazonia. *Environmental Conservation*, **17**, 213–226.

FSI (Forest Survey of India) (2000) *The State of Forest Report 1999*. Forest Survey of India, Dehradun, Ministry of Environment and Forests, Government of India.

FSI (Forest Survey of India) (2005) *The State of Forest Report 2003*. Forest Survey of India, Dehradun, Ministry of Environment and Forests, Government of India.

Gansser, A. (1964) *Geology of the Himalayas*. John Wiley & Sons, Inc., New York, NY.

Ghosh, P. K. (1994) *The Red Data Book on Indian Animals (Part I Vertebrata)*. Zoological Survey of India, Calcutta, India.

Ghosh, P. (2007) Urbanization: a potential threat to the fragile Himalayan environment. *Current Science*, **93**, 126–127.

Grumbine, R. E. and Pandit, M. K. (2013) Threats from India's Himalaya dams. *Science*, **339**, 36–37.

Guha, R. (2000) *The Unquiet Woods: Ecological Change and the Peasant Resistance in the Himalaya.*. Oxford University Press, Oxford.

Gup, T. (1994) Dammed from here to eternity: dams and biological integrity. *Trout*, **35**, 14–20.

Hewitt, G. M. (2000) The genetic legacy of the Quaternary ice ages. *Nature*, **405**, 907–913.

Jain, S. K. and Sastry, A. R. K. (1980) *Threatened Plants of India: A State of the Art Report*. Botanical Survey of India and Man and Biosphere Committee, New Delhi, India.

Jha, C. S., Dutt, C. B. S. and Bawa, K. S. (2000) Deforestation and land use changes in Western Ghats, India. *Current Science*, **79**, 231–238.

Jhingran, V. G. (1982) *Fish and Fisheries of India*. Hindustan Publishing Corporation, Delhi, India.

Joshi, R. and Dhyani, P. P. (2009) Environmental sustainability and tourism – implications of trend synergies of tourism in Sikkim Himalaya. *Current Science*, **97**, 33–41.

Joshi, P. K., Singh, S., Agarwal, S. and Roy, P. S. (2001) Forest cover assessment in western Himalayas, Himachal Pradesh using IRS 1C/1D WiFS data. *Current Science*, **80**, 941–947.

Kaiser, K., Miehe, G., Barthelmes, A., *et al.*, (2008) Turf-bearing topsoils on the central Tibetan Plateau, China: pedology, botany, geochronology. *Catena*, **73**, 300–311.

Kala, C. P. (2005) Indigenous uses, population density, and conservation of threatened medicinal plants in protected areas of the Indian Himalayas. *Conservation Biology*, **19**, 368–378.

Kawosa, M. A. (1988) *Remote Sensing of the Himalaya*. Natraj Publishers, Dehradun, India.

Knörzer, K.-H. (2000) 3000 years of agriculture in a valley of the High Himalayas. *Vegetation History and Archaeobotany*, **9**, 219–222.

Kumar, V. (1983) Pleistocene glaciation and evolutionary divergences in Himalayan rhododendrons, in *Proceedings XV International Congress of Genetics, New Delhi* (eds. M. S. Swaminathan, V. L. Chopra, B. C. Joshi, R. P. Sharma and H. C. Bansal), Oxford and IBH Publishing Co., New Delhi, India, p. 444.

Madan, S. and Rawat, L. (2000) The impacts of tourism on the environment of Mussoorie, Garhwal Himalaya, India. *The Environmentalist*, **20**, 249–255.

Manoharachary, C., Sridhar, K., Singh, R. (2005) Fungal biodiversity: Distribution, conservation and prospecting of fungi from India. *Current Science*, **89**, 58–71.

Menon, S. and Bawa, K. S. (1998) Deforestation in the tropics: reconciling disparities for estimates in India. *Ambio*, **27**, 576–577.

Miehe, G., Miehe, S. and Schlütz, F. (2009) Early human impact in the forest ecotone of southern High Asia (Hindu Kush, Himalaya). *Quaternary Research*, **71**, 255–265.

Mishra, C. (1997) Livestock depredation by large carnivores in the Indian trans-Himalaya: conflict perceptions and conservation prospects. *Environmental Conservation*, **24**, 338–343.

Mishra, C., van Wieren, S. E., Ketner, P., Heitkonig, I. M. A. and Prins, H. H. T. (2004) Competition between domestic livestock and wild bharal *Pseudois nayaur* in the Indian Trans-Himalaya. *Journal of Applied Ecology*, **41**, 344–354.

Mitsui, Y., Chen, S. T., Zhou, Z. K., Peng, C. I., Deng, Y. F.and Setoguchi, H. (2008) Phylogeny and biogeography of the genus *Ainsliaea* (Asteraceae) in the Sino-Japanese region based on nuclear rDNA and plastid DNA sequence data. *Annals of Botany*, **101**, 111–124.

Morrill, C., Overpeck, J. T. and Cole, J. E. (2003) A synthesis of abrupt changes in the Asian summer monsoon since the last deglaciation. *The Holocene*, **13**, 465–476.

Namgail, T. (2009) Mountain ungulates of the Trans-Himalayan region of Ladakh, India. *International Journal of Wilderness*, **15**, 35–40.

National Research Council (2012) *Himalayan Glaciers: Climate Change, Water Resources, and Water Security*. National Academy Press, Washington, D.C.

Nautiyal, D. C., Sharma, S. K. and Pandit, M. K. (2009) Notes on the taxonomic history, rediscovery and conservation status of two endangered species of Ceropegia (Asclepiadaceae) from Sikkim Hiamalaya. *Journal of Botanical Research Institute of Texas*, **3**, 815–822.

Nayar, M. P. (1996) *Hot Spots of Endemic Plants of India, Nepal, and Bhutan*. Tropical Botanic Garden and Research Institute, Palode, Thiruvananthapuram, India.

Nayar, M. P. and Sastry, A. R. K. (1987, 1988, 1990) *Red Data Book of Indian Plants*, Vols I–III. Botanical Survey of India, Calcutta, India.

Olsen, C. S. and Larsen, H. O. (2003) Alpine medicinal plant trade and Himalayan mountain livelihood strategies. *The Geographical Journal*, **169**, 243–254.

Pandit, M. K. (2009) Other factors at work in the melting Himalaya: follow-up to Xu *et al. Conservation Biology*, **23**, 1346–1347.

Pandit, M. K. and Babu, C. R. (1993) Cytology and taxonomy of *Coptis teeta* Wall. (Ranunculaceae). *Botanical Journal of the Linnean Society*, **111**, 371–378.

Pandit, M. K. and Grumbine, R. E. (2012) Potential effects of ongoing and proposed hydropower development on terrestrial biological diversity in the Indian Himalaya. *Conservation Biology*, **2**, 1061–1071.

Pandit, M. K., Sodhi, N. S., Koh, L. P., Bhaskar, A., Brook, B. W. (2007) Unreported yet massive deforestation driving loss of endemic biodiversity in Indian Himalaya. *Biodiversity and Conservation*, **16**,153–163.

Rangan, H. (1997) Property vs. control: the state and forest management in the Indian Himalaya. *Development and Change*, **28**, 71–94.

Raven, P. H. and Axelrod, D. I. (1974) Angiosperm biogeography and past continental movements. *Annals of the Missouri Botanical Garden*, **61**, 39–637.

Rowley, D. B. (1996) Age of initiation of collision between India and Asia: a review of stratigraphic data. *Earth and Planetary Science Letters*, **145**, 1–13.

Saxena, N. C. (2001) The new forest policy and joint forest management in India in *The Forests Handbook*, Vol. 2 (ed. J. Evans), Wiley-Blackwell, Oxford.

Sehgal, K. L. (1999) Coldwater fish and fisheries in the Indian Himalayas: rivers and streams in *Fish and Fisheries at Higher Altitudes: Asia* (ed. T. Petr). Food and Agricultural Organisation, Rome, Italy, pp. 41–63.

Sharma S. K. and Pandit, M. K. (2009) A new species of *Panax* L. (Araliaceae) from Sikkim Himalaya, India. *Systematic Botany*, **34**, 434–438.

Shepherd, C. R. and Nijman, V. (2008). The trade in bear parts from Myanmar: an illustration of the ineffectiveness of enforcement of international wildlife trade regulations. *Biodiversity and Conservation*, **17**, 35–42.

Singh, J. S. and Singh, S. P. (1987) Forest vegetation of the Himalaya. *Botanical Review*, **52**, 80–192.

Singh, S. P. (2007) *Himalayan Forest Ecosystem Services: Incorporating in National Accounting*. Central Himalayan Environment Association (CHEA), Nainital, India.

Singh, T. V. (1993) Development of tourism in the Himalayan environment: the problem of sustainability in *Himalaya: A Regional Perspective: Resources, Environment and Development* (ed. M. S. S. Rawat), Daya Publishing House, Delhi, India, pp. 63–78.

Telwala, Y., Brook, B. W., Manish, K. and Pandit, M. K. (2013) Climate-induced Elevational range shifts and increase in plant species richness in a Himalayan biodiversity epicentre. *PLoS ONE*, **8**(2), e57103. doi:10.1371/journal.pone.0057103.

Tikader, B. K. (1983) *Threatened Animals of India*. Zoological Survey of India, Calcutta, India.

Tiwari, P. (2008) Land use changes in Himalaya and their impacts on environment, society and economy: a study of the lake region in Kumaon Himalaya, India. *Advances in Atmospheric Sciences*, **25**, 1029–1042.

Terborgh, J. (1974) Preservation of natural diversity: the problem of extinction prone species. *BioScience*, **24**, 715–722.

Terborgh, J., Lopez, L., Nunez, P., Rao, M., Shahabuddin, G., Orihuela, G., Riveros, M., Ascanio, R., Adler, G. H., Lambert, T. D. and Balbas, L. (2001) Ecological meltdown in predator-free forest fragments. *Science*, **294**, 1923–1926.

Tucker, R. P. (1987) Dimensions of deforestation in the Himalaya: the historical setting. *Mountain Research and Development*, **7**, 328–331.

Vishnu-Mittre, (1984) Quaternary palaeobotany/palynology in the Himalaya: an overview. *Palaeobotanist*, **32**, 158–187.

Wakid, A. (2009) Status and distribution of the endangered Gangetic dolphin (*Platanista gangetica gangetica*) in the Brahmaputra River within India in 2005. *Current Science*, **97**, 1143–1151.

Wieringa, M. J. and Morton, A. G. (1996) Hydropower, adaptive management, and biodiversity. *Environmental Management*, **20**, 831–840.

WWF (World Wild Fund for Nature) (2009) The Eastern Himalayas: Where Worlds Collide: New Species Discoveries. Living Himalayas Network Initiatives.

Xu, J., Grumbine, E. R., Shreshtha, A., Eriksson, M., Yang, X., Wang, Y. and Wilkes, A. (2009) The melting Himalayas: cascading effects of climate change on water, biodiversity, and livelihoods. *Conservation Biology*, **23**, 520–530.

Yi-Ming L., Goa, Z. X., Li, X. H., Wang, S., Niemela, J. (2000) Illegal wildlife trade in the Himalayan region of China. *Biodiversity and Conservation*, **9**, 901–918.

Zhisheng, A., Kutzbach, J. E., Prell, W. L. and Porter, S. C. (2001) Evolution of Asian monsoons and phased uplift of the Himalaya–Tibetan plateau since Late Miocene times. *Nature*, **411**, 62–66.

CHAPTER 16

CONSERVATION CHALLENGES IN INDONESIA

Dewi M. Prawiradilaga[1] and Herwasono Soedjito[2]

[1]Division of Zoology, Research Centre for Biology, Indonesian Institute of Sciences, Cibinong-Bogor, Indonesia
[2]Division of Botany, Research Centre for Biology, Indonesian Institute of Sciences, Cibinong-Bogor, Indonesia

SUMMARY

Indonesia faces many challenges in conserving its biodiversity. The most important challenge is deforestation. Deforestation in Indonesia has mostly been associated with inappropriate policies and the misuse of natural resources. Experience has shown that here the decentralization of policies concerning resource extraction at local levels has had very damaging results. The existing economic and political systems tend to view natural resources, primarily forests, as sources of income, without considering sustainability. The situation was made worse because, even though forest concession companies obtained legal permits, they then proceeded to implement unsustainable forest management practices. The consequent high level of forest disturbance leads directly to a decrease in species diversity in the area, as well as to genetic erosion of those populations that manage to persist in the face of the destruction. Ironically, Indonesian biodiversity – despite being ranked as one of the highest in the world in terms of species richness and endemism – is poorly documented overall. As a consequence of this lack of knowledge, the country is on a path to losing a majority of its species and the associated resources without even realizing that they are present.

INTRODUCTION

Indonesia has long been recognized as a mega-diversity country – biologically and culturally. Mittermeier, Gil and Goettsch-Mittermeier, (1997) found that Indonesia has the second highest total biodiversity value, after Brazil, and the highest value for endemism. Plant diversity in Indonesia ranks fifth in the world, with more than 38,000 species, of which 55% are endemic. Palm diversity in Indonesia ranks the first in the world, with a total of 477 species, of which 225 species are endemic. At least 350 timber-producing tree species (mostly members of the family Dipterocarpaceae, see Figure 16.1) that have economic value are found in this country, 155 of them endemic to Borneo alone (Dephut, 1994; Newman, Burgess and Withmore, 1999). In terms of animal diversity, Indonesia hosts 12% of the world's mammal species, some 36% of them endemic, 17% of all bird species, 16% of all reptile and amphib-

Figure 16.1 View of low land Dipterocarp forest where most valuable tree exists. This forest ecosystem mostly under concessioners so that its conservation is a must. Photo © Herwasono Soedjito.

Table 16.1 Status of threatened species in Indonesia based on taxon group

Category	Animal (species)	Plant (species)
Extinct	3	2
Extinct in the wild	0	1
Critically endangered	44	114
Endangered	102	67
Vulnerable	242	203
Lower risk, conservation dependent	11	9
Lower risk, near threatened	294	72
Data deficient	93	40
Total	789	508

Source: BAPPENAS (2003)

ian species, 25% of all fish species, and an estimated 15% of all insect species (Indrawan, Primack and Supriatna, 2007). The country also has at least 47 distinct natural and man-made ecosystems such as botanic gardens, grand forests (*Taman Hutan Raya*), and community forest (agroforestry) (Sastrapradja *et al.*, 1989; BAPPENAS, 1993), which were later reclassified into about 90 ecosystem types (BAPPENAS, 2003). These ecosystems range from alpine grassland in the snowy mountains of West Papua (altitude >5,000 m above sea level), various tropical rainforest ecosystems from lowland to montane, shallow swamps to deep lakes, from mangroves to algal communities, sea grass beds, and coral reefs, as well as ocean ecosystems as deep as 8000 m below sea level (FAO, 2001).

Culturally, Indonesia is also exceptionally diverse. It has 336 cultural ethnic groups, the third highest cultural diversity in the world after Papua New Guinea and India (Mittermeier, Gil and Goettsch-Mittermeier, 1997). Correspondingly, Indonesia has 665 different local languages that are distributed in Papua (250), Mollucas (133), Sulawesi (105), Kalimantan (77), Lesser Sunda Islands (53), Sumatra (38), and Java-Bali (9). Some tribal communities in Papua, Kalimantan, and Sumatra still live in isolation and are traditionally dependent on forest products.

Although Indonesia is famous for its biological and cultural diversity, the country is facing extreme levels of habitat loss and associated threats to its biodiversity (Sodhi and Brook, 2006). The human population of Indonesia, currently 241 million, is projected to climb to about 310 million by mid-century (Population Reference Bureau, 2012). Coupled with a determination to achieve every higher levels of consumption while alleviating poverty, this growth will pose ever more severe challenges in the future for the attainment of sustain ability, as well as for the conservation of the rich biodiversity of the area. As in other countries, the most important threats to biodiversity in Indonesia are human activities, including habitat destruction, fragmentation, degradation (including pollution), global climate change, overexploitation, invasive species, disease, and synergies between these factors (Indrawan, Primack and Supriatna, 2007). Table 16.1 indicates the threatened status of Indonesian biodiversity based on taxonomic group.

In the face of the extreme threats faced by Indonesia's biodiversity, we attempt here to describe current conservation practices in Indonesia and the challenges that must be overcome to save as much as possible of the natural diversity in this globally important biodiversity hotspot.

CONSERVATION PRACTICES

Policy

The foundation of conservation policy is based on the Principal Law of the Republic of Indonesia, established in 1945 (*UUD 1945*), and its amendments. Article 33 section 1 of *UUD 1945* specifically states that forest ecosystem biodiversity belongs to the country and must be utilized for the wealth of the Indonesian people. This fundamental principle is supported by Act No. 5 in 1960 regarding the Basic Agrarian Law, and Act No. 5 in 1967 concerning the Basic Forestry Law. Act No. 5 in 1960 stated that the country owns all natural resources within the Indonesian border for the wealth of people, with the central government managing land use. Act No. 5 in 1967 established legally the definition of forest, including type of forest such as protected forest, wildlife reserve, and productive forest, and outlined the authority of the Minister of Forestry to manage all Indonesian forest including its biodiversity on behalf of the central government. These laws were strengthened by Law No. 5 in 1990, regarding conservation of biodiversity and ecosystems.

In 1999, there were 157 regulations concerning conservation, and this number increased over the next decade. This led to the promulgation of an additional 515 regulations dealing with the management of natural resources and establishing authority at the district level (Indrawan, Primack and Supriatna, 2007).

In the spirit of reform, Indonesia produced Law No. 32 in 2004, regarding autonomy (*UU Nomor 32 Tahun 2004 Tentang Pemerintahan Daerah*). This law gave even more authority in district level than had existed earlier. This decentralization policy allows the mayor of the district (*Bupati*) to determine the fate of the district's natural resources, and to issue permits for oil palm plantations and coal mining exploitation. Many of these environmentally destructive activities were implemented in forested areas that still hold high value for biodiversity, without considering specifically their effect on the nation's natural resources. Oil palm plantations and open pit coal mines have to date obtained permits to develop a total area of around 7 million hectares (ha); most of these areas have been completely deforested already.

At the international and regional level, Indonesia has ratified several conventions such as the Convention on International Trade in Endangered Species of Wild Fauna and Flora (CITES), RAMSAR (international convention on important wetlands) and the Convention on Biological Diversity (CBD). The CBD regulates intellectual property and benefit sharing from the development of biodiversity and regulates trade of protected species so that only the progeny of animals bred in captivity can be exported legally. Non-protected species can be exported with permits from the Ministry of Forestry for terrestrial or land species and from the Ministry of Marine Affairs and Fishery for fish or other fresh water and marine species.

Conservation areas

Protected areas in Indonesia include strict nature reserves and nature conservation areas as well as protected forests. Strict nature reserves – including nature reserves, wildlife sanctuaries, and biosphere reserves – have been established to protect biodiversity and ecosystem; they can be utilized only for limited activities such as research, education, and scientific development. In contrast, nature conservation areas – including national parks, nature recreational parks, grand forest parks and game reserves – allow a wider range of activities, provided that the area's natural resources are utilized sustainably. Finally, protected forests serve as sources of water, land, and habitats, and are expected to maintain and support healthy systems of living organisms.

Indonesia's conservation areas cover a total area of 28 million ha, consisting of 23 million ha of terrestrial and 5 million ha of marine protected areas (Sriyanto *et al.*, 2003). All terrestrial protected areas are managed under the Directorate General of Forest Protection and Nature Conservation, Ministry of Forestry, and all marine protected areas are under the Ministry of Marine Affairs and Fishery.

Species conservation and captive breeding efforts

Species conservation in Indonesia has been implemented through legal protection, limiting wildlife trade, and establishing various captive breeding programs oriented towards protection and sustainable use. Government Regulations No. 7 /1999 (*PP No. 7/1999*) protect endangered and important plant and animal species. Documents of Strategic and Action Plans for conservation of endangered species such as the Javan Hawk-eagle (*Spizaetus bartelsi*), Yellow-crested cockatoo (*Cacatua sulphurea*), Sumatran tiger (*Panthera tigris sumatranus*), Orangutans (*Pongo pygmaeus* and *Pongo abelii*), Javan gibbon (*Hylobates moloch*) and Anoa (*Bubalus depressicornis* and *B. quarlesi*) have been available. Those documents were prepared by related and concerned stakeholders and coordinated by the Directorate General of Forest Protection and Nature Conservation-Ministry of Forestry, which holds management authority. At least 538 species have been protected since 1993. To support in-situ conservation, ex-situ conservation has been established, including botanical parks, safari parks, zoos, and captive breeding facilities. In 1997, there were 23 zoos, 17 botanical parks, five arboreta, six elephant conservation centers, 31 animal captive breeding places, three safari parks, three bird parks and one reptile park (BAPPENAS, 2003; Indrawan, Primack and Supriatna, 2007). Captive breeding was first initiated in 1984 and has helped spur recoveries of 12 species including the Arowana fish from Kalimantan, Asian Box turtle (*Cuora amboinensis*), Salt water and New Guinea fresh water crocodiles (*Crocodylus porosus* and *C. novaeguineae*), Anoa from Sulawesi (*Bubalus* spp.), rusa deer, wild buffalo, banteng (*Bos javanicus*), Asian Two-horned rhinoceros (*Dicerorhinus sumatrensis*), and primates. Captive breeding of the Arowana fish, crocodile, and primate (only for long-tailed macaque (*Macaca fascicularis*)) is directed to supply the demand of wildlife trade.

Reference for biodiversity richness

Efforts to keep scientific collections of Indonesian biodiversity specimens began less than a century before the Indonesian independence. It was initiated by the Dutch biologists during Dutch colonization. The Herbarium Bogoriense (HB) was pioneered by Caspar Karl Reinwardt in 1841 and the Museum Zoologicum Bogoriense (MZB) was founded by J. C. Koningsberger in 1894. However, in further development, in 1982 Indonesian biologists founded the museum of ethnobotany to store Indonesian artefacts and, recently, LIPI Microbes and Culture (LIPIMC) to store important microbes and culture collections. All of these – HB, MZB, and LIPIMC – are under the Research Centre for Biology, Indonesian Institute of Sciences (LIPI), which holds national scientific authority for the Indonesian biodiversity. Currently the total number of scientific collections in HB, MZB, and LIPIMC have reached

Table 16.2 Herbarium collections at Herbarium Bogoriense, (LIPI)

Group	No. of specimens
Dicotyledonae	605,607
Monocotyledonae	97,140
Cryptogamae	55,968
Lichens	6,673
Algae	1,511
Musci	19,496
Hepaticae	16,876
Fungi	11,412
Pterydophyta	69,187
Gymnospermae	5,295
Fossil	166
Carpology	8,233
Wet collection	12,236
Type specimen	17,037
Total	870,869

Source: Widjaja *et al.* (2011)

870,869, 3,004,845, and 321 specimens, respectively (Widjaja *et al.*, 2011). An enumeration of herbaria and other museum collections is presented in Tables 16.2 and 16.3.

Local wisdom

Indonesian culture stores vast and diverse traditional knowledge and local wisdom (Soedjito and Sukara, 2006). Since the country spans thousands of large and small islands, it has many ethnic groups, with local cultures exceptionally diverse. These different local cultures hold their own knowledge systems on biodiversity utilization (see Soedjito, 2006; Soedjito, Purwanto and Sukara, 2009). Indeed, harmonized interaction between traditional societies and biodiversity in daily life produced dynamic knowledge systems that provide useful principles for subsisting on biological resources.

There are many examples of local wisdom in managing Indonesian biodiversity. In West Java, the Badui community has a taboo that prohibits people from overexploiting the forest. They successfully developed an agriculture system that involves conserving local rice varieties that are resistant to pests and diseases. In contrast, modern high-yield rice varieties may be more

Table 16.3 Fauna collections at the Museum Zoologicum Bogoriense, (LIPI)

Taxon	No. of specimens	No. of species	Type specimens No. of species	No. of specimens
Mammal	33,794	460	117	303
Bird	32,324	1,200	166	869
Insect	2,530,743	15,805	891	2,674
Ectoparasite (insect)	3,310	105	11	26
Ectoparasite (Acari)	7,073	97	38	308
Herpetofauna	24,988	800	105	618
Mollusc	201,420	3,007	165	1,053
Fish	144,516	1,300	250	536
Crustacea	37,060	270	109	167
Helminth	1,706	116	28	51

Source: Reproduced from Widjaja et al. (2011) Widjaja, E.A., Maryanto, I., Wowor, D. and Prijono, S.N. (eds.) (2011). Status Keanekaragaman Hayati Indonesia (Status of the Indonesian biodiversity). Jakarta: LIPI Press.

sensitive to pest and diseases because they carry the same cytoplasm (BAPPENAS, 2003; Hargrove, 1988). Traditional communities in eastern Indonesia developed a "SASI" system, forbidding people to enter or collect natural resources at specific areas for certain time periods. This system controls utilization of natural resources (forest and marine) by local people or newcomers. It was based on "adat" (local custom) and religion regulations (BAPPENAS, 2003).

CHALLENGES

There are many challenges to biodiversity conservation in Indonesia. In the following, we describe the most important or critical challenges, including a shortage of taxonomists, uncoordinated government regulations, deforestation, forest fires, the loss of traditional wisdom, and overexploitation.

Our inventory of Indonesian biodiversity is still limited, and a large part of this biodiversity remains unknown to science. Conservatively, probably more than 1.5 million species of eukaryotic organisms (all organisms except bacterial and viruses) exist in Indonesia, with fewer than 150,000 (10%) of them recorded scientifically (Raven, personal communication). This estimate is based on projections of the numbers of species in the better known groups, such as seed plants, vertebrates, and butterflies, and of the

numbers recorded for more poorly known ones such as many groups of insects, fungi, and nematodes. As a result of these projections, it can be calculated that, when tropical moist forest is destroyed for any reason, at most 1 in 10 of the species in it will be known scientifically before it is lost.

Taxonomy can help in managing and utilizing Indonesia's natural resources, and should be supported by other disciplines including morphology, genetics, anatomy, histology, physiology, biochemistry, and ecology. Because of the very large numbers involved, intelligent sampling systems would need to be devised for obtaining a picture of the total diversity present in the nation. In addition, very few species of organisms in Indonesia have been evaluated for human use. Thus, from around 38,000 plant species recorded in Indonesia, only about 6000 have been considered in this respect; 1000 animal species and 100 microbe species have been included in such surveys (KPPNN, 1992, cited by Indrawan, Primack and Supriatna, 2007). The Research Center for Biology of the Indonesian Institute of Sciences (LIPI) is the main institution where most of the taxonomists work. Universities could logically be developed as another resource to develop knowledge and increase the number of taxonomists in Indonesia, but only undertaking a major inventory effort in the country will increase the number of positions available for such workers once they are trained. International cooperation in the area of taxonomic inventory is also of prime importance, since no nation can within its own population support a sufficient number of experts to deal with the very large number of different groups of organisms that exist, many of them poorly know and with few specialists available even globally.

Governing conservation areas and protecting Indonesian biodiversity is complicated (Indrawan, Primack and Supriatna, 2007; Patlis, 2008). Although regulations and laws regarding biodiversity continued to be improved, there is still much confusion in the implementation of these laws because they span several different sectors such as forestry, fishery, and mining. A recent conflict between mining permit regulations and protected forest illustrates this confusion. Law No. 41 in 1999 prohibited mining activities in various forest types including conservation forests, protected forests, production forests, and parks. However, in 2004, the House of Representatives approved the government policy to give permits to mining companies to develop open-pit mines in protected forest areas, clearly contradicting Law No. 41. In order to legalize the approval, the House of Representatives revised Act No. 41 and adopted new legislation Act No. 19 in 2004, which allows mining activities in the protected forest. Certainly this policy threatens conservation in terms of sustaining biodiversity.

Indonesian forest resources and forest ecosystems are threatened by deforestation, fragmentation, and forest conversion. Deforestation increased dramatically in the beginning of the 1970s when large-scale commercial logging began. Since then, annual deforestation rates have increased from 1 million ha in the 1980s to 1.7 million ha in the 1990s to 2 million ha since 1996. These high rates of deforestation have reduced forest cover in Indonesia from about 193.7 million ha in the 1950s (Hannibal, 1950) to 119.7 million ha in 1985 and 100 million ha in 1997 (GOI/World Bank) to only 98 million ha in the 2000s (FWI/GFW 2001). About half of the remaining Indonesian forest has been fragmented by new networks of roads and other human land-use activities such as development of plantations, tree crops, and settlements. Recently, there has been an effort to design more comprehensive spatial plans for each province in Indonesia. It involves multidisciplines and stakeholders and will be used as the only reference for the development in the fields. In addition, as a result of lessons learned from Bogor Botanical Garden where hundreds of species of Indonesia endemic plants are conserved, more botanical gardens in district level (*Kebun Raya Daerah*) have

been established as ex-situ biodiversity conservation. About 17 of these ex-situ conservation areas ranging from hundreds to thousands ha have been developed, with potentially a substantial impact on the conservation of biodiversity.

Forest fires are another cause of the depletion and degradation of forest ecosystems, and in recent decades have increased in frequency and intensity due to increasing droughts and the conversion and degradation of forest habitats that leave them vulnerable to fires. In 1997–1998, forest and land fires burned no less than 9.75 million ha of forest in the five major islands of Indonesia. In Kalimantan and Sumatra, 3.1 million ha of lowland forest and 1.45 million ha of peat swamp forest (see Figure 16.2) were damaged by fire (BAPPENAS, 2003). Forest depletion and degradation threaten the integrity of forest ecosystems and the wildlife that live in them. In 2008, the Indonesian National Board for Disaster Management (Indonesian: *Badan Nasional Penanggulangan Bencana*/BNPB) was established and forest fire is included in its strategy. BNPB has provincial level authority and its main duty is to monitor weather and fire alert primarily during dry season.

Traditional knowledge is quickly disappearing as increasingly interconnected activities come into contact with isolated societies, threatening their very way of existence. Many young generations lack the interest to acquire this valuable traditional knowledge

Figure 16.2 Peat swamp forest, a unique ecosystem that has many endangered species and important value for climate change as carbon deposit site. Photo © Herwasono Soedjito.

from older generations. In addition, the destruction and degradation of forests threaten the very home of most of these cultural ethnic groups. The disappearance of ethnic groups together with their traditional wisdom is a serious problem for human beings generally, since the philosophical diversity inherent in such cultures is a treasure for us all, an important part of our common heritage. At the same time, this loss represents a threat to the future sustainability of biological diversity, because this traditional knowledge could be applied to help manage and sustain Indonesia's biodiversity. Through Man And the Biosphere (MAB) Program of UNESCO, MAB Indonesia encourages local government to establish biosphere reserve where biological conservation and economic development can go together to increase community livelihood, as well as nurturing cultural wisdom.

Finally, overexploitation is also threatening Indonesia's biodiversity as people increasingly obtain products that are made at the expense of natural forest habitats. This could be driven by high market demand, poor law enforcement, and poor awareness in local communities towards biodiversity laws (Sodhi and Brook, 2006) as well as an increase in human population. High rates of logging have threatened the survival of many members of economically important forest trees, such as Dipterocarpaceae, while overfishing has drastically reduced populations of certain marine fish and prawn species (BAPPENAS, 2003). Overexploitation has also threatened some critically endangered bird species such as the Sulphur-crested Cockatoo (*Cacatua sulphurea abbotti*), which had only eight individuals left on Masalembo Island in 2009 (Nandika and Agustina, personal communication), and the Bali starling (*Leucopsar rothschildi*) (Figure 16.3), which currently persists only because of individuals reintroduced to its natural habitat in Bali Barat National Park (Noerdjito, personal communication). Overexploitation is associated with greediness that seeks only short-term benefit. All individuals have values, attitudes, motivations, and judgments, and these are often based in and sanctified by religious beliefs (Awoyemi *et al.*, 2012). In recent years, there has been an increased emphasis on using religious principles as a basis for biological conservation (Arief, Prasetyo and Sari, 2003). Most religions emphasize the need to protect Creation, and many recognize sacred natural sites and religious-based systems that protect biodiversity in these and other areas (Dudley, Higgins-Zogib and Mansourian, 2005). Many sacred natural sites studied in various ethnic groups in Indo-

Figure 16.3 A pair of breeding Yellow-crested Cockatoo (*Cacatua sulphurea abbotti*) in Masalembo island. IUCN Status: Critically Endangered, CITES Appendix 1, Protected by Indonesian Laws. Photo © Dudi Nandika.

nesia effectively protect biodiversity (see Soedjito, Purwanto and Sukara, 2009).

In conclusion, Indonesia is facing major challenges and requires appropriate policy and immediate actions to conserve its mega-biodiversity. A comprehensive approach should be established including increasing the number of taxonomists concerned with national biological inventory and science-based spatial planning as guidance for development so that the great national asset represented by Indonesia's biological diversity can be preserved to the extent possible and provide a sustainable base for human benefit. Awareness that biological conservation is an asset for development is needed for policy makers. Conservation is a long-term work that requires commitment and persistent effort over decades and centuries to come.

ACKNOWLEDGMENTS

We are grateful to the late Professor Navjot S. Sodhi who invited us to contribute this chapter and Dr Siti N. Prijono, the Director of the Research Centre for Biology (LIPI) for her support. Dr Daisy Wowor, Dudi Nadika, Dwi Agustina and M. Noerdjito kindly shared their unpublished data.

REFERENCES

Arief, A. J., Prasetyo, E. B. and Sari, A. K. (2003) *Peran Agama dan Etika Dalam Konservasi Sumberdaya Alam Dan Lingkungan* (Role of Religion and Ethic in Conservation of Natural Resources and Environment). Pusat Penelitian Biologi – LIPI, Bogor, Indonesia.

Awoyemi, S. M., Gambrill, A., Ormsby, A., and Vyas, D. (2012) Global efforts to bridge religion and conservation: are they really working? In *Topics in Conservation Biology* (ed. T. Povilitis), InTech Publication, Rijeka, Croatia, pp. 97–110.

BAPPENAS (1993) *Biodiversity Action Plan for Indonesia*. Ministry of Development Planning Republik Indonesia, Jakarta, Indonesia.

BAPPENAS (2003) *Indonesian Biodiversity Strategy and Action Plan 2003–2020*. National Development Planning Agency (BAPPENAS), Jakarta, Indonesia.

Dephut (Departemen Kehutanan) (1994) *Pengelolaan Hutan Lestari* (Sustainable Forest Management). Ministry of Forestry, Jakarta, Indonesia.

Dudley, N., Higgins-Zogib, L. and Mansourian, S. (2005) *Beyond Belief: Linking Faiths and Protected Areas to Support Biodiversity Conservation*, WWF. http://wwf.panda.org/what_we_do/how_we_work/conservation/forests/publications/?uNewsID=58880 (accessed March 19, 2013).

FAO (Food and Agriculture Organızatıo) (2001) *Unusylva*, **205**, vol. 52.

FWI/GFW (2001) *Potret Keadaan Hutan Indonesia* (Picture of Indonesian Forest Condition). Forest Watch Indonesia, Bogor, Indonesia and Global Forest Watch, Washington, D.C.

Hannibal, L. W. (1950) Vegetation map of Indonesia. Planning Department, Forest Service, Jakarta, in *Forest Policies in Indonesia. The Sustainable Development of Forest Lands*. International Institute for Environment and Development and Government of Indonesia, Jakarta, Indonesia.

Hargrove, T.R. (1988) Twenty years of rice breeding: the role of semidwarf varieties in rice breeding for Asian farmers and the effects on cytoplasmic diversity. *BioScience*, **38**, 675–681.

Indrawan, M., Primack, R. B. and Supriatna, J. (2007) *Biologi Konservas, Edisi Revisi* (Conservation Biology, revised edition). Yayasan Obor Indonesia, Jakarta, Indonesia.

Mittermeir, R., Gil, P. and Goettsch-Mittermeier, C. (1997) *Megadiversity: Earth's Biologically Wealthiest Nations*. Cemex, Mexico.

Newman, M. F., Burgess, P. F. and Withmore, T. C. (1999) *Manual of Dipterocarps Series (Sumatra, Kalimantan, Jawa to Nugini)*. Prosea-Indonesia: Bogor, Indonesia.

Patlis, J. M. (2008) What protects the protected areas? Decentralization in Indonesia, the challenges facing its terrestrial and marine national parks and the rise of regional protected areas, in *Biodiversity and Human Livelihoods in Protected Areas* (eds N.S. Sodhi, G. Acciaioli, M. Erb and Tan, A. K-J.), Cambridge University Press, New York, NY, pp. 405–428.

Population Reference Bureau (2012) *World Population Data Sheet 2012*. http://www.prb.org/pdf12/2012-population-data-sheet_eng.pdf (accessed March 19, 2013.

Sastrapradja, D. S., Adisoemarto, S., Kartawinata, K., Sastrapradja, S. and Rifai, M. (1989) *Keanekaragaman Hayati Untuk Kelangsungan Hidup Bangsa* (Biodiversity for Sustainable Nation Life). Pusat Penelitian dan Pengembangan Bioteknologi-LIPI, Bogor, Indonesia.

Sodhi, N. S. and Brook, B. W. (2006). *Southeast Asian Biodiversity in Crisis*. Cambridge University Press, Cambridge.

Soedjito, H. (ed.) (2006) *Kearifan Tradisional dan Cagar Biosfer di Indonesia: Prosiding Piagam MAB 2005 untuk Peneliti Muda dan Praktisi Lingkungan di Indonesia* (Traditional Wisdom and Biosphere Reserves in Indonesia: Proceeding of MAB Awards 2005 for Young Scientists and Environment Practitioners in Indonesia). Komite Nasional MAB Indonesia – LIPI.

Soedjito, H. and Sukara, E. (2006). Mengilmiahkan pengetahuan tradisional: Sumber Ilmu masa depan Indonesia, in *Kearifan tradisional dan Cagar Biosfer di Indonesia: Prosiding Piagam MAB 2005 untuk Peneliti Muda dan Praktisi Lingkungan di Indonesia* (Traditional Wisdom and Biosphere Reserves in Indonesia: Proceeding of MAB Awards 2005 for Young Scientists and Environment Practitioners in Indonesia) (ed. S. Soedjito), National Committee of Man and the Biosphere (MAB) Indonesia–LIPI, Jakarta, Indonesia, pp. 1–17.

Soedjito, H., Purwanto, Y. and Sukara, E. (eds.) (2009) *Situs Keramat Alami Peran Budaya dalam Konservasi Keanekaragaman Hayati* (Sacred Natural Sites: The Role of Culture on Biodiversity Conservation). Yayasan Obor Indonesia, Komite Nasional MAB Indonesia, and Conservation International Indonesia, Jakarta, Indonesia.

Sriyanto, A., Wellesley, S., Suganda, D., Widjarnati, E. and Sutaryono, D. (2003). *Guidebook to 41 National Parks in Indonesia*. Ministry of Forestry–Republic Indonesia, UNESCO, and CIFOR, Jakarta, Indonesia.

Widjaja, E. A., Maryanto, I., Wowor, D. and Prijono, S. N. (eds) (2011) *Status Keanekaragaman Hayati Indonesia* (Status of the Indonesian Biodiversity). Jakarta: LIPI Press.

CHAPTER 17

SINGAPORE

Half Full or Half Empty?

Richard T. Corlett

Xishuangbanna Tropical Botanical Garden, Chinese Academy of Sciences, Menglun, Yunnan, China

SUMMARY

Singapore has had an influence on the development of tropical Asian biology over the past 200 years that is disproportionate to its size, through the many scientists who have been based there. Meanwhile, the island itself was subject to largely uncontrolled deforestation and exploitation, leaving only small areas of protected forest and a depleted native biota, while alien species dominate outside the nature reserves. Yet the biota is still very rich and new species continue to be found. Singapore has also proved immensely valuable as a relatively well-documented case study of extreme human impacts in the equatorial tropics.

INTRODUCTION

Singapore was once described by the Indonesian President Habibie as a "little red dot" and this epithet has subsequently been adopted by Singaporeans as a symbol of the nation's success, despite limitations of size ($704 \, km^2$), population (5.3 million), and natural resources (virtually none). That size is not everything is shown by Singapore's per capita gross domestic product, which ranks it with the richer countries of the European Union, and the global top three rankings for its port and airport. Less positively, a recent global review ranked Singapore worst for its environmental impact in relation to total resource availability (Bradshaw, Giam and Sodhi, 2010), although it can be argued that this is an inevitable consequence of the late twentieth-century model of development applied successfully to an island city state.

Biogeographically, in contrast, Singapore is nothing special, being separated from the southern tip of the Asian mainland by shallow straits less than a kilometer wide at the narrowest point. As one would expect, there are few endemic species and the local diversity in those groups of organisms for which comparable data are available is lower than at equivalent sites on the mainland or on the much larger neighboring islands of Borneo and Sumatra. Singapore's disproportionate influence on the development of biology in tropical Asia reflects not its own biota, therefore, but the efforts of the people who have worked there since the modern settlement was founded by Stamford Raffles in 1819.

NINETEENTH CENTURY: EXPLOITATION AND DEFORESTATION

Raffles himself was an enthusiastic amateur biologist, but the best description of Singapore in 1819 comes

---

from his assistant, the surgeon–botanist William Jack, in a letter to his family on June 20:

> It is impossible to conceive anything more beautiful than the approach to Singapore, through the archipelago of islands that lie at the extremity of the Straits of Malacca. Seas of glass wind among innumerable islets, clothed in all the luxuriance of tropical vegetation and basking in the full brilliance of a tropical sky . . . I have just arrived in time to explore the woods before they yield to the axe, and have made many interesting discoveries, particularly of two new and splendid species of pitcher-plant [*Nepenthes rafflesiana* and *Nepenthes ampullaria*], far surpassing any yet known in Europe.

As Jack foresaw, the primeval forest rapidly did "yield to the axe," but nineteenth-century Singapore, with its busy port, easily accessible habitats, and relatively comfortable living conditions, continued to attract a long list of famous – or soon to be famous – biological visitors and residents (Ng, Corlett and Tan, 2011). Most famous of all was Alfred Russel Wallace, who stayed in Singapore several times between 1854 and 1862 during his collecting trips throughout the Malay Archipelago. His local collections were mostly of beetles (Figure 17.1):

> In about two months, I obtained no less than 700 species of beetles, a large proportion of which were quite new. . . . Almost all these were collected in one patch of jungle, not more than a square mile in extent, and in all my subsequent travels in the East I rarely if ever met with so productive a spot. (Wallace, 1869)

This "patch of jungle" was Bukit Timah hill, at 164 m the highest point in Singapore. Wallace attributed the beetle diversity in a large part "to the labors of the Chinese wood-cutters" who had "furnished a continual supply of dry and dead and decaying leaves and bark, together with abundance of wood and sawdust, for the nourishment of insects and their larvae."

Uncontrolled deforestation for cash crops (gambier, pepper, and others) and exploitation of the remnants (for timber, rattans, firewood, and game) took its inevitable toll. By the time Nathaniel Cantley, Superintendent of the Singapore Botanic Gardens, published his *Report on the Forests of the Straits Settlements*, in 1884, there were serious concerns about the timber supply (Cantley, 1884). As Cantley reported:

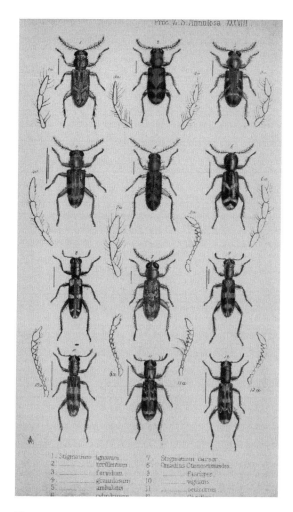

Figure 17.1 Twelve new species of checkered beetles (Cleridae) collected by Alfred Russel Wallace "within a few days after his arrival at Singapore" in 1854. Reproduced from Westwood (1855) Descriptions of some new species of Cleridae collected at Singapore by Mr. Wallace. Proceedings of the Zoological Society of London 23, 19–26.

> Such Crown forests as remain uncut are widely distributed in isolated patches over the island. These forest patches or clumps are of various sizes, from half an acre or so to about 25 acres [10 ha], and of no particular shape; their distance from each other may average a quarter of a mile though often exceeding a mile. The interspace is generally waste grassland, which supports, as a rule, only the strongly growing grass known locally as "lalang" [*Imperata cylindrica*]."

Extinction from isolated forest fragments can be a slow process and when H. N. Ridley, Cantley's successor in charge of the Botanic Gardens, described their fauna in 1895, they were still "haunted by numerous mammals" (Ridley, 1895). Ridley describes the scene at dusk when:

> [t]he wild pigs are making onslaughts on the pine-apple and tapioca fields, the deer come out to crop the shrubs on the edge of the woods, the tiger is moving quietly through the bushes in pursuit of the pigs and deer, . . . the large red flying squirrel is taking its surprising leaps from one lofty tree to another, . . . [and] the great fox-bats are flapping slowly overhead.

All these species were lost from the main island of Singapore over the coming century (although pigs have recently reinvaded).

This delay in extinction after the major period of nineteenth-century forest clearance is important because, with a few exceptions (such as the botanical records of Jack and Wallich, and the insects collected by Wallace), the period from 1870 to 1900, when >80% of the forest had already gone, marked the start of serious biological collections and thus the "baseline" on which later extinction estimates have been based. It is also important in predicting the ultimate fate of present-day examples of recently cleared tropical land-scapes, where the diversity of forest vertebrates can give a misleading impression of the long-term capacity of highly fragmented forests to support viable populations of forest animals.

TWENTIETH CENTURY: EXTINCTIONS, INVASIONS, AND CONSERVATION

Cantley's report led to the establishment of a system of forest reserves in Singapore. These eventually made up 10% of the land area, but they were by no means entirely forested. In any case, they were not taken seriously enough to apply much of a brake to the continued erosion of what little forest area remained. The biodiversity that had attracted so many biologists in the nineteenth century began to decline. When Ridley published his *Flora of Singapore* in 1900, he commented that some forest species recorded by Danish botanist Nathaniel Wallich in 1822 "appear to have quite vanished" while many open-country aliens had become

established over the same period (Ridley, 1900). The extinction of native, mostly forest-dependent, species and the establishment of alien, open-country species are recurring themes in natural history publications on Singapore throughout the twentieth century (Corlett, 1992).

Perhaps ironically, Ridley is most famous as the man responsible for establishing the rubber industry in Southeast Asia, and it was largely thanks to him that the patchwork of habitats described by Cantley in 1884 had been replaced by the 1930s, over 40% of Singapore, by a rubber monoculture. F. N. Chasen, who later became Director of the Raffles Museum, commented in the introduction to the first comprehensive checklist of the birds of Singapore that "rubber estates are notoriously unproductive from a naturalist's point of view," and went on to list bird species that had been lost since the nineteenth century, including all pheasants (although Chasen doubted the validity of the nineteenth-century records), hornbills, and trogons (Chasen, 1923). Much of the unprotected forest that still survived was on swampy ground, but most of this freshwater swamp forest was cleared in the 1920s and 1930s. We would know nothing of its botanical composition were it not for the efforts of E. J. H. Corner, Assistant Director of the Singapore Botanic Gardens, who used these fellings as an opportunity to make comprehensive collections of the trees, climbers, and epiphytes (Corner, 1978).

The fact that any primary forest at all survived to the present day is largely due to the efforts of Corner's boss, the Director of the Gardens, Eric Holttum. In 1936, when the forest reserves established by Cantley were abolished (after a brief and damaging period without protection), he managed to get Bukit Timah regazetted "on grounds of amenity . . . and botanical interest": still a forest reserve in name, but a nature reserve in practice. Then, after World War II, Holttum again intervened to stop the increasing threat from encroaching granite quarries and the resulting enquiry resulted in the Nature Reserves Act of 1951 and the formal creation of Bukit Timah Nature Reserve (BTNR). BTNR and the adjacent Central Catchment Nature Reserve form the non-coastal component of the present nature reserve system, protecting all the surviving primary forest fragments (totaling circa 200 ha) and most of the native secondary forest (circa 1600 ha). The other reserves are Sungei Buloh Wetland Reserve (130 ha), consisting largely of secondary coastal wetlands, and Labrador Nature Reserve (10 ha), which contains secondary dry coastal forest and cliff vegetation.

When Navjot Sodhi arrived in Singapore in 1995, he became interested in the island's potential as a well-documented "worst-case scenario" from which general conservation principles could be derived. He started with an analysis of bird extinctions (65 more species had gone since Chasen's 1923 list) (Castelletta, Sodhi and Subaraj, 2000). Over the next decade, he and his students looked at bees (Liow, Sodhi and Elmqvist, 2001), butterflies (Koh, Sodhi and Brook, 2004), dung beetles (Lee *et al.*, 2009), and angiosperms (Sodhi *et al.*, 2008). Data on decapods, phasmids, fishes, amphibians, reptiles, and mammals were added in the classic paper, "Catastrophic extinctions follow deforestation in Singapore" (Brook, Sodhi and Ng, 2003), which summarized the lessons learned from Singapore. These lessons included the high overall extinction rates (28–73%, depending on assumptions made about extinctions before the first reliable species records in the 1870s), the great variation between taxonomic groups (5–43%, for observed extinctions), the greater vulnerability of forest than non-forest species (33% versus 7% for observed extinctions), and the concentration of the survivors (>50%) in the tiny nature reserves. The paper then goes on to infer likely extinctions resulting from deforestation elsewhere in Southeast Asia over the next century.

TWENTY-FIRST CENTURY SINGAPORE: HALF FULL OR HALF EMPTY?

The pessimistic (half-empty) case for Singapore biodiversity is easy to make. Most of the people mentioned earlier made their names for work on a larger regional stage. Singapore was important as a base and, in some cases, a training ground, but much less so as a study area. Singapore's biota was never particularly rich by regional standards and around half of it has subsequently been lost. Moreover, approximately half of the surviving species in well-studied groups are nationally endangered (Davison, Ng and Ho, 2008). Deforested areas were initially occupied by tolerant native plants and animals, mostly of coastal origin, in the nineteenth and early twentieth centuries, but these have now been largely displaced by aliens, which dominate most habitats outside the nature reserves. It is not yet clear if this alien dominance can ever be reversed, or, conversely, if it threatens the last areas of native forest (Corlett, 2010). Climate change, in the form of a predicted 3–4°C warming by the end of this century, is an addi-

tional threat that we are a long way from being able to quantify (Corlett, 2011, 2012). The nature reserves are well protected, but occupy less than 5% of Singapore's land area and have no marine counterparts, so the future for Singapore's native biodiversity does not look bright.

But there are other ways of looking at Singapore. First, it is still stunningly rich in native species for a tiny (704 km²), largely urbanized, island city state. There are still around 1600 extant species of native vascular plants, 350 birds, 26 mammals, 117 reptiles, 25 amphibians, 35 freshwater fish, 300 butterflies, 124 dragonflies, and so on (Ng, Corlett and Tan, 2011). With these numbers, it is not hard to see Singapore as half full. Moreover, species new to science are discovered almost every month, by both resident and visiting scientists. More than a hundred such species have been described in the last few years, including new species of mosses, fungi, lichens, fishes, nematodes, spiders, mites, harvestmen, wasps, beetles, bugs, flies, shrimps, barnacles, and crabs (Ng and Corlett, 2011).

Second, its value as a case study – as recognized by Navjot Sodhi – has grown with time. Singapore may be the worst-case scenario for tropical continental biodiversity, but the fact that so much has survived – so far – suggests that extinction is at least slower than we had feared. Attempts to use Singapore to identify traits that predict vulnerability to extinction have had varied degrees of success (Castelletta, Sodhi and Subaraj, 2000; Koh, Sodhi and Brook, 2004; Sodhi *et al.* 2008), but this is a useful lesson, suggesting either that we are investigating the wrong traits, or that extinction can sometimes be trait-neutral. Moreover, Singapore's urbanization has interest beyond its importance as a driver of extinction. By 2030, the percentage of the total population that is urban in the Asian, African, and American tropics is projected to be 52%, 59%, and 84%, respectively (Montgomery, 2008; DeFries *et al.*, 2010). Indeed, most global population growth is now in tropical cities and lessons learned in Singapore (e.g., on invasive urban birds; Lim *et al.*, 2003) can have wide application.

Despite these factors that highlight the value of Singapore's natural assets, Singapore has been uncharacteristically timid when it comes to biodiversity conservation, so that many tools in the modern conservation toolbox remain unused. Aliens have had a free run outside the nature reserves, ecological restoration has been ad hoc and small scale, and there has been no systematic attempt to reintroduce species lost from Singapore that still survive in the immediately adjacent areas of Malaysia and Indonesia. A serious attempt to expand the area of native-dominated forest, perhaps to 10% of Singapore, to control the most invasive of the aliens, and to reintroduce those species that succumbed to threats, such as hunting, which have now been controlled, could transform the situation. We need to learn how to do these things in the lowland tropics and Singapore is an excellent place to start.

Climate change may yet overshadow all other threats to lowland tropical biodiversity, in Singapore and elsewhere (Corlett, 2011, 2012), but this again is an opportunity for the world's only all-equatorial, all-lowland nation to make a globally significant contribution. The close proximity of lowland rainforest fragments and world-class laboratories makes Singapore a logical center for research into the potential impacts of climate change. Rigorous, quantitative, long-term monitoring of tree growth, survival, and fecundity would be a good start (Clark and Clark, 2011), but the biggest gap in our current knowledge is information on thermal tolerances and acclimation capacity needed to predict the future of equatorial biodiversity in a warming world (Corlett, 2011).

Finally, the lowland ecosystems of Malaysia and Indonesia have experienced catastrophic human impacts over recent decades, with vast areas now under industrial crop monocultures or expanding urban areas (Miettinen, Shi and Liew, 2011). It is no longer safe to assume that a non-endemic species lost from Singapore still persists elsewhere in the region. Singapore's nature reserves are tiny and fragmented, but they are safe from conversion, logging, and hunting. If Singapore plays its cards right, it will still be half full a century from now.

ACKNOWLEDGMENTS

This essay benefited greatly from many years of discussions with Navjot Sodhi, and also with Hugh Tan, Peter Ng, and many other friends and colleagues in Singapore.

REFERENCES

Bradshaw, C. J. A., Giam, X. and Sodhi, N. S. (2010) Evaluating the relative environmental impact of countries. *PLoS One*, **5**, e10440.

Brook, B. W., Sodhi, N. S. and Ng, P. K. L. (2003) Catastrophic extinctions follow deforestation in Singapore. *Nature*, **424**, 420–423.

Cantley, N. (1884) *Report on the Forests of the Straits Settlement*. Singapore Printing Office, Singapore.

Castelletta, M., Sodhi, N. S. and Subaraj, R. (2000) Heavy extinctions of forest-dependent avifauna in Singapore: lessons for biodiversity conservation in Southeast Asia. *Conservation Biology*, **14**, 1870–1880.

Chasen, F. N. (1923) An introduction to the birds of Singapore Island. *Singapore Naturalist*, **2**, 87–112.

Clark, D.A. and Clark, D.B. (2011) Assessing tropical forests' climatic sensitivities with long-term data. *Biotropica*, **43**, 31–40.

Corlett, R. T. (1992) The ecological transformation of Singapore: 1819–1990. *Journal of Biogeography*, **19**, 411–420.

Corlett, R. T. (2010) Invasive aliens on tropical East Asian Islands. *Biodiversity and Conservation*, **19**, 411–423.

Corlett, R. T. (2011) Impacts of warming on tropical lowland rainforests. *Trends in Ecology and Evolution*, **26**, 606–613.

Corlett, R. T. (2012) Climate change in the tropics: the end of the world as we know it? *Biological Conservation*, **151**, 22–25.

Corner, E. J. H. (1978) The freshwater swamp-forest of South Johore and Singapore. *Gardens' Bulletin, Singapore*, Supplement 1, 1–266.

Davison, G. W. H., Ng, P. K. L. and Ho, H. H. (2008) *The Singapore Red Data Book: Threatened Plants and Animals of Singapore*, 2nd edn. Nature Society, Singapore.

DeFries, R., Rudel, T. K., Uriarte, M. and Hansen, M. (2010) Deforestation driven by urban population growth and agricultural trade in the twenty-first century. *Nature Geoscience*, **3**, 178–181.

Koh, L.P., Sodhi, N. S. and Brook, B. W. (2004) Coextinction of tropical butterflies and their host plants. *Biotropica*, **36**, 272–274.

Lee, J. S. H., Lee, I. Q. W., Lim, S. L.-H., Huijbregts, J. and Sodhi, N. S. (2009) Changes in dung beetle communities and associated dung removal services along a gradient of tropical forest disturbance in South-East Asia. *Journal of Tropical Ecology*, **25**, 677–680.

Lim, H. C., Sodhi, N. S., Brook, B. W. and Soh, M. C. K. (2003) Undesirable aliens: factors determining the distribution of three invasive bird species in Singapore. *Journal of Tropical Ecology*, **19**, 685–695.

Liow, L. H., Sodhi, N. S. and Elmqvist, T. (2001) Bee diversity along a disturbance gradient in tropical lowland forests of Southeast Asia. *Journal of Applied Ecology*, **38**, 180–192.

Miettinen, J., Shi, C. and Liew, S. C. (2011) Deforestation rates in insular Southeast Asia between 2000 and 2010. *Global Change Biology*, **17**, 2261–2270.

Montgomery, M. R. (2008) The urban transformation of the developing world. *Science*, **19**, 761–764.

Ng, P. K. L. and Corlett, R. T. (2011) Biodiversity in Singapore: an overview, in *Singapore Biodiversity: An Encyclopedia of the*

Natural Environment and Sustainable Development (eds P. K. L. Ng, R. T. Corlett and H. T. W. Tan), Editions Didier Millet, Singapore, pp. 18–36.

Ng, P. K. L., Corlett, R. T. and Tan, H. T. W. (eds) (2011) *Singapore Biodiversity: An Encyclopedia of the Natural Environment and Sustainable Development*. Editions Didier Millet, Singapore.

Ridley, H. N. (1895) The mammals of the Malay Peninsula. *Natural Science*, **6**, 23–29, 89–96, 161–166.

Ridley, H. N. (1900) Flora of Singapore. *Journal of the Straits Branch of the Royal Asiatic Society*, **33**, 27–196.

Sodhi, N. S., Koh, L. P., Peh, K. S.-H., Tan, H. T. W., Chazdon, R. L., Corlett, R. T., Lee, T. M., Colwell, R. K., Brook, B. W., Sekercioglu, C. H. and Bradshaw, C. J. A. (2008) Correlates of extinction proneness in tropical angiosperms. *Diversity and Distributions*, **14**, 1–10.

Wallace, A. R. (1869) *The Malay Archipelago*. Macmillan, London, pp. 37–38.

Westwood, J. O. (1855) Descriptions of some new species of Cleridae collected at Singapore by Mr. Wallace. *Proceedings of the Zoological Society of London*, **23**, 19–26.

CHAPTER 18

WANT TO AVERT EXTINCTIONS IN SRI LANKA?

Empower the Citizenry!

Rohan Pethiyagoda

Ichthyology Section, Australian Museum, Sydney, NSW, Australia

SUMMARY

With dozens of species already declared extinct and hundreds more threatened with imminent extinction, Sri Lanka's biodiversity is in trouble. The country has hitherto relied on a failed paradigm: that government is the sole custodian and protector of biodiversity. It asserts that protective legislation and effective enforcement alone will save threatened species and ecosystems. However, the management interventions that are needed to save the species most at risk (namely, conservation) demand a more participatory approach. I argue that the adversarial relationship between paternalistic government and a distanced civil society is inimical to conservation. Government, the scientific community, and the public need to acknowledge a shared ownership of – and responsibility for – biodiversity: only imaginative, participatory, science-based partnerships can deliver positive conservation outcomes. This calls for a fundamental reform of conservation policy and the enactment of enabling legislation that encourages trust and engenders engagement between government and civil society.

INTRODUCTION

For a continental island of its size (65,000 km^2), Sri Lanka hosts remarkable biodiversity and, together with the Western Ghats, forms a Global Biodiversity Hotspot (Myers *et al.*, 2000; Kumar, Pethiyagoda and Mudappa, 2004; Gunawardene *et al.* 2007). A total of 4086 species of vascular plants are indigenous to the island, and nearly a quarter (986) of them are endemic (Dassanayake and Fosberg, 1980–2006). Endemism is high in many groups of terrestrial animals too: 120 of the island's 210 reptile species occur nowhere else, as do 95 of 111 amphibians, 40 of 79 freshwater fishes, 204 of 246 land snails and all 50 species of freshwater crabs (Gunawardene *et al.*, 2007; Gunatilleke, Pethiyagoda and Gunatilleke, 2008; Green *et al.*, 2009; Beenaerts *et al.*, 2010; Pethiyagoda and Meegaskumbura, in press). Endemism within more vagile groups is lower: e.g., only 18 of 91 mammal species and 33 of 482 birds are endemic to the island. Projecting from the global estimates for all groups of eukaryotic organisms, at least 1% of all species of eukaryotic organisms (estimated 12 million species) could occur in Sri Lanka, with a majority of them yet to be discovered (Raven, personal communication). Considering that half or more of these species may be endemic, the very high value of the island's ecosystem cannot be doubted.

Within the island, the evergreen forests of the southwestern "wet zone" (rainfall >2.5 m/year) lowlands harbor the greatest proportion of endemic plants and animals. For example, while 219 flowering-plant species are restricted to the ~10,000 km^2 wet zone, only 30 species are restricted to the 55,000 km^2 dry and intermediate zones (rainfall <2.5 m/yr; data from Dassanayake and Fosberg, 1980–2006). This pattern is repeated across other floral and faunal groups (Gunatilleke, Pethiyagoda and Gunatilleke, 2008).

Sri Lanka's population stands at 21.2 million (2.5 times that at independence in 1948), a density of 323 people per square kilometer, by far the highest among the global biodiversity hotspots (Cincotta, Wisnewski and Engelman, 2000; Population Reference Bureau, 2012). Population density in the wet zone, however, is approximately double the national average, exerting substantial pressure on land and other natural resources. Fortunately, the population of Sri Lanka is projected to grow relatively slowly to an estimated 25.7 million by 2050 (Population Reference Bureau, 2012), but levels of consumption will likely rise hugely, adding a great deal of pressure on the natural environment.

Although almost 30% of Sri Lanka is forested, near-primary forest persists on only 1.2% of the island, covering 800 km^2 (8%) of the biodiversity-rich wet zone (FPU, 1995). Even this modest extent of forest exists as several fragments, all but three of them less than 500 hectares (ha) in size. Addressing the deleterious effects of fragmentation is thus among the country's main conservation challenges. The wet zone additionally contains some 1180 km^2 of secondary and partly harvested forest, largely at lower elevations (Perera, 2001).

EXTINCTION AND ENDANGERMENT

In this chapter, I present a brief review of the conservation status of plants, amphibians, freshwater fish, freshwater crabs, and the Asian elephant in Sri Lanka.

These taxa were selected because they were recently surveyed during conservation assessments.

Plants

Although only a single plant species has formally been declared extinct in Sri Lanka (IUCN, 2011), Dassanayake and Fosberg (1980–2006) reported that as many as 61 endemic flowering-plant species (including 23 tree species) had not been collected anywhere on the island in the preceding 50 years. While three tree species (*Albizia lankaensis*, *Crudia zeylanica*, and *Diospyros albiflora*) are now likely extinct (the last known individuals have disappeared), Gunatilleke, Pethiyagoda and Gunatilleke (2008) reported that, of the 1099 species of angiosperms assessed, 42 endemic species were tentatively considered extinct. They also listed 81 narrowly restricted endemics (including 26 tree species) that are each restricted to a single site or locality. Even among plants that are more widely distributed, there are several that deserve priority whether because they are endemic genera that are monotypic (e.g., *Cyphostigma*, *Davidsea*, *Dicellostyles*, *Farmeria*, *Leucocodon*, *Loxococcus*, *Nargedia*, *Schizostigma*), contain only a few species (e.g. *Phoenicanthus*, *Schumacheria*, *Scyphostachys*), are biogeographical curiosities with disjunct distributions (e.g., *Axinandra*, *Cephalocroton*, *Hortonia*, *Ptychopyxis*), or are wild relatives of commercially important species (e.g., the six species of cinnamon endemic to Sri Lanka).

Amphibians

Recent research has shown that 21 of Sri Lanka's 111-species amphibian inventory are apparently extinct (IUCN, 2011). Of the 90 surviving amphibians, 11 are assessed as critically endangered and 36 as endangered, clearly indicative of a fauna in trouble. Yet, chytridiomycosis, the fungal disease implicated in many amphibian population declines elsewhere (Stuart *et al.*, 2008,), has not yet been recorded from Sri Lanka. Rather, the island's amphibian fauna appears to be a victim primarily of habitat loss and degradation. With endemic amphibians preponderantly restricted to the wet zone, 92% of which has been stripped of its primary forest, further extinctions must be expected. But as with plants, the conservation situation is made much more difficult by the extremely limited ranges of many species: as many as 20 of the 90 extant amphibian species fall in this category, being micro-endemics, restricted to areas of occupancy of less than 10km^2 (Figure 18.1). Even highly localized disturbance could, in such cases, lead to a species' demise. For example, Morningside, a 1100-m high plateau of some 10km^2, is home to 17 amphibian species endemic to Sri Lanka, eight of them restricted to this one site. Though much larger, the Knuckles Hills, with 22 endemic species, is an important area of localized amphibian diversity. In both areas, much of the forest understory is being cleared for cardamom cultivation, which endangers the long-term viability of the forest itself and destroys many of its original habitats.

Freshwater fishes

With about 80 species (40 endemic), Sri Lanka's freshwater fish fauna is unexceptional compared with that of southern India. Nevertheless, since the 1930s, several of the smaller and more colorful fishes have been part of the ornamental fish trade, becoming well known internationally. Although the fishes remain to be assessed for conservation purposes, several species are already known to be in trouble (Pethiyagoda, 2006), including the local endemic species *Devario pathirana*, *Puntius asoka*, *P. bandula* and *P. srilankensis*. The ranges of these fishes lie within agricultural or suburban landscapes outside the protected areas network: their survival must therefore depend on the informed goodwill of local communities.

The populations of two formerly abundant and widespread endemic fishes, *Labeo lankae* and *Macrognathus pentophthalmos*, crashed precipitously between 1980 and 1990 (Pethiyagoda, 1994). Neither showed a gradual decline, and, while there exists a single informal record of the former from 2009, the latter has not been reported during the past two decades, despite extensive searches. *Labeo lankae*, sufficiently common in dry-zone reservoirs to warrant a separate fishery (Senanayake, 1980), might have fallen victim to competition from exotic carp introduced by fisheries authorities. *Macrognathus pentophthalmos* was once common even in rice fields; its demise remains cryptic. Sri Lanka's aquatic biodiversity is also threatened by several dozen alien species of fishes and mollusks imported by the ornamental-fish industry that are now naturalized; their impact remains to be assessed.

Freshwater crabs

Sri Lanka is the beneficiary of a remarkable radiation of freshwater crabs, a fauna that includes five endemic genera and 50 endemic species (Bahir *et al.*, 2005; Beenaerts *et al.*, 2010). This fauna has benefited from a complete conservation assessment (IUCN, 2011), which shows that 23 species are critically endangered and a further eight endangered, 26 of them because they are so limited in range. While no extinctions are known (likely because the historical baseline provided is poor), many of the species occur outside the protected areas network and therefore could fall victim to hydrological alterations and other disturbances.

Elephants

The Asian elephant is awarded a section of its own here because, despite being a single species that is also widely distributed outside Sri Lanka and "merely" endangered (IUCN, 2011), it is the country's conservation flagship, absorbing almost the entirety of species-specific conservation investment. Domestic elephants are part of Sri Lanka's cultural and religious heritage, adding to their importance in the island. Despite this, about three elephants and one farmer, on average, die each week as a result of the conflict that has arisen between the two groups (Bandara and Tisdell, 2001; Fernando *et al.*, 2005, 2011).

The Sri Lankan elephant is now in effect a synanthropic species. With some 10,000 dry-zone reservoirs offering a perennial supply of water and abundant grass on fallow lands on which swidden cultivation is practiced seasonally, elephants have come to benefit from – and depend on – a human-modified habitat. Given that Sri Lanka remains primarily an agricultural society based around a farming peasantry, and with a human-population growth of around 160,000 people per year (Population Reference Bureau, 2011), the conflict is set to intensify. Worse, with ever more land being set aside for permanent crops such as sugarcane (another favorite of elephants), leading also to a decline in the secondary vegetation elephants depend on for food, even more trouble can be expected. Despite substantial numbers of elephant deaths being reported annually, it is moot whether its population in Sri Lanka is declining: indeed, successive censuses suggest an increasing trend: 1600 (McKay, 1973),

SRI LANKA

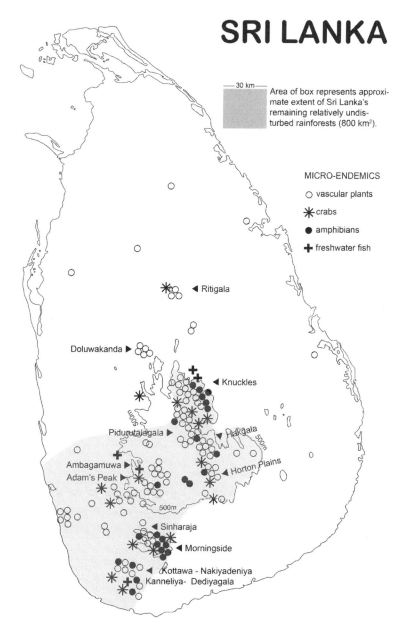

—30 km— Area of box represents approximate extent of Sri Lanka's remaining relatively undisturbed rainforests (800 km²).

MICRO-ENDEMICS

○ vascular plants

✳ crabs

● amphibians

✚ freshwater fish

◀ Ritigala

Doluwakanda ▶

◀ Knuckles

Pidurutalagala ▶   ◀ Hakgala

Ambagamuwa ▶

Adam's Peak ▶   ◀ Horton Plains

500m

◀ Sinharaja

◀ Morningside

◀ Kottawa - Nakiyadeniya
Kanneliya- Dediyagala

Figure 18.1 Distribution of micro-endemic vascular plants, amphibians, freshwater fish and freshwater crabs in Sri Lanka, based on distribution records in Dassanayake and Fosberg (1980–2006), Manamendra-Arachchi and Pethiyagoda (2007), Pethiyagoda (1994) and Bahir *et al.* (2005), respectively. The wet zone (shaded) and the montane zone are especially rich in restricted-range species; some key sites are labeled. Note: symbols have been spread out to avoid overlap; most sites are substantially smaller than suggested by the distributions shown.

1967 (Hendavitharana *et al.*, 1994), 4500 (de Silva and de Silva, 2007) and 7379 (Rodrigo, 2011).

WHERE MUST THERE BE CHANGE?

Protection versus conservation

Sri Lanka's governing law for biodiversity conservation is the Fauna and Flora Protection Ordinance, a poorly drafted and deeply flawed catalogue of prohibitions and penalties (e.g., even the noctuid and pyralid moths that are common pests of rice are strictly protected on pain of five years' imprisonment, despite pesticides that specifically target them being legally available and in widespread use). By seeking to assert sovereignty over biodiversity, as bestowed by the 1992 Convention on Biological Diversity, the law has unwittingly served to distance civil society, especially the scientific community, from engagement in conservation activity including research (Pethiyagoda, 2004; Pethiyagoda *et al.*, 2007). Besides, while the ordinance affords protection (from exploitation) to species and protected areas, it says almost nothing about conservation, or even preventing extinctions, "the most important problem conservation science must address" (Pimm and Jenkins, 2010). Most importantly, it does not require the government to actually do anything other than enforce the various prohibitions – certainly not to undertake conservation actions, even to the extent of identifying species and ecosystems at risk, or drafting and implementing recovery plans.

Biodiversity conservation in Sri Lanka falls within the remits of the Departments of Wildlife Conservation and Forests, which together manage some 30% of the country's land area. The two departments have a staff of 3900, which does not, however, include any significant capacity for conservation biology (Pethiyagoda *et al.*, 2007). Their annual recurrent-expenditure budget in 2010 was US $10.5 million, of which 82% was for personal emoluments (Anonymous, 2011). Of the US $7.4 million capital-expenditure budget, 84% was for the construction or maintenance of buildings. There was no provision for conservation interventions save an allocation of US$2.7 million for "buildings and structures" (i.e., electric fencing) for the "conservation of elephants." No funds were available for monitoring threatened-species populations, designing and implementing recovery plans, undertaking research, or devising conservation partnerships with civil society.

Indeed, there is no record of such activities even being proposed or budgeted.

So draconian is the law as it stands that the public is prohibited from contact with wildlife under any circumstances, as evidenced by the recent successful prosecution of a man who rescued a common toad from a saltwater swimming pool into which it had fallen, and placed it in a bucket of freshwater in which to recover (Wright, 2012). Conversely, however, the law turns a blind eye toward the maintenance in Buddhist temples of hundreds of Asian elephants (an endangered species), for work, purposes of ritual, mascots, or simply as pets.

Fragmentation

The earlier review shows, however, that enforcement alone is insufficient – and sometimes inimical – to delivering positive conservation outcomes in Sri Lanka. If it is accepted that biodiversity conservation should aim not just for the short-term preservation of species but for the long-term sustenance of evolutionary processes, immense challenges must be overcome. Prime among these is the establishment of habitat connectivity between isolated forest fragments of the wet zone, whose small size would otherwise see inevitable attrition from island effects (e.g., Bierregaard *et al.*, 2001; Bennett and Saunders, 2010). Restoration at the landscape level can pose enormous challenges (Pethiyagoda and Nanayakkara, 2011), especially when an elevation range from sea level to 2500 m is involved. While Sri Lanka's contemporary demography might not allow for land reform on the required scale (only 15% of the population lives in urban areas, Anonymous, 2010), continuing industrialization is likely to accelerate urbanization, easing demand for agricultural land. Given that restoration initiatives need to be planned at least in the decadal time frame, much research and experimentation needs to be done urgently, but none is.

Local Endemics

Conserving the hundreds of locally highly restricted endemic species is also a formidable challenge that remains to be addressed. Two opportunities, however, deserve to be tested. In the case of plants, surviving populations need to be located, monitored, and propagated; this could be a potential role for the National

Botanic Gardens (the public is precluded from engagement because the species are, nominally and regardless of whether or not they actually exist, strictly protected). Indeed, one species, *Alphonsea hortensis*, was in 1985 known only from a single tree in the Peradeniya Gardens (Dassanayake and Fosberg, 1980–2006). It may well be commercially possible for such species to be propagated for use as garden plants, which in turn will generate public interest while yielding a conservation dividend. Such an initiative would offer an opportunity for the public to become involved in a native-plant restoration project as, for example, in Australia (see www.anpsa.org.au).

The monitoring and management of micro-endemic species that occur outside of the protected areas network is impossible without the active participation of civil society. Enforcement alone offers no benefits to taxa such as crabs, which have no commercial value. Similarly, in-situ conservation of micro-endemic fishes such as *Devario pathirana*, *Puntius asoka*, *P. bandula* and *P. srilankensis* that occur outside protected areas can succeed only through participation by the local communities who would need to improve agricultural practices, preserve riparian vegetation, desist from fishing in the river, and ensure that no pollutants enter it. Such engagement could be enhanced through community-led on-site captive breeding programs that could arguably be successful and financially sustainable. No legal provision exists, however, for community participation in conservation initiatives of this kind.

Aquatic species

Given the high degree of hydrological disturbance and the large number of potentially harmful alien fishes in Sri Lanka's inland waters, only ex-situ measures are likely to assure the conservation of threatened aquatic species in the short term, while longer term measures are being designed and implemented. Thanks to the aquarium fish industry, methods to breed most of the fishes in danger have already been developed, though urgent research is needed for recalcitrant species such as *Puntius asoka*. The sad fate of *Labeo lankae* and *Macrognathus pentophthalmos* recalls the "Frankenstein effect" (Moyle, Li and Barton, 1986), in which freshwater-fish extinctions occur more rapidly than data on population declines can be acquired. A precautionary ex-situ approach might have saved these species.

Elephants

The management of the wild elephant population is central to the national conservation strategy, arguably to the detriment of other species in more urgent need of attention. Scientists, conservation managers, and lay conservation enthusiasts form three distinct and sometimes mutually antagonistic constituencies that have their own subset of ideas on how the problem should best be tackled. Working essentially with the same dataset, each advocates a distinct set of remedies stemming from different priorities and ideologies.

What, for example, should be the fate of elephants hemmed within purely agricultural landscapes, which are in conflict with farmers and have no prospect of successful in-situ conservation in the long term? Translocation is the only option that has been tested, but translocated elephants tend to return to their natal territory, apparently preferring their natal area because of their intimate knowledge of its food, water, and other resources (Fernando *et al.*, 2012). Besides, translocations tend to remove mostly the relatively trouble-free herds of females and juveniles, not the troublesome sexually mature males. Conservation managers have invested heavily in the construction of electric fences to keep elephants within protected areas such as Yala National Park, but this, too, has its drawbacks: the more experienced elephants tend to escape with ease, while others, especially vulnerable juveniles, starve because of limited food and water. At the same time, capture and domestication, though relatively easy, yields no conservation benefit. Put bluntly, Sri Lanka has more elephants than the available habitat can support, leading to a conflict that leads to the deaths of tens of people and hundreds of elephants annually. The unwillingness of the public to acknowledge this fact has resulted in officials resorting to expensive measures that merely palliate and do not cure. Interventions such as electric fences and elephant translocations are not only (arguably) futile: they absorb the entire budget, leaving no resources for the conservation of biodiversity at large.

A PATHWAY TO REFORM

It is important to recognize that Red Lists need to be more than just paper. Every threatened species and ecosystem demands (after prioritization through a process of triage) a scientifically valid recovery plan with an appropriate slice of the conservation budget. The state alone cannot possibly implement every one of the hundreds of recovery plans that are needed: it needs to involve non-government entities and academia, admittedly within an effective regulatory framework. Also important is a transparent system of monitoring progress and accountability for failure.

Meanwhile, institutions such as the National Botanic Gardens, National Zoological Gardens and competent non-governmental organizations (NGOs) must be mandated to institute both in- and ex-situ conservation programs for the most vulnerable species including micro-endemics. These institutions and their activities are of obvious appeal to nature tourism and could be financially viable. Likewise, a drive to identify, locate, and propagate threatened native plants (an activity presently prohibited), even at the domestic-garden level, is likely to generate public enthusiasm and engender participation and enhanced awareness.

It is important also to recognize that threatened species inhabit not just the protected areas but the landscape at large. Thus, for example, the patchwork of "home gardens" that characterize the wet zone's rural landscape offers rich opportunities for conservation. Despite being dominated by alien plant species, these secondary forests create novel ecosystems (see Marris, 2009) that can serve as important refugia for animals. Raheem et al. (2008) found them to be an important resource for endemic land snails, while Pethiyagoda and Manamendra-Arachchi (2012), found healthy populations of eight endangered amphibian species to occur in a 25-ha montane secondary forest dominated by two alien invasive plant species at Agrapatana, in the central highlands. Thus, rather than being dismissed as being of negligible conservation value, secondary and plantation forests, even if poor in native plant species, should be viewed as a conservation opportunity rather than a threat, with incentives for owners to develop them also for their value to threatened biodiversity.

Recovery plans for many less well-known species will call for research, which should be prioritized and farmed out to competent government and civil-society agencies. Likewise, competent researchers need to be commissioned to study and monitor general threats such as climate change, invasive species, and montane-forest dieback (see Schaefer, 1998; Chandrajith et al., 2009; Gunasekera, 2009).

Given the threats to the wet zone's biodiversity from fragmentation effects, also urgent is the re-designing of the protected areas network so as to benefit the species and ecosystems most at risk, establishing habitat corridors to link proximal forest fragments. This is admittedly a long-term endeavor that calls for land reform, and economic and demographic changes, but, if built into the development-planning process, could be implemented as circumstances permit. Ecosystem restoration, a complex process, will call for much experimentation and advance research, and this could commence sooner. Other biodiversity-relevant economic changes could be implemented more quickly: e.g., given that almost half of Sri Lanka's energy comes from biomass combustion (mostly harvested from the wild), fossil fuels should be made more accessible to rural communities living close to forests, thereby reducing demand for firewood.

From where, then, will the money for all this come? Understandably, the government's investment priorities focus on economic development. However, the generation of a few million dollars for on-ground conservation initiatives annually calls for only marginal budgetary adjustment. For example, Sri Lanka generates some 40 GWH of hydroelectric energy annually, with a retail value of approximately US $250 million. Given that this industrial benefit is entirely derived from an ecosystem service, a tax of e.g., 5% could not only generate substantial funds for conservation programs, but also be tolerated by consumers. Likewise, a tax of just 1% on imported plastic raw materials could yield an annual revenue of US$3 million for conservation (indeed, in 2003 the plastics industry endorsed just such a proposal). Likewise, engagement with the 18 publicly quoted companies that control most of the 350,000 ha of tea and rubber plantations in the wet zone (together with substantial extents of forest, forestry plantations, and "waste land") could yield a significant biodiversity dividend – e.g., through a system of habitat corridors, stepping stones, and set asides.

None of this, however, can happen unless the state takes the people into its confidence and fosters genuine partnerships with the private sector, the academic community, conservation NGOs, and the conservation-minded public. Almost all the actions discussed earlier require the adoption of a new legislative paradigm that enables and empowers civil society rather than treating it as an adversary (while recognizing that enforcement, too, is important). A high literacy rate and system of universal free education, together with the conservation ethos imbued by the reverence for all life embedded in Buddhism, also stands in Sri Lanka's favor in facing the future challenges of biodiversity conservation.

CONCLUSION

With dozens of plant and animal species already extinct, and its rainforest cover reduced to 8% of its pre-colonial extent, Sri Lanka's biodiversity is in trouble. The official response to this crisis has been to introduce a draconian regulatory regime that distances the people from their country's biodiversity, seeking to (a) prevent exploitation and (b) assert sovereign ownership of this resource as envisaged by the Convention on Biological Diversity. Much of Sri Lanka's threatened biodiversity, however, lies outside the government-owned protected areas network, whether in plantations, home gardens, or abandoned farmland (secondary forest). Its persistence can be assured only if the government takes the people into its confidence and develops meaningful public-private partnerships within an enlightened framework of regulations and incentives. While protection (i.e., regulation, policing) is primarily a function of government, conservation (i.e., the design and implementation of scientific management interventions wherever threatened biodiversity occurs) is an endeavor that necessarily calls for the public – including scientists, NGOs, rural communities, and civil society at large – to be full partners. Unless such partnerships occur in the near future, the long-term prospects for Sri Lanka's biodiversity are bleak.

ACKNOWLEDGMENTS

This essay benefited from the discussions with a number of Sri Lankan biodiversity scientists, among whom it is a pleasure to thank especially Prithiviraj Fernando, Nimal Gunatilleke, Savitri Gunatilleke, Sarath Kotagama, Kalana Maduwage, Kelum Manamendra-Arachchi, Madhava Meegaskumbura, Anjana Silva, and Ruchira Somaweera. I am grateful to Rohan S. Pethiyagoda for several useful suggestions made in the course of a critical review of an earlier draft of this article.

REFERENCES

Anonymous (2010) *Appropriation Act, No. 7 of 2010. Gazette of the Democratic Socialist Republic of Sri Lanka, Supplement II*, Colombo, Sri Lanka. http://www.treasury.gov.lk/BOM/nbd/pdfdocs/act/appropriationact2010.pdf (accessed March 19, 2013).

Anonymous (2011) *Population and Housing*. Department of Census and Statistics, Colombo, Sri Lanka. http://www.statistics.gov.lk/page.asp?page=Population%20and%20Housing (accessed March 19, 2013).

Bahir, M. M., Ng, P. K. L., Crandall, K. and Pethiyagoda, R. (2005) A conservation assessment of the freshwater crabs of Sri Lanka. *The Raffles Bulletin of Zoology Supplement*, **12**, 351–380.

Bandara, R. and Tisdell, C. (2001) Conserving Asian elephants: economic issues illustrated by Sri Lankan concerns. Working paper 59, Department of Economics, University of Queensland, Brisbane, Australia.

Beenaerts, N., Pethiyagoda, R., Ng, P. K. L., Yeo, D. C. J., Bex, G. J., Bahir, M. M. and Artois, T. (2010) Phylogenetic diversity of Sri Lankan freshwater crabs and its implications for conservation. *Molecular Ecology*, **19**, 183–196.

Bennett, A. F. and Saunders, D. A. (2010) Habitat fragmentation and landscape change, in *Conservation Biology for All* (eds N. S. Sodhi and P. R. Erlich), Oxford University Press, Oxford, pp. 88–106.

Bierregaard, R. O. Jr, Laurance, W. F., Gascon, C., Benitex-Malvido, J., Fearnside, P. M., Fonseca, C. R., Ganade, G., Malcolm, J. R., Martins, M. B., Mori, S., Oliveira, M., Rankin-de Mérona, J., Scariot, A., Spironello, W. and Williamson, B. (2001) Principles of forest fragmentation and conservation in the Amazon, *Lessons from Amazonia: the Ecology and Conservation of a Fragmented Forest* (eds R. O. Bierregaard, Jr, C. Gascon, T. E. Lovejoy, and R. Mesquita), Yale University Press, New Haven, CT, pp. 371–385.

Chandrajith, R., Koralegedara, N., Ranawana, K. B., Tobschall, H. J. and Dissanayake, C. B. (2009) Major and trace elements in plants and soils in Horton Plains National Park, Sri Lanka: an approach to explain forest die back. *Environmental Geology*, **57**, 17–28.

Cincotta, R. P., Wisnewski, J. and Engelman, R. (2000) Human population in the biodiversity hotspots. *Nature*, **404**, 990–992.

Dassanayake, M. D. and Fosberg, F. R. (eds) (1980–2006) *A Revised Handbook to the Flora of Ceylon*, 16 vols, Oxford and IBH Publishing Co., New Delhi, India.

de Silva, M. and de Silva, P. K. (2007) *The Sri Lankan Elephant: Its Evolution, Ecology and Conservation*. WHT Publications, Colombo, Sri Lanka.

FPU (Forestry Planning Unit) (1995) *Sri Lanka Forestry Sector Master Plan*. Forestry Planning Unit, Ministry of Agriculture, Lands and Forestry, Battaramulla, Sri Lanka.

Fernando, P., Jayewardene, J., Prasad, T., Hendavitharana, W. and Pastorini, J. (2011) Current status of Asian elephants in Sri Lanka. *Gajah*, **35**, 93–103.

Fernando P., Leimgruber P., Prasad T., Pastorini J. (2012) Problem-elephant translocation: translocating the problem and the elephant? *PLoS ONE*, **7**, e50917. doi:10.1371/journal.pone.0050917

Fernando, P., Wikramanayake, E., Weerakoon, D., Jayasinghe, L. K. A., Gunawardene, M. and Janaka, H. K. (2005) Perceptions and patterns of human-elephant conflict in old and new settlements in Sri Lanka: insights for mitigation and management. *Biodiversity and Conservation*, **14**, 2465–2481.

Green, M. J. B., How, R., Padmalal, U. K. G. K. and Dissanayake, S. R. B. (2009) The importance of monitoring biological diversity and its application to Sri Lanka. *Tropical Ecology*, **50**, 41–56.

Gunasekera, L. (2009) *Invasive Plants: A Guide to the Identification of the Most Invasive Plants in Sri Lanka*. The author, Colombo, Sri Lanka.

Gunatilleke, N., Pethiyagoda, R. and Gunatilleke, S. (2008) Biodiversity of Sri Lanka. *Journal of the National Science Foundation of Sri Lanka*, **36**, 25–62.

Gunawardene, N. R., Daniels, A. E. D., Gunatilleke, I. A. U. N., Gunatilleke, C. V. S., Karunakaran, P. V., Nayak, K. G., Prasad, S., Puyravaud, P., Ramesh, B. R., Subramanian, K. A. and Vasanthy, G. (2007) A brief overview of the Western Ghats – Sri Lanka biodiversity hotspot. *Current Science*, **93**, 1567–1572.

Hendavitharana, W., Dissanayake, S., de Silva, M. and Santiapillai, C. (1994) The survey of elephants in Sri Lanka. *Gajah*, **12**, 1–30.

IUCN (The International Union for Conservation of Nature and Natural Resources) (2011) *The IUCN Red List of Threatened Species*. http://www.iucnredlist.org (accessed March 19, 2013).

Kumar, A., Pethiyagoda, R. and Mudappa, D. (2004) Western Ghats and Sri Lanka, in *Hotspots Revisited: Earth's Richest and Most Endangered Terrestrial Ecoregions* (eds R. A. Mittermeier, P. R. Gill, M. Hoffmann, J. Pilgrim, T. Brooks, C. G. Mittermeier, J. Lamoreux and G. A. B. Da Fonseca), Cemex, Mexico, pp. 152–158.

Manamendra-Arachchi, K. and Pethiyagoda, R. (2007) *Sri Lankawe ubhayajeevin* ["The am¬phibian fauna of Sri Lanka"] (in Sinhala). WHT Publications, Colombo. 440 pp., 88 pl.

Marris, E. (2009) Ragamuffin Earth. *Nature*, **460**, 450–453.

McKay, G. M. (1973) Behavior and ecology of the Asiatic elephant in South-eastern Ceylon. *Smithsonian Contributions to Zoology*, **125**, 1–113.

Moyle, P. B., Li, H. W. and Barton, B. A. (1986) The Frankenstein effect. The impact of introduced fishes on native fishes of North America, in *Fish Culture in Fisheries Management* (ed. R. H. Stroud), American Fisheries Society, Bethesda, MD, pp. 415–426.

Myers, N., Mittermeier, R. A., Mittermeier, C. G., da Fonseca, G. A. B. and Kent, J. (2000) Biodiversity hotspots for conservation priorities. *Nature*, **403**, 853–859.

Perera, G. A. D. (2001) The secondary forest situation in Sri Lanka: a review. *Journal of Tropical Forest Science*, **13**, 768–785.

Pethiyagoda, R. (1994) Threats to the indigenous freshwater fishes of Sri Lanka and remarks on their conservation. *Hydrobiologia*, **285**, 189–201.

Pethiyagoda, R. (2004) Biodiversity law has had some unintended effects. *Nature*, **429**, 129.

Pethiyagoda, R. (2006) Conservation of freshwater fishes, in *The Fauna of Sri Lanka: Status of Taxonomy, Research and Conservation* (ed. C. N. B. Bamabaradeniya), The World Conservation Union, Colombo, Sri Lanka, pp. 103–112.

Pethiyagoda, R. and Meegaskumbura, M. (in press) Sri Lankan amphibians: extinctions and endangerment, in *Amphibian Biology, Vol. 11: Status of Decline of Amphibians: Eastern Hemisphere, Part 5: Afghanistan, Pakistan, India, Sri Lanka* (ed. I. Das), Surrey Beatty & Sons, Bauklham Hills, Australia.

Pethiyagoda, R., Gunatilleke, M., De Silva, M., Kotagama, S., Gunatilleke, S., de Silva, P., Meegaskumbura, M., Fernando, P., Ratnayeke, S., Jayewardene, J., Raheem, D., Benjamin, S. and Ilangakoon, A. (2007) Science and biodiversity: the predicament of Sri Lanka. *Current Science*, **92**, 426–427.

Pethiyagoda, R. S. and Manamendra-Arachchi, K. (2012) Endangered anurans in a novel forest in the highlands of Sri Lanka. *Wildlife Research*, **39**, http://dx.doi.org/10.1071/WR12079

Pethiyagoda, R. S. and Nanayakkara, S. (2011) Invasion by *Austroeupatorium inulifolium* (Asteraceae) arrests succession following tea cultivation in the highlands of Sri Lanka. *Ceylon Journal of Science*, **40**, 175–181.

Pimm, S. L. and Jenkins, C. N. (2010) Extinctions and the practice of preventing them, in *Conservation Biology for All* (eds N. S. Sodhi and P. R. Erlich), Oxford University Press, Oxford, pp. 181–198.

Population Reference Bureau (2011) *2011 World Population Data Sheet*. http://www.prb.org/pdf11/2011population-data-sheet_eng.pdf (accessed March 19, 2013).

Population Reference Bureau (2012) *World Population Data Sheet 2012*. http://www.prb.org/pdf12/2012-population-data-sheet_eng.pdf (accessed March 19, 2013.

Raheem, D. C., Naggs, F., Preece, R. C., Mapatuna, Y., Kariyawasam, L. and Eggleton, P. (2008) Structure and conservation of Sri Lankan land-snail assemblages in fragmented lowland rainforest and village home gardens. *Journal of Applied Ecology*, **45**, 1019–1028.

Rodrigo, M. (2011) Sri Lanka has 7,370 elephants. *Down to Earth*. http://www.downtoearth.org.in/content/sri-lanka-has-7370-elephants (accessed March 19, 2013).

Schaefer, D. (1998) Climate change in Sri Lanka? Statistical analyses of long-term temperature and rainfall records, in *Sri Lanka: Past and Present: Arachaeology, Geography, Economics: Selected Papers on German Research* (eds M. Domroes and H. Roth), Margraf Verlag, Weikersheim, Germany, pp. 103–117.

Senanayake, F. R. (1980) The biogeography and ecology of the inland fishes of Sri Lanka. PhD thesis, Department of Wildlife and Fisheries Biology, University of California, Davis, CA.

Stuart, S., Hoffmann, M., Chanson, J. S., Cox, N. A., Berridge, R. J., Ramani, P. and Young, B. E. (eds) (2008) *Threatened Amphibians of the World*. Lynx Ediciones, Barcelona, Spain.

Wright, A. (2012). Zoo man Damian Goodall fined over frog in Sri Lanka. *Herald Sun*, 2 March. http://www.news.com.au/zoo-man-fined-over-frog/story-fn7x8me2-1226286684771 (accessed March 19, 2013).

CHAPTER 19

CONSERVATION OF HORNBILLS IN THAILAND

Pilai Poonswad[1], Vijak Chimchome[2], Narong Mahannop and Sittichai Mudsri[3]

[1]Department of Microbiology, Faculty of Science, Mahidol University, Bangkok, Thailand
[2]Department of Forest Biology, Faculty of Forestry, Kasetsart University, Bangkok, Thailand
[3]Department of National Parks, Wildlife and Plant Conservation, Bangkok, Thailand

INTRODUCTION

In Thailand, as in other developing countries in Asia, the state of conservation has not kept pace with development. The management of natural resources is the responsibility of several government agencies, but they are often in conflict. Partly in consequence of this division of responsibility, Thailand has lost more than 40% of its forested areas, which are the richest terrestrial habitats in biodiversity within the past five decades (1961–2010), partly because of poor coordination between and execution of policies in the areas of economic, social, and political development (FAO, 2010). In 1951, Dr Boonsong Lekagul and his colleagues founded the Association for Conservation of Wildlife, the first non-governmental organization established in Thailand for conservation. Through its immense struggle to attract attention from the government, this group helped to establish laws concerning the conservation of wildlife and their habitats, including the Wild Animals Reservation and Protection Act (1960) and the National Park Act (1961). For his efforts, Dr Boonsong was named "Father of Conservation" in Thailand. The protection and conservation of natural resources in Thailand really only began with the formation of the Association in 1951.

CONSERVATION ISSUES IN THAILAND

Over the past century, the natural resources of Thailand have been depleted continuously and rapidly, driven by such pressures as population growth, poverty, and globalization, including economic expansion that places overriding importance on the size of the country's gross domestic product. By 1985, the government had set a national forest policy target whereby no less than 40% of the country's area was to be protected, 15% as conservation forests and 25% as economic forests (RFD, 1985). In the Seventh National Economic and Social Development Plan (1993–1996), the target for conservation of national forest was changed to 25% for conservation forest and 15% for economic forest (NESDB, 2008), still a total of 40% of the total area. By 1989, however, only 28% of the total land area remained forested.

To sustain Thailand's exceptional species diversity, the government has put a major effort into protecting the forest and its animal inhabitants by various conservation measures. Wild animals and their habitats are now completely protected by the National Park Acts (1961) and the revised Wild Animals Reservation and Protection Act (1992). A total of 123 national parks and 58 wildlife sanctuaries have been established in the

Conservation Biology: Voices from the Tropics, First Edition. Navjot S. Sodhi, Luke Gibson, and Peter H. Raven.
© 2013 John Wiley & Sons, Ltd. Published 2013 by John Wiley & Sons, Ltd.

past 50 years, but they cover only 15% of the country's area, still well short of the goals stated through the years. Thailand has also ratified some international conventions related to natural resource conservation and management, including the Convention on the International Trade in Endangered Species of Wild Fauna and Flora (CITES) in 1983 and the Convention on Biological Diversity (CBD) in 2003 (ICEM, 2003). All of these treaties support wildlife conservation not only in Thailand but also in trans-boundary areas and at a global scale. However, despite the existence of good laws, enforcement mechanisms do not always seem to work, so that there is often further depletion of primary forest. In an evaluation of forest resources in 2004, only about 18% out of the target of 25% of total land areas were protected (Trisurat, 2007).

The protected areas in Thailand are highly fragmented, but they can be grouped into 19 complexes that include 17 forest areas and two marine and coastal habitats. To connect these fragmented areas, a Biodiversity Conservation Corridors Initiative (BCI; Phase 1: 2006–2008) was set up, with a pilot project, "Wildlife and its Habitat Assessment in the Corridor Zone in the Tenasserim Western Forest Complex (WEFCOM), Thailand" run by the Wildlife Conservation Society (Thailand). Seven landscape-wide species, including tiger, elephant, gaur, sambar deer, barking deer, serow, and great hornbill, were monitored in the complex. In Phase II (2009–2011), the Department of National Parks, Wildlife and Plant Conservation has raised the corridor program to the national level so as to evaluate and prioritize potential areas for connectivity within and between complexes. This program will be not only a response to the CBD, but will also serve the goal of reducing biodiversity loss from its 2000 level, as declared by the World Biodiversity Summit 2002 in Johannesburg, South Africa. Although the poaching of wildlife and collection of wild plants for illegal trade violate both national and international laws, the cost of this trade is still severe: between 2005 and 2008, it was estimated at more than US$1 million.

WHY CONSERVE HORNBILLS?

This question is the one asked most frequently by people living in concrete habitats! The answer to this question is not easy, but it is challenging to help people living in cities, who are familiar only with pigeons, sparrows, and crows, to understand the uniqueness

and intrinsic value of these ancient birds and the threats they are facing.

Hornbills originated at least 50 million years ago in the Eocene Period. The living species are unique and attractive birds, not only in their appearance but also in their intriguing nesting habits. Thai hornbills are very large (body mass 680–3,400 g) and very noisy, which makes them conspicuous in the forest. Although the birds are omnivorous, fruits are the main component in their diet (Poonswad, Tsuji and Jirawatkavi, 2004); therefore, hornbills are characterized as large frugivorous birds. To satisfy their needs, they require intact primary forest that provides suitable nest sites and sufficient food resources.

Hornbills have intimate relationships with forest plants as food, and probably coevolved with many of these plants as seed-dispersal agents of primary importance. In ecological terms, hornbills are often considered "keystone" species whose essential service is to move seeds away from parent trees and spread them over a larger area. By doing this they help to regenerate the forest and maintain the diversity of plants within their habitat. Past studies have shown the significance of hornbills in dispersing seeds, particularly of those plants with large seeds (>25 mm) that rely almost entirely on the large bills of hornbills. By combining existing information on their home range, number of fruit species consumed, and their habits of regurgitating a few seeds while perching and in flight, some estimates are possible. Great Hornbills (*Buceros bicornis*) move seeds around their home range of 30 km^2 and over a distance of 15 km daily, while Wreathed Hornbill (*Rhyticeros undulatus*) move around over 35 km^2 (Poonswad and Tsuji, 1994). Hornbills are undoubtedly important in forest regeneration, particularly to establish connections between forest patches along their flyways or within their nomadic range (Holbrook, Smith and Hardesty, 2002; Kinnaird and O'Brien, 2007). It is no exaggeration that Kinnaird and O'Brien (2007) name hornbills as "farmers of the forest."

THREATS TO AND THE CONSERVATION STATUS OF HORNBILLS

Since hornbills rely entirely on the availability of natural tree cavities for reproduction and depend very much on fruit as their food resource, any loss of habitat means loss not only of breeding sites but also of food resources. Because of their rigid requirement for breeding sites, hornbills are inevitably threatened by any human activities within forests, besides such natural phenomena as storm and decomposition processes that are also considered important threats to hornbill populations (Table 19.1).

Human activities have altered the environment of other vertebrate species to an immense degree. Population growth coupled with rapid economic growth is profoundly important as a driver in the exploitation of natural resources, particularly forest resources. Unlimited demand, poor planning, and unsustainable use of this valuable and natural capital resource have resulted in a great depletion of forest resources and hence of hornbill habitat. The depletion of forest resources results from a range of activities, the most obvious of which are now discussed.

Deforestation and logging

Deforestation poses the most serious threat to hornbills. The 13 species of hornbills that occur in Thailand inhabit various types of forests, the most important being evergreen forest, which ranges from lowland plains up to 1500 m above sea level and which exists as dry evergreen, semi-evergreen (Santisuk, 2007; Corlett, 2009), and hill evergreen forest subtypes. Evergreen forests dominate the forested area in Thailand (43%) but are also the most threatened forest type. Cul-

tivation is the major human activity altering these forests, which destroys primary forest in Thailand at an estimated rate of 0.7% annually (FAO, 2005). The most obvious consequence of evergreen forest destruction is the loss of dominant large trees, particularly trees in the genera *Dipterocarpus*, *Hopea* and *Shorea* (Dipterocarpaceae). Losses of primary forests have impacts on hornbill populations by reducing their potential breeding sites and depleting their food resources. Dipterocarp trees are the predominant hornbill nest trees, accounting for 40% of nest sites (Poonswad, 1995). Deforestation has already extirpated three sympatric hornbill species, the Great, Wreathed and Rufousnecked *Aceros nipalensis* Hornbills from parts of their range in northern Thailand (Poonswad, 1993).

Hunting

Hunting is another major threat to wildlife, including hornbills. Although hornbills are canopy-living species, their large size and noisy behavior make them conspicuous so that they make an easy target for hunters. Even their secretive nesting behavior cannot elude hunters. The purposes for hunting hornbills include for food and the pet trade. Hunting them depends on the existence of areas where laws are not effective, with the hunting often linked to tribes and villagers who live in or near the forest. Hill tribes in northern and western

Table 19.1 Nest cavity information (a) and chick production (b) at Khao Yai National Park (KY) and Budo Mountain (Budo), with national status of hornbill species

(a)

	KY (1981–2008)		Budo (1994–2008)	
	Total	Annual	Total	Annual
No. recorded nest trees	226	8.7	188	12.5
No. nest loss	95	4.1	44	2.9
% nest loss	42.0	4.9	23.4	2.7
No. bad cavities	123	5.1	25	1.7
% bad cavities	54.4	6.2	10.6	13.6
No. repair/improved (1994-2008)	77	4.7	20	–
No. cavities available	120	70.8	119	90.3

(Continued)

Table 19.1 (*Continued*)

(b)

	KY (1994–2008)			Budo (1994–2008)		
	Total	Annual	From repair	Total	Annual	From repair
Great Hornbill (near threatened)**						
No. sealed	312	20.8	134	263	17.5	–
% success	77.9	78.4	42.9	83.6	84.2	–
No. chicks (x1/pair)	243	16.2	134	220	14.6	–
Rhinoceros Hornbill (endangered)**						
No. sealed	–	–	–	158	10.5	–
% success	–	–	–	72.8	71.8	–
No. chicks (x1/pair)	–	–	–	115	7.7	–
Helmeted Hornbill (endangered)**						
No. sealed	–	–	–	43	2.8	–
% success	–	–	–	75	82.1	–
No. chicks (x1/pair)	–	–	–	32	2.3	–
Wreathed Hornbill (near threatened)**						
No. sealed	148	9.9	56	55	4.2	–
% success	89.7	87.9	33.6	73.5	72.2	–
No. chicks (x1/pair)	133	8.9	49	40	3.0	–
Brown Hornbill (Vulnerable)**						
No. sealed	119	7.9	36	–	–	–
% success	90.6	87.3	27.4	–	–	–
No. chicks (x2.3/pair)*	248	16.6	74	–	–	–
Bushy-crested Hornbill (near threatened)**						
No. sealed	–	–	–	35	2.5	–
% success	–	–	–	65.5	67.9	–
No. chicks (x2.3/pair)*	–	–	–	53	3.5	–
White-crowned Hornbill (endangered)**						
No. sealed	–	–	–	12	1.7	–
% success	–	–	–	80	–	–
No. chicks (x1.5/pair)*	–	–	–	15	1.4	–
Oriental Pied Hornbill (not determined)**						
No. sealed	371	24.7	53	–	–	–
% success	92.2	91.8	12.4	–	–	–
No. chicks (x1.5/pair)*	513	34.2	69	–	–	–
All species						
No. sealed	950	63.3	312	566	37.7	–
% success	86.9	86.5	27.6	77.8	77.8	–
No. chicks	1,137	75.8	326	440	29.3	–

*Average number of chicks per breeding pair derived from Poonswad (1993)
**National status from Sanguansombat (2005) and ONEP (2007)
Source: (a and b) Based on data from Poonswad (1993) Sanguansombat (2005) and ONEP (2007).

Thailand hunt hornbills mainly for food, and this hunting might have an important impact on the resilience of certain species, such as Tickell's Brown Hornbill (*Ptilolaemus tickelli*), particularly when the hunting pressure is exacerbated locally by deforestation.

STATUS OF HORNBILLS IN THAILAND

Evaluation and revision of the conservation status of Thai flora and fauna by various criteria have been made in the past few decades. Lekagul and Round (1991) determined the status of Thai birds based on abundance and habitat restriction. More recently, the Office of Natural Resource and Environmental Policy and Planning (ONEP, 2007) revised the Thailand Red Data status for vertebrates, including birds, based on the IUCN criteria in their 2001 Red List, version 3.1. The Red Data status for the 13 hornbill species found in Thailand suggest this group requires urgent conservation actions (Figure 19.1): two species, the Black *Anthracoceros malayanus* and Wrinkled *Rhyticeros corrugatus* Hornbills, are critically endangered; four species, the Rufous-necked, Plain-pouched *Rhyticeros subruficollis*, Helmeted *Rhinoplax vigil* and Rhinoceros *Buceros rhinoceros* Hornbills, are endangered; six more species, the White-crowned Hornbill *Berenicornis comatus*, White-throated Brown *Ptilolaemus austeni*, Tickell's Brown Hornbills, the Bushy-crested *Anorrhinus galeritus*, Great and Wreathed Hornbills are vulnerable; only one species, the Oriental Pied *Anthracoceros albirostris*, is of least concern. The evaluation of status is important, providing warning information and basic guidelines for setting the priority and degree of intensiveness of conservation activities. Since Thailand is situated on the Asian mainland and extends onto the Thai-Malay Peninsula, it shares hornbill species with other geographical areas where the conditions for effective conservation may differ. Working toward the conservation of Thai hornbills for their long-term existence, we need to take serious consideration of our national status level and that of our neighbors.

RECOMMENDATIONS

Loss of forest means loss of habitat for wildlife, particularly for hornbills, which require large trees for both nesting holes and for food resources. Due to this, intensive survey and study of threatened and endangered

species are required to ascertain the actual status, density, and potential of the remaining habitats for target species. Twelve out of 19 protected area complexes have potential as hornbill habitats. The remaining seven complexes are too small and severely fragmented, and no hornbills have been recorded in them for the past two decades (Poonswad, 1993). Intensive surveys using point-count transects in three complexes of different sizes in which extensive research and conservation activities are continuously conducted (WEFCOM, Dong Phayayen-Khao Yai, and Hala-Bala) indicate that these complexes still support viable populations of Great, Wreathed, Rufous-necked, Rhinoceros and Oriental-Pied Hornbills, while other species do not have adequate data for analysis (Thailand Hornbill Project, unpublished data; see also Table 19.1). Increasing connectivity of suitable habitat between and within complexes to facilitate hornbill movement, as being practiced for conservation planning in WEFCOM, is a promising approach to maintain population viability of hornbills (Trisurat *et al.*, 2010).

The degree of threats mentioned earlier may differ by area or region. The goal of the conservation efforts is to increase hornbill populations to minimum viable sizes and so sustain them for long-term survival. To achieve this goal, clear identification of threats or problems in each area or region is very important in order to implement the most suitable strategy. According to our knowledge and experience, we recommend our two most successful strategies: research- and community-based conservation.

Research-based conservation

Initiation of hornbill research project

Some 30 years ago, knowledge on the biology of Thai hornbills was limited to general information on distribution, habitat, and anecdotes about food and behavior (Lekagul and Cronin, 1974). Given the very large size of the birds, one could imagine the magnitude of requirements – they need large nest cavities and large amounts of food. But, how large must the cavities be? Being secondary-cavity nesters that are unable to excavate their own nests, hornbills do not have much choice. Finding a suitable cavity is a principal factor that limits hornbill reproduction, but what attracts them in seeking out a nest cavity? Ground-breaking research to reveal the basic requirements of four

Male	Female	National status*	Male	Female	National status*
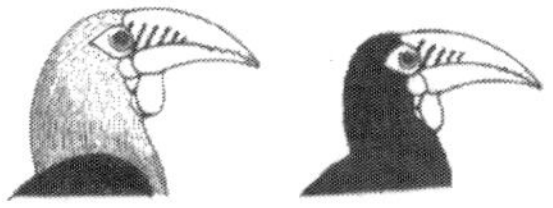 Rufous-necked Hornbill (*Aceros nipalensis*)		Endangered	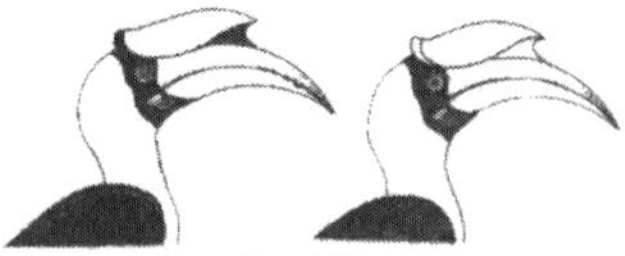 Great Hornbill (*Buceros bicornis*)		Vulnerable
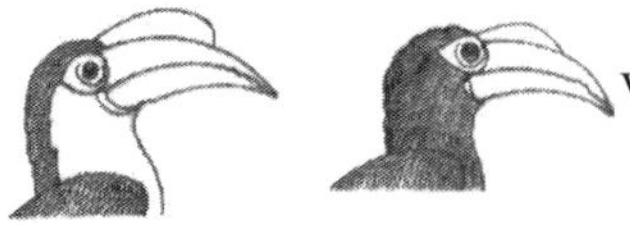 White-throated Brown Hornbill (*Anorrhinus austeni*)		Vulnerable	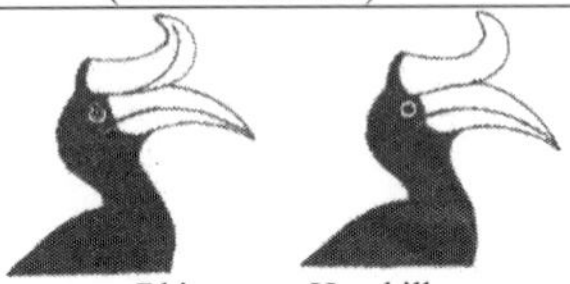 Rhinoceros Hornbill (*Buceros rhinoceros*)		Endangered
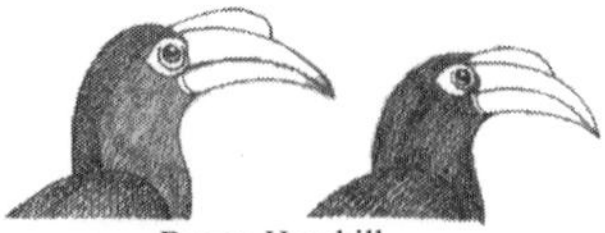 Brown Hornbill (*Anorrhinus tickelli*)		Vulnerable	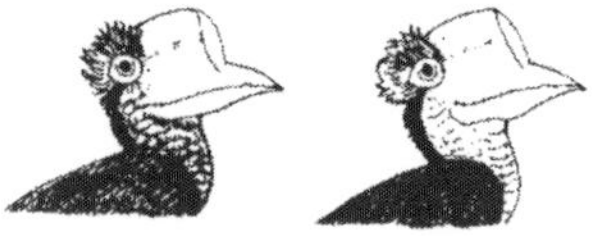 Helmeted Hornbill (*Rhinoplax vigil*)		Endangered
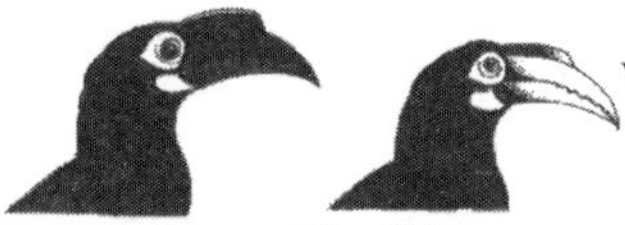 Bushy-crested Hornbill (*Anorrhinus galeritus*)		Vulnerable	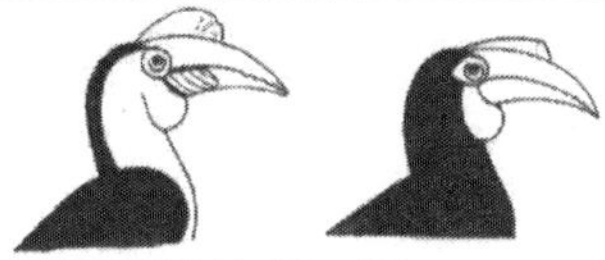 Wrinkled Hornbill (*Rhyticeros corrugatus*)		Critically endangered
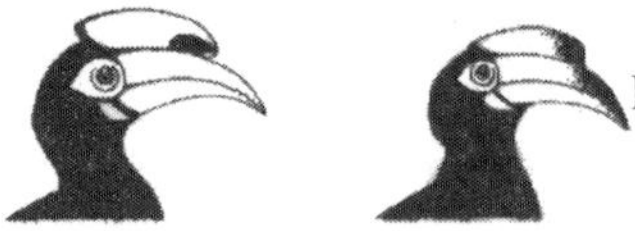 Oriental Pied Hornbill (*Anthracoceros albirostris*)		Least concern	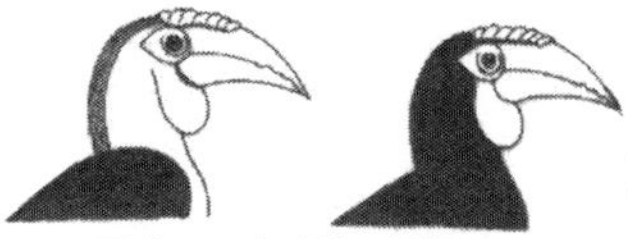 Plain-pouched Hornbill (*Rhyticeros subruficollis*)		Endangered
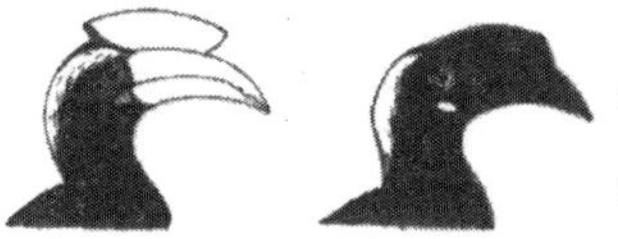 Black Hornbill (*Anthracoceros malayanus*)		Critically endangered	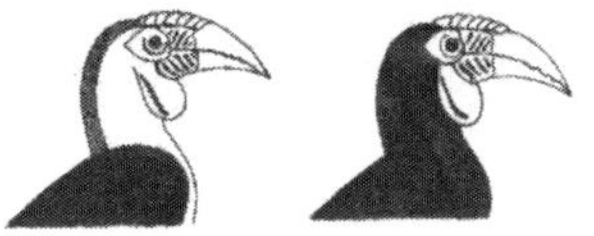 Wreathed Hornbill (*Rhyticeros undulatus*)		Vulnerable
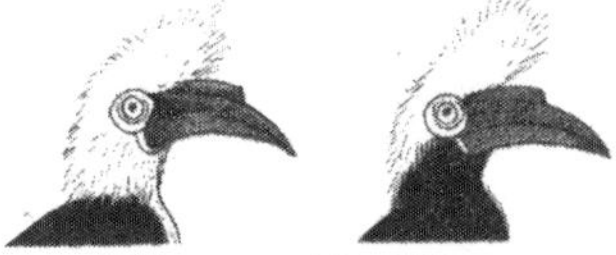 White-crowned Hornbill (*Berenicornis comatus*)		Vulnerable			

Figure 19.1 Sketches of 13 hornbill species and their conservation status.
*National status from Sanguansombat (2005) and ONEP (2007).
Source: Based on Sanguansombat (2005) and ONEP (2007).

sympatric species, Great, Wreathed, White-throated Brown and Oriental Pied Hornbills, for breeding at Khao Yai National Park (KY, 2,168 km^2) was begun in 1981 on a 150 km^2 study area within semi-evergreen forest and continues to the present day.

Implementation of knowledge

From long-term research at KY, we have amassed a great sample of nests under observation, and have proved the success of implementation of field knowledge to refurbish nesting cavities. Between 1981 and 2008, we located a total of 226 different nest trees (Table 19.1 and Figure 19.2a). Hornbills used mainly trees of the genera *Dipterocarpus* of the family Dipterocarpaceae (40%) followed by *Syzygium* of the Myrtaceae (20%). Nest trees are very large, with a diameter at breast height (dbh) of, on average, over 100 cm; tall and emergent above the forest canopy; and hence mainly aged or over-mature trees. These trees are prone to damage by wind storms that cause irreversible breakage to this rare resource. Over 26 years, 42% of nest trees were lost to strong winds (Table 19.1). Nest cavities that form in these aged trees are also subject to gradual decay by rot fungi and 50% of those we examined had become unsuitable for this reason (Table 19.1). Losses of nest trees and poor cavity condition can be a natural threat to breeding hornbills and their populations.

A consequence of the nest losses is competition for good cavities, which may subsequently cause nest abandonment. Competition among animals over a limited resource is normal and may be enhanced by a range of factors, such as safe location. In the case of hornbills at KY, the annual nest abandonment is 36% of nest cavities (1981–2008), and competition over nest cavities is as high as 40% (Poonswad *et al.*, 2005), with 53% of disputed cavities being abandoned (Poonswad *et al.*, 1999). These two aspects are good clues to tell us the situation of cavities; we used them when we began to inspect the condition of the cavities (Table 19.1). Of 152 cavities inspected, 123 (80.9%) were unsuitable and 77 (62.6%) were repaired. The most serious problems were a sunken nest floor (deeper than 15 cm, 50%) and a closed or narrowed entrance (less than 10 cm wide, 40%). The repair was done prior to the breeding season, and was a simple operation: soil filling for cavities with deep floor, and enlarging the entrance by chisel for narrowed or closed entrance.

Realizing the shortage of good cavities, the THP team has improved those natural cavities that have the potential to be nests and so increase the breeding opportunities for hornbills. Over 15 years (1994–2008), an overall breeding success rate of 86% produced 1137 chicks of four species, of which about 30% successfully fledged from repaired and improved cavities (Table 19.1). Among these, the Great Hornbill, the largest species, benefited most by producing 243

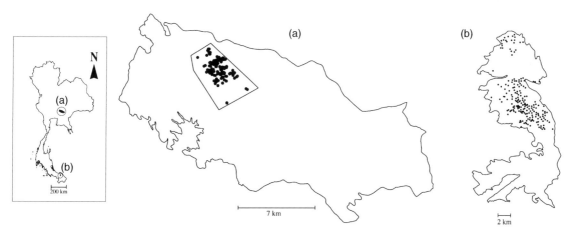

Figure 19.2 Study areas and location of nests at (a) Khao Yai National Park and (b) Budo Mountain. Permission from the Department of National Parks, Wildlife and Plant Conservation (DNP).

chicks, of which 134 (55%) successfully fledged (Table 19.1). Without cavity monitoring and management, hornbills at KY, particularly Great Hornbills, would undoubtedly have declined.

Community-based conservation

The significance of Budo Mountain

Unlike KY, which is much larger in size and well protected and managed, Budo Sungai Padi National Park (BSNP: only 340 km^2) was designated as a national park only in 1999, following 12 years of preparation. It comprises Budo and Sungai Padi, two isolated mountains situated in peninsular Thailand. These mountains are covered with tropical lowland rainforest, encompassing the rich biodiversity of fauna and flora of the Thai-Malay Archipelago, a Sundaic habitat that is of limited extent in Thailand and has been much reduced in Malaysia. The Park incorporates areas in three provinces where social unrest has continued since 2003. In effect, these two mountains are discrete and separate islands of rainforest within a "sea" of human-modified habitats that are occupied mainly by villagers and farmers of Muslim faith.

Amazingly, the forest of Budo mountain (190 km^2, about 90 km^2 of primary forest and 100 km^2 of disturbed forest, rubber plantations, and fruit orchards), supports six species of hornbills. Some of these species are of conservation concern internationally, including the Rhinoceros, Helmeted, and White-crowned Hornbills. They are also of high concern nationally, particularly the Rhinoceros Hornbill that was previously thought to have become extinct in Thailand. During the period 1994–2008, 188 nest trees of these six hornbill species were located by ex-poachers and ex-illegal loggers (villagers hereafter) (Table 19.1, Figure 19.2). Budo has been under military operations for most of the years during our research and conservation practices.

Initiation of community-based conservation

Before 1994, hornbills at Budo were severely poached for the pet trade and food, and were in jeopardy of extinction. The discovery of poaching of hornbill chicks from their nests and their illegal sale into aviculture and the pet trade was an important source of supplementary income in the villagers' lives, given that the most desired species could fetch up to US$750 per bird. Early in 1994, the immediate goal of stopping poaching and increasing hornbill populations was established as Phase I. Intensive efforts were undertaken promptly to convince and persuade villagers to participate in research and conservation activities, wherein they became guides and later research assistants.

The project did not develop without some unexpected problems. At first, some villagers had reservations about dealing with non-Muslim strangers, but these were overcome through person-to-person contact, ethical project operation and principles, and development of mutual respect. In 1997 there was a severe economic crisis in Thailand, which reduced many local corporate sources of funds and the resumption of poaching became a risk. Withdrawal of the project would have been considered a failure, implying lack of determination of our part and potentially impairing the trust we had gained. It was imperative to keep the project running, despite the economic crisis.

Hornbill family adoption

In response to the crisis, fundraising to run the Budo hornbill conservation program through a program of "Hornbill Family Adoption" was initiated in 1997. This successful effort has become recognized as a win-win program more widely. The program encouraged people outside BSNP, particularly those living in urban areas, to participate by making an annual donation for each hornbill nest that was adopted. All donations went into training and hiring villagers as guides and assistants, since overheads for the project were borne internally by the THP. In return, the adopter(s), who chose the species they preferred, gained knowledge by receiving an annual report, prepared by THP's staff, with details of the nest tree, breeding cycle, breeding success, food, and feeding behavior. They also received pictures of hornbills at their nests, hornbill foods, and the villager(s) who protected and collected data at their nests. Those who wished to visit the species they had adopted were welcomed and guided by the villagers, so that the parties met and conversed. In this way, the practice led to the initiation of ecotourism related to the project.

On achievement of Phase I, a further goal to increase and sustain the hornbill populations to reach minimum viable population size was set as Phase II. An intensive and continuous community-collaborative campaign of research and conservation has proven highly successful in eradicating poaching and significantly increasing hornbill populations, particularly those of two of the endangered species (Table 19.1). Over 16 years, under this community-based conservation, there were no less than 440 hornbill chicks of six species fledged, 50% being Great and 26% Rhinoceros Hornbills (Table 19.1).

Despite the poor economic conditions, exacerbated by regional unrest, which has a pervasive effect on community lives and livelihoods around Budo, the hornbills remain under this form of community care. Around Budo, the annual outreach campaign among school children and teachers about hornbills and nature conservation shows a progressive increase in numbers from 200 individuals in 2006 to 1600 in 2010. Even though a nest cavity is an important factor for hornbills to reproduce, at Budo the future of hornbills relies on community concern as the prime factor. With limiting natural resources in a conservation context, it is a challenge to transform an economically based community into an altruistic community. If a community does realize the intrinsic value of their natural heritage with pride, as a gift to future generations, as a responsibility, rather than focusing on economic benefit, then conservation in these forests will truly bloom. Partnerships with outsiders who may help to support the effort financially can also be extremely helpful.

ACKNOWLEDGMENTS

We are deeply indebted to the late Professor Sodhi, who inspired and kindly included a part of our research in this valuable book. We would like to thank the Department of National Parks, Wildlife and Plant Conservation for granting permission to conduct research in Khao Yai and Budo Sungai Padi National Parks. We extend our thanks to park superintendents, park staff, THP staff, and villagers of Budo Mountain for data collection and excellent cooperation. We are grateful to Dr Alan Kemp for his unlimited assistance. Our special thanks go to Preeda Thiensongrusme for illustrations of hornbills and to Porntip Poolswat for preparation of this manuscript. Research and conservation activities are supported by various organizations and private sectors: Thailand, National Center of Genetic Engineering and Biotechnology, Hornbill Research Foundation, the Siam Cement Group, I.C.C International Public Company Limited, PTT Exploration and Production Public Company Limited, Mahidol and Kasetsart Universities; USA, Woodland Park Zoo, Seattle, WA, American Association of Zoo Keepers (AAZK); UK, Chester Zoo, Cheshire.

REFERENCES

Corlett, R.T. (2009) *The Ecology of Tropical East Asia*. Oxford University Press, Oxford.

FAO (2005) *State of the World's Forests 2005*. FAO, Rome, Italy.

FAO (2010) *Global Forest Resources Assessment 2010*. FAO, Rome, Italy.

Holbrook, K. M., Smith, T. B. and Hardesty, B. D. (2002) Implication of long-distance movements of frugivorous rain forest hornbills. *Ecography*, **25**, 745–750.

ICEM (International Center for Environmental Management) (2003) *Thailand National Report on Protected Areas and Development. Review of Protected Areas and Development in the Lower Mekong River Region*. Indooroopilly, Queensland, Australia.

Kinnaird M. F. and O'Brien, T. G. (2007) *The Ecology and Conservation of Asian Hornbills: Farmers of the Forest*. The University of Chicago Press, Chicago, IL.

Lekagul, B. and Cronin, P. (1974) *Bird Guide of Thailand*, 2nd edn. Kurusapa Ladprao Press, Bangkok, Thailand.

Lekagul, B. and Round, P. D. (1991) *A Guide to the Birds of Thailand*. Saha Karn Bhaet Co., Ltd., Bangkok, Thailand.

NESDB (Office of the National Economic and Social Development Board) (2008) *The Seventh National Economic and Social Development Plan (1993–1996)*. www.nesdb.go.th (accessed March 19, 2013).

ONEP (Office of Natural Resources and Environmental Policy and Planning) (2007) *Thailand Red Data: Vertebrates*. Ministry of Natural Resources and Environment, Bangkok, Thailand.

Poonswad, P. (1993) Current status and distribution of hornbills and their habitats in Thailand, in *Manual to the Conservation of Asian Hornbills* (eds P. Poonswad and A. C. Kemp). Sirivatana Interprint Co., Ltd., Bangkok, Thailand, pp. 436–475.

Poonswad, P. (1995) Nest site characteristics of four sympatric species of hornbills in Khao Yai National Park, Thailand. *Ibis*, **137**, 183–191.

Poonswad, P. and Tsuji, A. (1994) Ranges of males of the Great Hornbill *Buceros bicornis*, Brown Hornbill *Ptilolaemus tickelli* and Wreathed Hornbill *Rhyticeros undulatus* in Khao Yai National Park, Thailand. *Ibis*, **136**, 79–86.

Poonswad, P., Tsuji, A. and Jirawatkavi, N. (2004) Estimation of nutrients delivered to nest inmates by four sympatric species of hornbills in Khao Yai National Park, Thailand. *Ornithological Science*, **3**, 99–112.

Poonswad, P., Chimchome, V., Plongmai, K. and Chuailua, P. (1999) Factors influencing the reproduction of Asian hornbills, in *Proceedings of 22nd International Ornithological Congress, Durban* (eds N. J. Adams and R. H. Slotow), 1740–1755. BirdLife South Africa, Johannesburg, South Africa.

Poonswad, P., Sukkasem, C., Phataramata, S., Hayeemuida, S., Plongmai, K., Chuailua, P., Thiensongrusame, P. and Jirawatkavi, N. (2005) Comparison of cavity modification and community involvement as strategies for hornbill conservation in Thailand. *Biological Conservation*, **122**, 385–393.

RFD (Royal Forest Department) (1985) *National Forest Policy.* Royal Forest Department, Bangkok, Thailand.

Sanguansombat, W. (2005) *Thailand Red Data: Birds.* Office of Natural Resources and Environmental Policy and Planning (ONEP), Bangkok, Thailand.

Santisuk, T. (2007) *Forests of Thailand.* Department of National Parks, Wildlife and Plants Conservation, Bangkok, Thailand.

Trisurat, Y. (2007) Applying gap analysis and comparison index to evaluate protected areas in Thailand. *Environmental Management*, **39**(2), 235–245.

Trisurat, Y., Pattanavibool, A., Gale, G. A. and Reed, D. H. (2010) Improving the viability of large mammal populations using landscape indices for conservation planning. *Wildlife Research*, **36**, 401–412.

Section 4: Oceania

CHAPTER 20

TIPPING POINTS AND THE VULNERABILITY OF AUSTRALIA'S TROPICAL ECOSYSTEMS

William F. Laurance

Centre for Tropical Environmental and Sustainability Science (TESS) and School of Marine and Tropical Biology, James Cook University, Queensland, Australia

SUMMARY

I identify the major tropical ecosystems in Australia that are most vulnerable to tipping points, in which modest environmental changes can cause disproportionately large changes in ecosystem properties. To accomplish this, I surveyed 24 researchers in Australia to produce a list of candidate ecosystems, which was then refined during a workshop in late 2010. The list includes (1) montane rainforests; (2) tropical savannas; (3) tropical floodplains and wetlands; (4) coral reefs; (5) drier rainforests; (6) tropical islands; and (7) salt marshes and mangroves. Some of these ecosystems are vulnerable to widespread phase-changes that could fundamentally alter ecosystem properties such as habitat structure, species composition, fire regimes, or carbon storage. Others appear susceptible to major changes across only part of their geographic range, whereas yet others are susceptible to a large-scale decline of key biotic components, such as small mammals or stream-dwelling amphibians. For each ecosystem I consider the intrinsic features and external drivers that render it susceptible to tipping points, and identify subtypes of the ecosystem that appear especially vulnerable.

INTRODUCTION

Various vulnerability assessments have been carried out for Australian terrestrial and marine ecosystems. Some have focused on identifying vulnerable ecological communities (e.g., EPBC, 1999) or species (e.g., Watson *et al.*, 2011), whereas others have assessed particular environmental threats, such as climatic change and its potential impacts on biodiversity (Hennessy *et al.*, 2007; Johnson and Marshall, 2007; Steffen *et al.*, 2009) and ecosystem function (Hughes, 2003; Murphy, Russell-Smith and Prior, 2010).

In concert with two dozen colleagues, I recently led an effort to assess the potential vulnerability of Australian ecosystems to tipping points (Laurance *et al.*, 2011), in which modest environmental changes can cause disproportionately large changes in ecosystem properties. Such an exercise is important because Australian ecosystems will face important environmental challenges in the future (Beeton *et al.*, 2006). Current projections of climate change, for instance, suggest that minimum and maximum temperatures will continue to increase whereas precipitation will become more seasonal and sporadic across large swaths of the Australian continent (CSIRO-Australian Bureau of

Figure 20.1 Striking contrast between a natural tropical savanna-woodland near Bachelor, Northern Territory, Australia and similar habitat 300 m away that is heavily invaded by Gamba grass (*Andropogon gayanus*), an exotic species. The grass promotes high-intensity fires that dramatically transform the ecosystem.
Photos courtesy of © S. Setterfield

Meteorology, 2007). By the end of this century, arid and semi-arid zones of northern Australia could experience more heat waves (Tebaldi *et al.*, 2006). Large expanses of the Australian continent are likely experiencing fire regimes for which their ecosystems are poorly adapted (Ward, Lamont and Burrows, 2001; Mooney *et al.*, 2010; Setterfield *et al.*, 2010). In the surrounding oceans, sea levels are rising while sea-surface temperatures and acidity are both increasing (De'ath, Lough and Fabricius, 2009; Hughes *et al.*, 2010). Habitat loss and degradation continue apace in parts of the continent, and many ecosystems are suffering seriously from invasions of non-native plants and animals (Rea and Storrs, 1999; Rossiter-Rachor *et al.*, 2009; Setterfield *et al.*, 2010) or from emerging pests and pathogens (Laurance, McDonald and Speare, 1996; Cahill *et al.*, 2008). Key components of the native biota have been lost, and continue to be lost, from many Australian ecosystems (Hero *et al.*, 2006; Jones *et al.*, 2007; Burbidge *et al.*, 2009; AWC, 2011; Woinarski *et al.*, 2010).

Much of Australia lies in the tropics or subtropics. Here I identify tropical and subtropical ecosystems in Australia that appear vulnerable to tipping points, based largely on a recent Australia-wide assessment (Laurance *et al.*, 2011). My colleagues and I defined a tipping point rather loosely as a circumstance by which

a relatively modest change in an environmental driver or perturbation can cause a major shift in key ecosystem properties (Figure 20.1), such as habitat structure, species composition, community dynamics, fire regimes, carbon storage, or other important functions. The tipping point is an ecological threshold beyond which major change becomes inevitable and is often very difficult to reverse. Because of ecological feedbacks, many ecosystems seem relatively stable as they approach a tipping point, but then shift abruptly to an alternative state once they reach it (see Washington-Allen *et al.*, 2009; Hughes *et al.*, 2010, and references therein).

In conducting our recent analysis (Laurance *et al.*, 2011), my colleagues and I found it useful to distinguish among three broad categories of ecosystems that vary in their geographic extent and severity of their tipping points. "Tipping" ecosystems are likely to experience profound regime changes across most or all of their geographic range, whereas "dipping" ecosystems experience similarly profound changes, but these are restricted geographically, affecting only a portion of the entire ecosystem. Finally, "stripping" ecosystems are being stripped of important ecosystem components, such as their small mammal, amphibian, or large predator fauna, but such changes are more insidious and less visually apparent than major regime changes, at least at present.

Here I highlight vulnerable terrestrial and near-coastal marine ecosystems in tropical and subtropical Australia. For each I highlight some of the intrinsic features and external drivers that render it susceptible to tipping points, and identify subtypes of the ecosystem that we consider especially vulnerable. I emphasize that this exercise is exploratory and thought provoking, not definitive. My goal is to stimulate critical thinking about tipping points while highlighting Australian ecosystems that could – in the absence of effective conservation or management interventions – change dramatically in the future.

METHODS

The analysis on which this chapter is largely based (Laurance *et al.*, 2011) was conducted in two phases. In early October 2010, my 24 coauthors and I each submitted independent lists of major terrestrial and marine ecosystem types in Australia that we considered vulnerable to tipping points, along with potential intrinsic characteristics or external threats that were thought to render each nominated ecosystem vulnerable. I compiled these data into a preliminary list, with the nominated ecosystems ranked by the number of investigators who considered them vulnerable.

In late October 2010, we met in Cairns, Queensland, for an intensive two-day workshop in which we discussed and refined the initial list. We had five goals: (1) to identify the "top 10" major Australian ecosystems vulnerable to tipping points; (2) to highlight key subtypes of each ecosystem type currently at critical risk; (3) to identify the intrinsic features of each ecosystem that predisposed it to tipping points; (4) to identify major external threats to each ecosystem; and (5) to cross-tabulate the intrinsic features and external threats across all 10 vulnerable ecosystems to identify any general attributes that render them vulnerable to tipping points. To achieve aims (3) and (4), we devised general schemes to categorize intrinsic ecosystem features (Table 20.1) and external threats (Table 20.2) that predispose ecosystems to tipping points. For all

Table 20.1 Intrinsic features of seven Australian tropical and subtropical ecosystems that can render them vulnerable to tipping points, as perceived by 25 environmental experts. For each ecosystem type, the most important feature is numbered 1 with those of lesser importance numbered subsequently

Intrinsic feature	Montane rainforests	Tropical savannas	Coastal wetlands	Coral reefs	Drier rainforests	Islands	Estuarine wetlands
Narrow environmental envelope	1		4	1	1	2	1
Near threshold	3			3			
Geographically restricted	2		1		2	1	2
History of fragmentation			2		3		4
Reliance on ecosystem engineers		3					
Reliance on framework species					2		6
Reliance on predators or keystone mutualists*							
Positive feedback		1		4	4		
Proximity to humans			3	5	5		3
Social vulnerability		2					5

*No figures shown as this criteria was not considered to rank among the top threats.
Reproduced with permission from Laurance *et al.* (2011) The 10 Australian ecosystems most vulnerable to tipping points. *Biological Conservation* 144, 1472–1480. Elsevier.

Table 20.2 Environmental threats to seven Australian tropical and subtropical ecosystems that render them vulnerable to tipping points, as perceived by 25 environmental experts. For each ecosystem type, the most important threat is numbered 1 with those of lesser importance numbered subsequently

Environmental threat	Montane rainforests	Tropical savannas	Coastal wetlands	Coral reefs	Drier rainforests	Islands	Estuarine wetlands
Increased temperatures	1			1	2	6	
Changes in water balance and hydrology	2		3		3		3
Extreme weather events	3	3	2	2		2	1
Ocean acidification				3			
Sea-level rise			1			3	2
Changed fire regimes	8	2	8		1		
Habitat reduction	5		5	5	5	4	4
Habitat fragmentation	6	4	6	6	6	5	
Invasives	4	1	4		4	1	
Pests and pathogens	7					7	
Salinization				4			
Pollution			7				5
Overexploitation		5		7	7		

Reproduced with pemission from Laurance *et al.*, (2011) The 10 Australian ecosystems most vulnerable to tipping points. *Biological Conservation* 144, 1472–1480. Elsevier.

analyses, we reached a final consensus via a combination of discussion, debate, and formal voting.

RESULTS: VULNERABLE ECOSYSTEMS

To produce this chapter, I focused the analysis only on ecosystems occurring in tropical or subtropical regions of Australia. I recast our earlier analyses (Laurance *et al.*, 2011) in this light, beginning with the ecosystems for which consensus among the original panel of experts was strongest.

Montane rainforests

Montane rainforests in Australia occur almost exclusively in the Great Dividing Range, which skirts the country's eastern seaboard, ranging from northern New South Wales northward to the Cape York Peninsula in northern Queensland. These forests range from about 300–1700 m in elevation. The most vulnerable rainforests are those that rely substantially on cloud stripping for moisture inputs during the drier months (Hutley *et al.* 1997; McJannet, Wallace and Reddell, 2007) or sustain high numbers of restricted endemic species (Williams, Pearson and Walsh, 1996; Hoskin, 2004).

Montane rainforests are considered inherently vulnerable because of their often-narrow environmental envelopes, their geographically restricted distribution, and the fact that many appear to be near climatic thresholds (Table 20.1). We regard global warming (Williams, Bolitho and Fox, 2003), potential changes in moisture inputs and a rising cloud base (Pounds, Fogden and Campbell, 1999; Still, Foster and Schneider, 1999), and extreme weather events (Tebaldi et al., 2006) as the most serious future threats (Table 20.2). Further perils include invasive plants and fauna, habitat loss and fragmentation (Laurance, 1991), and new pests and pathogens, such as the chytrid fungus that has decimated many stream-dwelling amphibian populations (Skerratt et al., 2007).

Tropical savannas

Tropical savanna-woodlands are one of the most extensive environments in Australia, spanning much of the northern third of the continent (Mackey et al., 2007). This system is experiencing severe regime changes in only parts of its geographic range – and hence is a "dipping" ecosystem. Invasive weeds and animals (Setterfield et al., 2010; Woinarski et al., 2010), changing fire regimes (Midgley, Lawes and Chamaillé-Jammes, 2010; Prior, Williams and Bowman, 2010;), and extreme weather events are seen as the major threats, with habitat fragmentation and overgrazing by livestock (Kutt and Woinarski, 2007) being further perils (Table 20.2). In addition, this ecosystem is currently experiencing an apparently widespread decline of its small mammal fauna – a feature of a "stripping" ecosystem – for reasons that remain uncertain (AWC, 2011; Woinarski et al., 2010).

A key reason for the high vulnerability of tropical savannas is massive weed invasions (Figure 20.1) that profoundly alter fire regimes and other fundamental ecosystem attributes such as carbon storage and nitrogen cycling (Rea and Storrs, 1999; Rossiter-Rachor et al., 2009; Setterfield et al., 2010) (Table 20.1). We believe that sandstone savannas and heaths, which have an endemic flora (Woinarski et al., 2006) and fauna and a highly restricted geographic range, are especially vulnerable habitats, with increasing fire incidence their principal threat (Russell-Smith, Ryan and Cheal, 2001; Sharp and Bowman, 2004).

Coastal floodplains and wetlands

Coastal floodplains and wetlands are freshwater (or only slightly brackish) ecosystems in coastal areas throughout Australia (Adam, 1992; Kingsford et al., 2004), including extensive areas of the tropics and subtropics. They are most widespread in the vast tropical floodplains of the Northern Territory (Cowie, Short and Osterkamp Madsen, 2000), Queensland, and Western Australia. Principal threats to these systems are rising sea levels caused by global warming, extreme weather events (such as storm surges that cause major saltwater incursions inland), and massive plant invasions (Table 20.2). Hydrological changes, habitat loss and fragmentation, pollution, and changing fire regimes are seen as important localized threats (Table 20.2).

In general, coastal floodplains and wetlands are vulnerable to tipping points because of their restricted and naturally fragmented geographic distribution, narrow environmental envelopes, and frequently close proximity to land-use pressures in coastal areas (Table 20.1). Many sustain sensitive wildlife. We believe the most susceptible habitats are relatively flat, topographically restricted wetlands, especially those trapped between habitat conversion or topography on the inland side and rising sea levels on the seaward side. Wetlands adjoining coastal areas with high tidal amplitudes (5–13 m), which have more physical energy to drive seawater inland, are also highly vulnerable. They are often connected, at least intermittently, to intertidal wetlands, making them vulnerable to saltwater intrusions both at the surface and via groundwater. Salinity is toxic to amphibians and demonstrably alters fish populations (Sheaves and Johnston, 2008; Sheaves, 2009).

Coral reefs

Coral reefs occur in warm, shallow seas along much of northeastern Australia with smaller, scattered reefs along the Western Australian coast. These reefs are considered vulnerable to tipping points because of their narrow thermal and water-quality tolerances, heavy reliance on key "framework" species (reef-building corals), and high susceptibility to nutrient run-off and eutrophication (Johnson and Marshall, 2007; Hughes et al., 2010). In our view, the most vulnerable reefs are those near rivers carrying heavy nutrient loads from

Figure 20.2 Bleaching of tropical corals in the eastern Pacific during a major heat wave in 2005. Darker, unaffected corals are visible in the background. Photo courtesy of © W. Chaffey

nearby farmlands, and those at near-equatorial latitudes off Cape York Peninsula and northern Western Australia (Table 20.1), which are susceptible to coral bleaching (Figure 20.2) associated with global warming. Isolated reefs, such as Ningaloo Reef in Western Australia, are also vulnerable because local species declines are not as easily offset by immigration as occurs in less-isolated reefs (e.g., Underwood, 2009).

The greatest threat to coral reefs in Australian waters is probably rising sea temperatures, followed by extreme weather events (especially heat waves and destructive storms), ocean acidification, and pollution. Reef destruction and overharvesting of fish, crustaceans, gastropods, and other reef species are ancillary threats (Table 20.2), but are lesser problems in Australia than elsewhere in the tropics.

Drier rainforests

Relatively dry rainforest types, including vine thickets, monsoonal vine-thickets, and semi-deciduous rainforest types, such as Mabi forest in far north Queensland, occur in moist, comparatively fire-proof refugia scattered across much of northern Australia (Russell-Smith, 1991; Bowman, 2000). Shifts in fire regime, rising temperatures, changing rainfall regimes, and extreme weather events (especially droughts and heat waves) are considered their greatest threats, although many sites are also heavily invaded by lantana (*Lantana*

camara), rubber vine (*Cryptostegia grandiflora*), and other tropical weeds that can suppress tree recruitment, provide fuel for destructive surface fires (Humphries, Groves and Mitchell, 1991; Russell-Smith and Bowman, 1992; Fensham, 1994), and render the habitat unsuitable for some native species (e.g., Valentine, Roberts and Schwarzkopf, 2007). Some are also being degraded by human habitat disruption and overgrazing by livestock (Table 20.2).

In broad terms, drier rainforest types are vulnerable to tipping points because of their narrow environmental tolerances, their highly restricted and patchy distributions (Bowman and Woinarski, 1994; Price *et al.*, 1999), and the destabilizing positive feedbacks that occur when heavy weed invasions increase fire incidence, which in turn opens up the forest and makes it more prone to further weed invasions and fire (Table 20.1). We believe that forest patches that are small, near human settlements, in frequently burned areas, and in low-lying areas prone to rising sea levels are especially vulnerable.

Offshore islands

Australia has over 8000 offshore islands (Ecosure, 2009), many of which occur at tropical or subtropical latitudes. In Australia, as elsewhere, islands are considered vulnerable to dramatic changes because of their restricted size, physical isolation, often-narrow environmental envelopes, and relatively limited (yet often highly endemic) biodiversity that may facilitate species invasions (Table 20.1) (Burbidge and Manly, 2002; Ecosure, 2009). We believe the most vulnerable are small, relatively species-poor islands with many vacant ecological niches, which are prone to species invasions; those with large human populations or visitation; those near ocean-circulation boundaries or with many species that depend on upwelling; and low-lying islands susceptible to rising sea levels. Not all Australian islands have suffered invasions; some have provided important refugia for native wildlife that have been extirpated elsewhere by introduced predators and competitors (Morton, Short and Barker, 1995; Burbidge, 1999).

The chief threats to Australia's tropical and subtropical islands are myriad invading species such as rats, mice, pigs, cats, toads, and fire ants (Burbidge and Manly, 2002); extreme weather events such as intense storms or droughts that can have disproportionately large impacts on insular ecosystems; rising sea levels; habitat loss and degradation; rising sea-surface temperatures that might affect oceanic circulation and the upwelling of nutrient-rich waters; and emerging pathogens and pests (Table 20.2).

Estuarine wetlands (salt marshes and mangroves)

Salt marshes and mangroves are estuarine ecosystems that are frequently found in the tropics and subtropics and that play many important environmental roles. These roles include stabilizing coastal sediments, acting as nutrient and pollution traps, providing protection from storm surges and tsunamis, sustaining wildlife populations, and functioning as vital "nurseries" for breeding fish and crustaceans (Beck *et al.*, 2009). Estuarine wetlands are vulnerable for several reasons: their narrow environmental tolerances, geographically restricted nature, proximity to dense human populations in coastal regions, patchy and fragmented distribution (Duke *et al.*, 2007), and reliance on a few key framework species (Table 20.1). We believe that salt marshes and coastal-fringe mangroves (those in narrow strips along coastlines rather than in estuarine areas) are especially susceptible, particularly those in densely populated areas.

In the future, increasing storm intensity could be a serious threat to salt marshes and particularly to mangroves at the seaward edge (e.g., Cahoon *et al.*, 2003). They also are increasingly likely to be squeezed between human land uses or topography on the landward side and rising sea levels on the seaward side (Eslami-Andargoli *et al.*, 2010). Furthermore, water pollution and small changes in salinity and hydrology can cause dramatic changes in estuarine communities (Table 20.2).

DISCUSSION

Predisposing factors and key drivers

The tropical ecosystems identified in this analysis are deemed at risk of changing dramatically in the near future. Why are they intrinsically susceptible? I can draw tentative conclusions by evaluating the most important features (those ranked 1–3 by my original panel of experts; Laurance *et al.*, 2011) across the

seven vulnerable ecosystem types (Table 20.1). The most frequently cited features of vulnerable ecosystems are a narrow environmental envelope, rendering them sensitive to even relatively modest changes in environmental conditions, and a restricted (or highly patchy) geographic range, which limits their capacity to withstand anthropogenic pressures simply by persisting in places where such pressures are absent. Montane rainforests, drier rainforests, oceanic islands, and estuarine ecosystems are all considered vulnerable for these reasons.

Which environmental drivers are most likely to threaten these ecosystems? This analysis is based on ranking the relative importance of 13 environmental drivers for each of the seven vulnerable ecosystems (Table 20.2). As before, my focus is on the drivers that were regarded as most important (those ranked 1–3 for each ecosystem by my panel of experts; Laurance *et al.*, 2011). At the outset, I note that the anthropogenic threats identified here may well differ from those that have altered Australian ecosystems in the past (see Flannery, 1994; Johnson, 2006).

The two most important of the top-ranked drivers, extreme weather events and changes in water balance and hydrology, were each considered important for 4–6 ecosystems (Table 20.2). Extreme weather events include severe, short-term phenomena such as heat waves, droughts, and intense storms. The Australian continent, whose precipitation and hydrology are strongly influenced by the El Niño-Southern Oscillation (Nicholls, Drosdowsky and Lavery, 1997; Chiew *et al.*, 1998), whose ancient, relatively flat land surface is poor at capturing rainfall, and which is dominated by strongly seasonal environments at tropical and subtropical latitudes, may be particularly susceptible to such events. Changes in water balance and hydrology usually arise from changes in moisture inputs, a phenomenon that under plausible scenarios of future climate change could imperil montane ecosystems that rely on orographic rainfall and/or cloud stripping (Still, Foster and Schneider, 1999; Bradley *et al.*, 2006).

Some ecosystems are also vulnerable to rising sea levels or rising temperatures (Table 20.2), both of which relate directly to global warming. Among the myriad ways in which global change phenomena could affect Australian ecosystems, and one of the potentially most important, is by altering fire regimes (Bradstock, 2010). Fire regimes are largely determined by weather and fuel loads. Increasing atmospheric CO_2 could potentially increase fuel loads via enhanced primary productivity (Donohue, McVicar and Roderick, 2009; Sun et al., 2010), but this effect could be magnified or diminished by changes in available moisture, depending on the location. In some ecosystems, serious weed invasions are profoundly altering fire regimes (Figure 20.1). Fire-promoting invaders can dramatically transform ecosystems, usually favoring short-lived annuals and exotics at the expense of long-lived trees. Closed-canopy rainforests generally seem less vulnerable to weed invasions than are more open ecosystems such as savannas, wetlands, and drier (seasonally deciduous) rainforests, in which light-loving weeds can become established more easily (Laurance et al., 2002).

Although factors relating to climatic change are likely to play a key role in predisposing Australian ecosystems to tipping points, all of the vulnerable ecosystems are being influenced by multiple drivers (Table 20.2). This reinforces a general view that synergisms among different environmental drivers can be extremely important, predisposing species and ecosystems to serious environmental changes (Laurance and Cochrane, 2001; Brook, Sodhi and Bradshaw, 2008; Laurance and Useche, 2009). Examples of such synergisms are pervasive in the Australian tropics – for example, between weed invasions and fire, between land-use change and climatic change, between anthropogenic activities and introduced pathogens, and between coastal land-use pressures and rising sea levels. For the Australian environment, as elsewhere, combinations of environmental perils may be the death knell for many ecosystems.

Conservation actions to avoid tipping points

The threats facing vulnerable ecosystems in Australia are often multifaceted and, at least for some perils such as global climate change, rising ocean acidity, and the continued spread of certain invasive species and pathogens, largely beyond the control of Australian resource managers. In practical terms, this limits the tools that can be applied to mitigate these pressures. Rather than preaching despair, however, I believe much can be done to limit the further decline of vulnerable Australian ecosystems.

A key priority is to identify likely or imminent changes in vulnerable ecosystems and taxa (e.g., Hughes et al., 2010; Woinarski et al., 2010). A full discussion of this concept is beyond the scope of this chapter, but I note here two key points. First, the best

approach for judging whether an ecosystem is approaching a tipping point may be to examine key ecological processes involved in proper ecosystem functioning and integrity (Dunning, Danielson and Pulliam, 1992; Didham et al., 1996), rather than biodiversity indicators (such as species richness) that can have delayed responses to disturbance effects (Loehle and Li, 1996; Vellend et al., 2006). Second, a key harbinger of tipping points may be a "critical slowing" of ecosystem dynamics. This can include slower recovery from disturbances, increased variance in ecosystem dynamics, and increased auto-correlation in ecosystem properties as the tipping point is approached (see van Nes and Scheffer, 2007; Biggs, Carpenter and Brock, 2009; Scheffer et al., 2009; Drake and Griffen, 2010; Scheffer, 2010, for discussion). Further, phenomena such as an increased variance and spatial autocorrelation might be detectable from spatial patterns in vegetation (Bailey, 2011), potentially allowing ecosystem vulnerability to be evaluated via remote sensing, rather than requiring detailed field studies. Such approaches might provide important insights into the status and vulnerability of particular ecosystems.

In addition, on-the-ground conservation and management actions can often have a profound impact on ecosystem resilience. In broad terms, concrete steps such as increasing the size and number of protected areas, limiting external disturbances such as habitat conversion and new roads (Goosem, 2007; Laurance, Goosem and Laurance, 2009), creating buffer zones and wildlife corridors, restoring key habitats and landscape linkages (Shoo et al., 2011), and designing and locating nature reserves to maximize their resilience to climate change (Hannah et al., 2007; Loarie et al., 2009; Shoo et al., 2011) can play vital roles in maintaining ecosystem viability. Key phenomena such as fire regimes can often be managed via steps such as prescriptive burning, silviculture, livestock grazing, fire suppression, and controlling human ignition sources (Yibarbuk et al., 2001; Murphy et al., 2009; Russell-Smith, Price and Murphy, 2010).

Managing natural and semi-natural ecosystems in a world that is continually in flux is a great challenge, but societies are adapting to these realities. Environmental regulations and policies are changing profoundly in an effort to address complex and multi-faceted environmental challenges (Lockwood et al., 2010). Conservation efforts are increasingly being integrated across institutions and among public, private, and civil sectors to address uncertainty and "wicked" environmental problems (Holling, 1978; Robinson *et al.*, 2009) in an adaptive and flexible manner (Dietz, Ostrom and Stern, 2003; Armitage *et al.*, 2009). Environmental "horizon scanning" is being used to anticipate new threats (Laurance and Peres, 2006; Sutherland and Woodroof, 2009). Great challenges lie ahead for Australian tropical ecosystems, as elsewhere, but much can still be done to address them.

ACKNOWLEDGMENTS

I thank my collaborators in Australia (see Laurance *et al.*, 2011) for much input. This chapter is a product of the Innovative Research Universities of Australia (www.iru.edu.au).

REFERENCES

Adam, P. (1992) Wetlands and wetland boundaries: problems, expectations, perceptions and reality. *Wetlands*, **10**, 60–67.

Armitage, D., Plummer, R., Berkes, F., Arthur, R., Charles, A., Davidson-Hunt, I., Diduck, A., Doubleday, N., Johnson, D., Marschke, M., McConney, P., Pinkerton, E. and Wollenberg, E. (2009) Adaptive co-management for social–ecological complexity. *Frontiers in Ecology and the Environment*, **7**, 95–102.

AWC (2010) Where Have All the Mammals Gone? *Australian Wildlife Conservancy*. http://www.australianwildlife.org/images/file/Northern_Mammal_WM_Winter_2010_-lowres.pdf (accessed March 19, 2013).

Bailey, R. M. (2011) Spatial and temporal signatures of fragility and threshold proximity in modelled semi arid vegetation. *Proceedings of the Royal Society B*, **278**, 1064–1071.

Beck, M. W., Heck, K. L., Able, K. W., Childers, D. L., Eggleston, D. B., Gillanders, B. M., Halpern, B., Hays, C. G., Hoshino, K., Minello, T. J., Orth, R. J., Sheridan, P. F. and Weinstein, M. P. (2009) The identification, conservation, and management of estuarine and marine nurseries for fish and invertebrates. *BioScience*, **51**, 633–641.

Beeton, R. J. S., Buckley, K. I., Jones, G. J., Morgan, D., Reichelt, R. E. and Trewin, D. (2006) *Australia State of the Environment 2006*. Department of Environment and Heritage, Canberra, Australia.

Biggs, R., Carpenter, S. R. and Brock, W. A. (2009) Turning back from the brink: detecting an impending regime shift in time to avert it. *Proceedings of the National Academy of Sciences of the USA*, **106**, 826–831.

Bowman, D. M. J. S. (2000) *Australian Rainforest: Island of Green in a Land of Fire*. Cambridge University Press, Cambridge.

Bowman, D. M. J. S. and Woinarski, J. C. Z. (1994) Biogeography of Australian monsoon rainforest mammals: implications for the conservation of rainforest mammals. *Pacific Conservation Biology*, **1**, 98–106.

Bradley, R. S., Vuille, M., Diaz, H. F. and Vergara, W. (2006) Threats to water supplies in the tropical Andes. *Science*, **312**, 1755–1756.

Bradstock, R. A. (2010) A biogeographic model of fire regimes in Australia: current and future implications. *Global Ecology and Biogeography*, **19**, 145–158.

Brook, B. W., Sodhi, N. S. and Bradshaw, C. J. A. (2008) Synergisms among extinction drivers under global change. *Trends in Ecology and Evolution*, **23**, 453–460.

Burbidge, A. A. (1999) Conservation values and management of Australian islands for non-volant mammal conservation. *Australian Mammalogy*, **21**, 67–71.

Burbidge, A. A. and Manly, B. F. J. (2002) Mammal extinctions on Australian islands: causes and conservation implications. *Journal of Biogeography*, **29**, 465–475.

Burbidge, A. A., McKenzie, N. L., Brennan, K. E. C., Woinarski, J. C. Z., Dickman, C. R., Baynes, A., Gordon, G., Menkhorst, P. W. and Robinson, A. C. (2009) Conservation status and biogeography of Australia's terrestrial mammals. *Australian Journal of Zoology*, **56**, 411–422.

Cahill, D. M., Rookes, J. E., Wilson, B. A., Gibson, L. and McDougall, K. L. (2008) *Phytophthora cinnamomi* and Australia's biodiversity: impacts, predictions and progress towards control. *Australian Journal of Botany*, **56**, 279–310.

Cahoon, D. R., Hensel, P., Rybczyk, J., McKee, K. L., Proffitt, C. E. and Perez, B. C. (2003) Mass tree mortality leads to mangrove peat collapse at Bay Islands, Honduras after Hurricane Mitch. *Journal of Ecology*, **91**, 1093–1105.

Chiew, F. H. S., Piechota, T. C., Dracup, J. A. and McMahon, T. A. (1998) El Niño/Southern Oscillation and Australian rainfall, streamflow and drought: links and potential for forecasting. *Journal of Hydrology*, **204**, 138–149.

Cowie, I. D., Short, P. and Osterkamp Madsen, M. (2000) *Floodplain Flora: A Flora of Coastal Floodplains of the Northern Territory, Australia*. Australian Biological Resources Study, Canberra, Australia.

CSIRO-Australian Bureau of Meteorology (2007) *Climate Change in Australia: Technical Report 2007*. CSIRO, Canberra, Australia.

De'ath, G., Lough, J. M. and Fabricius, J. E. (2009) Declining coral calcification on the Great Barrier Reef. *Science*, **323**, 116–119.

Didham, R. K., Ghazoul, J., Stork, N. E. and Davis, A. J. (1996) Insects in fragmented forests: a functional approach. *Trends in Ecology and Evolution*, **11**, 255–260.

Dietz, T., Ostrom, E. and Stern, P. (2003) The struggle to govern the commons. *Science*, **302**, 1907–1912.

Donohue, R. J., McVicar, T. R. and Roderick, M. L. (2009) Climate-related trends in Australian vegetation cover as inferred from satellite observations, 1981–2006. *Global Change Biology*, **15**, 1025–1039.

Drake, J. M. and Griffen, B. D. (2010) Early warning signals of extinction in deteriorating environments. *Nature*, **467**, 456–459.

Duke, N. C., Meynecke, J. O., Dittmann, S., Ellison, A. M., Anger, K., Berger, U., Cannicci, S., Diele, K., Ewel, K. C., Field, C. D., Koedam, N., Lee, S. Y., Marchand, C., Nordhaus, I. and Dahdouh-Guebas, F. (2007) A world without mangroves? *Science*, **317**, 41–42.

Dunning, J. B., Danielson, B. J. and Pulliam, H. R. (1992) Ecological processes that affect populations in complex landscapes. *Oikos*, **65**, 169–175.

Ecosure (2009) *Prioritisation of High Conservation Status Offshore Islands*. Department of Environment, Water, Heritage and the Arts, Canberra, Australia.

EPBC (1999) *Environmental Protection and Biodiversity Conservation (EPBC) Act List of Threatened Ecological Communities*. Australian Government, Canberra. http://www.environment. gov.au/cgi-bin/sprat/public/ publiclookupcommunities.pl (accessed March 19, 2013).

Eslami-Andargoli, L., Dale, P. E. R., Sipe, N. and Chaseling, J. (2010) Local and landscape effects on spatial patterns of mangrove forest during wetter and drier periods: Moreton Bay, Southeast Queensland, Australia. *Estuarine and Coastal Shelf Science*, **89**, 53–61.

Fensham, R. J. (1994) The invasion of *Lantana camara* L. in Forty Mile Scrub National Park, north Queensland. *Australian Journal of Ecology*, **19**, 297–305.

Flannery, T. (1994) *The Future Eaters: An Ecological History of the Australasian Lands and People*. Grove Press, New York, NY.

Goosem, M. (2007) Fragmentation impacts caused by roads through rainforests. *Current Science*, **93**, 1587–1595.

Hannah, L., Midgley, G., Andelman, S., Araújo, M., Hughes, G., Martinez-Meyer, E., Pearson, R. G. and Williams, P. (2007) Protected area needs in a changing climate. *Frontiers in Ecology and the Environment*, **5**, 131–138.

Hennessy, K., Fitzharris, B., Bates, B. C., Harvey, N., Howden, M., Hughes, L., Salinger, J. and Warrick, R. (2007) Australia and New Zealand, in *Climate Change 2007: Impacts, Adaptation and Vulnerability* (eds M. L. Parry, O. Canziani, J. Palutikof, van P. der Linden and C. Hanson), Cambridge University Press, Cambridge, pp. 507–540.

Hero, J.-M., Morrison, C., Gillespie, G., Roberts, J. D., Newell, D., Meyer, E., McDonald, K., Lemckert, F., Mahony, M., Osborne, W., Hines, H., Richards, S., Hoskin, C., Clarke, J., Doak N. and Shoo, L. (2006) Overview of the conservation status of Australian frogs. *Pacific Conservation Biology*, **12**, 313–320.

Holling, C. (1978) *Adaptive Environmental Assessment and Management*. John Wiley & Sons, London.

Hoskin, C. J. (2004) Australian microhylid frogs (*Cophixalus* and *Austrochaperina*): phylogeny, taxonomy, calls, distributions and breeding biology. *Australian Journal of Zoology*, **52**, 237–269.

Hughes, L. (2003) Climate change and Australia: trends, projections and impacts. *Austral Ecology*, **28**, 423–443.

Hughes, T. P., Graham, N. A. J., Jackson, J. B. C., Mumby, P. J. and Steneck, R. S. (2010) Rising to the challenge of sustaining coral reef resilience. *Trends in Ecology and Evolution*, **25**, 633–642.

Humphries, S. E., Groves, R. H. and Mitchell, D. S. (1991) *Plant Invasions of Australian Ecosystems: Kowari 2*. Australian National Parks and Wildlife Service: Canberra, Australia.

Hutley, L. B., Doley, D., Yates, D. J. and Boonsaner, A. (1997) Water balance of an Australian sub-tropical rainforest at altitude: the ecological and physiological significance of intercepted cloud and fog. *Australian Journal of Botany*, **45**, 311–329.

Johnson, C. N. (2006) *Australia's Mammal Extinctions: A 50,000 Year History*. Cambridge University Press, Melbourne, Australia.

Johnson, J. E. and Marshall, P. E. (eds) (2007) *Climate Change and the Great Barrier Reef: A Vulnerability Assessment*. Great Barrier Reef Marine Park Authority, Townsville, Australia.

Jones, M., Jarman, P., Lees, C., Hesterman, H., Hamede, R., Mooney, N., Mann, D., Pukk, C., Bergfeld, J. and McCallum, H. (2007) Conservation management of Tasmanian devils in the context of an emerging, extinction-threatening disease: Devil Facial Tumor Disease. *EcoHealth*, **4**, 326–337.

Kingsford, R. T., Brandis, K., Thomas, R. F., Crighton, P., Knowles, E. and Gale, E. (2004) Classifying landform at broad spatial scales: the distribution and conservation of wetlands in NSW, Australia. *Marine and Freshwater Research*, **55**, 17–31.

Kutt, A. and Woinarski, J. C. Z. (2007) The effects of grazing and fire on vegetation and the vertebrate assemblage in a tropical savanna woodland in north-eastern Australia. *Journal of Tropical Ecology*, **23**, 95–106.

Laurance, W. F. (1991) Ecological correlates of extinction proneness in Australian tropical rainforest mammals. *Conservation Biology*, **5**, 79–89.

Laurance, W. F. and Cochrane, M. A. (2001) Synergistic effects in fragmented landscapes. *Conservation Biology*, **15**, 1488–1489.

Laurance, W. F. and Peres, C. A. (eds.) (2006) *Emerging Threats to Tropical Forests*. University of Chicago Press, Chicago, IL.

Laurance, W. F. and Useche, D. C. (2009) Environmental synergisms and extinctions of tropical species. *Conservation Biology*, **23**, 1427–1437.

Laurance, W. F., Goosem, M. and Laurance, S.G. (2009) Impacts of roads and linear clearings on tropical forests. *Trends in Ecology and Evolution*, **24**, 659–669.

Laurance, W. F., McDonald, K. R. and Speare, R. (1996) Epidemic disease and the catastrophic decline of Australian rain forest frogs. *Conservation Biology*, **10**, 406–413.

Laurance, W. F., Lovejoy, T., Vasconcelos, H., Bruna, E., Didham, R., Stouffer, P., Gascon, C., Bierregaard, R., Laurance, S. and Sampiao, E. (2002) Ecosystem decay of Amazonian forest fragments: a 22-year investigation. *Conservation Biology*, **16**, 605–618.

Laurance, W. F., Dell, B., Turton, S., Lawes, M., Hutley, L., McCallum, H., Dale, P., Bird, M., Hardy, G., Prideaux, G., Gawne, B., McMahon, C., Yu, R., Hero, J.-M., Schwarzkopf, L., Krockenberger, A., Setterfield, S., Douglas, M., Silvester, E., Mahony, M., Vella, K., Saikia, U., Wahren, C.-H., Xu, Z., Smith, B. and Cocklin, C. (2011) The 10 Australian ecosystems most vulnerable to tipping points. *Biological Conservation*, **144**, 1472–1480.

Loarie, S. R., Duffy, P. B., Hamilton, H., Asner, G. P., Field, C. B. and Ackerly, D. D. (2009) The velocity of climate change. *Nature*, **462**, 1052–1055.

Lockwood, M., Davidson, J., Curtis, A., Stratford, E. and Griffith, R. (2010) Governance principles for natural resource management. *Society and Natural Resources*, **23**, 986–1001.

Loehle, C. and Li, B.-L. (1996) Habitat destruction and the extinction debt revisited. *Ecological Applications*, **6**, 784–749.

Mackey, B. G, Woinarski, J. C. Z., Nix, H. and Trail, B. (2007) *The Nature of Northern Australia: Its Natural Values, Ecology, and Future Prospects*. ANU Electronic Press, Canberra, Australia.

McJannet, D. L., Wallace, J. S. and Reddell, P. (2007) Precipitation interception in Australian tropical rainforests: II. Altitudinal gradient of cloud interception, stemflow, throughfall and interception. *Hydrological Processes*, **21**, 1703–1718.

Midgley, J. J., Lawes, M. J. and Chamaillé-Jammes, S. (2010) Savanna woody plant dynamics; the role of fire and herbivory, separately and synergistically. *Australian Journal of Botany*, **58**, 1–11.

Mooney, S. D., Harrison, S. P., Bartlein, P. J., Daniau, A., Stevenson, J., Brownlie, K., Buckman, S., Cupper, M., Luly, J., Black, M., Colhoun, E., D'Costa, D., Dodson, J., Haberle, S., Hope, G., Kershaw, P., Kenyon, C., McKenzie, M. and Williams, N. (2010) Late Quaternary fire regimes of Australia. *Quaternary Science Reviews*, **30**, 28–46.

Morton, S. R., Short, J. and Barker, R. D. (1995) *Refugia for Biological Diversity in Arid and Semi arid Australia*. Biodiversity Series Paper 4, Department of the Environment, Sport and Territories, Canberra, Australia.

Murphy, B. P., Russell-Smith, J. and Prior, L. D. (2010) Frequent fires reduce tree growth in northern Australian savannas: implications for tree demography and carbon sequestration. *Global Change Biology*, **16**, 331–343.

Murphy, B. P., Russell-Smith, J., Watt, F. A. and Cook, G. D. (2009) Fire management and woody biomass carbon stocks in mesic savannas, in *Culture, Ecology and Economy of Fire Management in North Australian Savannas: Rekindling the Wurrk Tradition* (eds J. Russell-Smith, P. J. Whitehead and P. Cooke), CSIRO Publishing, Collingwood, Victoria, Australia, pp. 361–387.

Nicholls, N., Drosdowsky, W. and Lavery, B. (1997) Australian rainfall variability and change. *Weather*, **52**, 66–71.

Pounds, J. A., Fogden, M. and Campbell, J. (1999) Biological response to climate change on a tropical mountain. *Nature*, **398**, 611–615.

Price, O. F., Woinarski, J. C. Z. and Robinson, D. (1999) Very large area requirements for frugivorous birds in monsoon rainforests of the Northern Territory, Australia. *Biological Conservation*, **91**, 169–180.

Prior, L. D., Williams, R. J. and Bowman, D. M. J. S. (2010) Experimental evidence that fire causes a tree recruitment bottleneck in an Australian tropical savanna. *Journal of Tropical Ecology*, **26**, 595–603.

Rea, N. and Storrs, M. J. (1999) Weed invasions in wetlands of Australia's Top End: reasons and solutions. *Wetlands Ecology and Management*, **7**, 47–62.

Robinson, C. J., Eberhard, R., Wallington, T. and Lane, M. (2009) *Institutional Collaboration for Effective Environmental Governance in Australia's Great Barrier Reef*. CSIRO Water for Healthy Country and MTSRF Technical Report, Brisbane, Australia.

Rossiter-Rachor, N. A., Setterfield, S. A., Douglas, M. M., Hutley, L. B., Cook, G. D. and Schmidt, S. (2009) Invasive *Andropogon gayanus* (gamba grass) is an ecosystem transformer of nitrogen relations in Australian savanna. *Ecological Applications*, **19**, 1546–1560.

Russell-Smith, J. (1991) Classification, species richness, and environmental relations of monsoon rain forest in northern Australia. *Journal of Vegetation Science*, **2**, 259–278.

Russell-Smith, J. and Bowman, D. M. J. S. (1992) Conservation of monsoon rainforest isolates in the Northern Territory, Australia. *Biological Conservation*, **59**, 51–63.

Russell-Smith, J., Price, O. F. and Murphy, B. P. (2010) Managing the matrix: decadal responses of eucalypt-dominated mesic savanna to ambient fire regimes in three north Australian conservation reserves. *Ecological Applications*, **20**, 1615–1632.

Russell-Smith, J., Ryan, P. G. and Cheal, D. (2001) Fire regimes and the conservation of sandstone heath in monsoonal northern Australia: frequency, interval, patchiness. *Biological Conservation*, **104**, 91–106.

Scheffer, M. (2010) Complex systems: foreseeing tipping points. *Nature*, **467**, 411–412.

Scheffer, M., Bascompte, J., Brock, W. A., Brovkin, V., Carpenter, S. R., Dakos, V., Held, H., van Nes, E. H., Rietkerk, M. and Sugihara, G. (2009) Early-warning signals for critical transitions. *Nature*, **461**, 53–59.

Setterfield, S. A., Rossiter-Rachor, N. A., Hutley, L. B., Douglas, M. M. and Williams, R. J. (2010) Turning up the heat: the impacts of *Andropogon gayanus* (gamba grass) invasion on fire behaviour in northern Australian savannas. *Diversity and Distributions*, **16**, 854–861.

Sharp, B. R. and Bowman, D. (2004) Patterns of long-term woody vegetation change in a sandstone-plateau savanna woodland, Northern Territory, Australia. *Journal of Tropical Ecology*, **20**, 259–270.

Sheaves, M. (2009) Consequences of ecological connectivity: the coastal ecosystem mosaic. *Marine Ecology Progress Series*, **391**, 107–115.

Sheaves, M. and Johnston, R. (2008) Influence of marine and freshwater connectivity on the dynamics of subtropical estuarine wetland fish metapopulations. *Marine Ecology Progress Series*, **357**, 225–243.

Shoo, L. P., Storlie, C., Vanderwal, J., Little, J. and Williams. S. E. (2011) Targeted protection and restoration to conserve tropical biodiversity in a warming world. *Global Change Biology*, **17**, 186–193.

Skerratt, L. F., Berger, L., Speare, R., Cashins, S., McDonald, K., Phillott, A., Hines, H. and Kenyon, N. (2007) Spread of chytridiomycosis has caused the rapid global decline and extinction of frogs. *Ecohealth* **4**, 125–134.

Steffen, W., Burbidge, A. A., Hughes, L., Kitching, R., Lindenmayer, D., Musgrave, W., Stafford Smith, M. and Werner, P. A. (2009) *Australia's Biodiversity and Climate Change*. Natural Resource Management Ministerial Council, Canberra, Australia.

Still, C. J., Foster, P. N. and Schneider, S. H. (1999) Simulating the effects of climate change on tropical montane cloud forests. *Nature*, **398**, 608–610.

Sun, F. F., Kuang, Y. W., Wen, D. Z., Xu, Z. H., Li, J. L., Zuo, W. D. and Hou, E. Q. (2010) Long-term tree growth rate, water use efficiency, and tree ring nitrogen isotope composition of *Pinus massoniana* L. in response to global climate change and local nitrogen deposition in southern China. *Journal of Soils and Sediments*, **10**, 1453–1465.

Sutherland, W. J. and Woodroof, H. J. (2009) The need for environmental horizon scanning. *Trends in Ecology and Evolution*, **24**, 523–527.

Tebaldi, C., Hayhoe, K., Arblaster, J. M. and Meehl, G. A. (2006) Going to extremes: an intercomparison of model-simulated historical and future changes in extreme events. *Climatic Change*, **79**, 185–211.

Underwood, J. N. (2009) Genetic diversity and divergence among coastal and offshore reefs in a hard coral depend on geographic discontinuity and oceanic currents. *Evolutionary Applications*, **2**, 222–233.

Valentine, L. E., Roberts, B. and Schwarzkopf, L. (2007) Mechanisms driving avoidance of non-native plants by native lizards. *Journal of Applied Ecology*, **44**, 228–237.

van Nes, E. H. and Scheffer, M. (2007) Slow recovery from perturbations as a generic indicator of a nearby catastrophic shift. *American Naturalist* **169**, 738–747.

Vellend, M., Verheyen, K., Jacquemyn, H., Kolb, A., van Calster, H., Peterken, G., and Hermy, M. (2006) Extinction debt of forest plants persists for more than a century following habitat fragmentation. *Ecology*, **87**, 542–548.

Ward, D. J., Lamont, B. B. and Burrows, C. L. (2001) Grasstrees reveal contrasting fire regimes in eucalypt forest before and after European settlement of southwestern Australia. *Forest Ecology and Management*, **150**, 323–329.

Washington-Allen, R. A., Briske, D. D., Shugart, H. H. and Salo, L. F. (2009) Introduction to special feature on catastrophic thresholds, perspectives, definitions, and applications. *Ecology and Society*, **15**, 38. http://www.ecologyandsociety.org/vol15/iss3/art38 (accessed March 19, 2013).

Watson, J. E. M., Evans, M. C., Carwardine, J., Fuller, R. A., Joseph, L. N., Segan, D. B., Taylor, M. F. J., Fensham, R. J. and Possingham, H. P. (2011) The capacity of Australia's protected-area system to represent threatened species. *Conservation Biology*, **25**, 324–332.

Williams, S. E., Bolitho, E. E. and Fox, S. (2003) Climate change in Australian tropical rainforests: an impending environmental catastrophe. *Proceedings of the Royal Society B*, **270**, 1887–1892.

Williams, S. E., Pearson, R. G. and Walsh, P. J. (1996) Distributions and biodiversity of the terrestrial vertebrates of Australia's Wet Tropics: a review of current knowledge. *Pacific Conservation Biology*, **2**, 327–362.

Woinarski, J. C. Z., Hempel, C., Cowie, I., Brennan, K., Kerrigan, R., Leach, G. and Russell-Smith, J. (2006) Distributional pattern of plant species endemic to the Northern Territory, Australia. *Australian Journal of Botany*, **54**, 627–640.

Woinarski, J. C. Z., Armstrong, M., Brennan, K., Fisher, A., Griffiths, A. D., Hill, B., Milne, D. J., Palmer, C., Ward, S., Watson, M., Winderlich, S. and Young, S. (2010) Monitoring indicates rapid and severe decline of native small mammals in Kakadu National Park, northern Australia. *Wildlife Research*, **37**, 116–126.

Yibarbuk, D., Whitehead, P. J., Russell-Smith, J., Jackson, D., Godjuwa, C., Fisher, A., Cooke, P., Choquenot, D. and Bowman, D. (2001) Fire ecology and Aboriginal land management in central Arnhem Land, northern Australia: a tradition of ecosystem management. *Journal of Biogeography*, **28**, 325–343.

CHAPTER 21

BIODIVERSITY AND CONSERVATION IN THE PACIFIC ISLANDS: WHY ARE WE NOT SUCCEEDING?

Gilianne Brodie[1], Patrick Pikacha[2] and Marika Tuiwawa[3]

[1]School of Biological and Chemical Sciences, University of the South Pacific, Suva, Fiji Islands
[2]Solomon Islands Community Conservation Partnership, Honiara, Solomon Islands
[3]South Pacific Regional Herbarium, University of the South Pacific, Suva, Fiji Islands

SUMMARY

There are more than 25,000 relatively small islands located in the Pacific region. The flora and fauna of these islands are highly diverse, and many of the species that occur in the region have limited ranges and are not found elsewhere in the world. By nature of their small size, the majority of Pacific Islands are ecologically fragile and particularly vulnerable to climate change, overexploitation of natural resources, and invasive species. Many of the human-induced ecosystem changes currently occurring on these fragile islands are irreversible; they often relate to changes in community values and beliefs as well as the growing desire for income generation. In light of the continuing loss of traditional knowledge and practices, we are in dire need of strong, creative, ethical leaders who are not afraid to think outside the box. The education of our Pacific Island youth (emerging leaders) and a considerable strengthening of our human resource development are essential foundations for us to have any chance of making the necessary changes in human behavior needed to achieve long-term conservation success and sustainable environmental practices that will enhance the future health of Pacific Island communities.

INTRODUCTION

The Pacific Ocean contains approximately 25,000 islands of variable size and topography with a total coastline of 135,663 kilometers and an ocean surface area of 165 million square kilometers (IUCN, 2009a). This area is larger than the world's total land surface, with the islands divided among 56 different Pacific Island and Rim countries and territories. Within the Pacific Ocean lies the region referred to as Oceania, which includes the islands of the tropical Pacific Ocean. The terrestrial diversity and endemism per unit area in Oceania are among the highest in the world, with more than half the diversity in independent, developing island nations (Keppel et al., 2012).

Globally, biologists have identified up to 34 biodiversity "hotspots" that are extremely rich in endemic species and considered highly threatened by human activities (Myers *et al.*, 2000; Mittermeier *et al.*, 2005). These hotspots include the Oceania island sub-regions of Polynesia, Micronesia, and Melanesia. Six major threatening processes are driving biodiversity decline in Oceania: habitat loss and degradation, invasive species, climate change, overexploitation, pollution and disease, and implementation capacity (see Lees and Siwatibau, 2007; COS, 2008; Kingsford *et al.*, 2009). Good science is essential to achieve successful conservation outcomes in relation to these threats, since sound management and decision making cannot occur without adequate background information and an adequate understanding of the ecosystems and biodiversity we wish to conserve.

Simply knowing the facts will not lead to the conservation of biotas, however. What we need is a change in human behavior at the community, business, and political level. There is general agreement in the Pacific Island conservation sector that our biodiversity, and the essential ecosystem services it provides for human livelihoods, is in crisis. In contrast, the average person on the street sees this crisis as either not relevant to him or her or as a problem that someone else (e.g., village chief, local council, national government, or God) should and will solve. In other words, the local people do not think of themselves as either responsible for generating the problem or contributing to its solution.

ISLAND ECOSYSTEM FRAGILITY

The islands of Oceania are not only numerous but highly variable in size. The Solomon Islands, for example, consist of approximately 900 islands, but only seven have an area greater than 2000 km^2. The majority of Oceanic islands are relatively small, with fragile ecosystems that are functionally vulnerable to change.

In fragile island ecosystems, even a seemingly small change can have far-reaching effects, and the impact of a threat such as the introduction of an invasive species is greatly magnified as compared with that of a comparable introduction into a more robust mainland community. Often such changes can severely and irreversibly alter ecosystem function (Veitch, Clout and Towns, 2011). Examples of these irreversible changes to islands in the region are numerous. Well-documented accounts include (1) the impact of the alien crazy ant (*Anoplolepis gracilipes*) on Christmas Island, where a

rapid and catastrophic shift in the functioning of the rain forest ecosystem has occurred (O'Dowd, Green and Lake, 2003); and (2) the accidental introduction of the brown tree snake (*Boiga irregularis*) to the island of Guam, where most native vertebrates are now either endangered or extinct (see Rodda, Fritts and Conry, 1992; Wiles *et al.*, 2003). Similarly, the introduction of the cane toad (*Bufo marinus*) to areas like the Solomon Islands has had an impact on the native goanna (*Varanus indicus*), reducing their populations in the coastal forests on the main islands.

Once the ecosystems of small islands are irreversibly altered, there is no going back. The loss of ecosystem function and biodiversity often results in diminished contributions to human livelihoods (Daily, 1997).

LOSS OF TRADITIONAL KNOWLEDGE AND PRACTICES

In the past, sustainable resource use practices on and around islands occurred routinely and worked well (Johannes, 1978; Morauta, Pernetta and Heaney, 1980). However, much of this traditional knowledge is fast becoming lost or forgotten or is now underutilized (Thaman, 2010). Sadly, we are facing a major breakdown and continued decline in the transfer of traditional knowledge and the acceptance of its value to finding solutions for the future (Johannes, 1978; Berkes, Colding and Folke, 2000; Laird, 2002). Traditional biodiversity knowledge, and a clear understanding of the fact that such knowledge can often provide the foundation for the sustainability of island lifestyles, seems well recognized only by our active environmental youth groups, a relatively small number of government staff, and experienced regional conservation practitioners.

However, several regional programs (e.g., Secretariat of the Pacific Community Land Resources Division) have been working for many years to retain and revive past knowledge of the use of biological diversity in our historically diverse gardens, with their numerous crop varieties. Such knowledge can play an important role – for example, in adequately addressing and mitigating island climate change impacts.

In more "natural" systems, the gradual loss of traditional knowledge is evident by the way many communities currently perceive and manage their forest resources. Traditional societies held a comprehensive understanding of rainforest plants, distinguishing edible plants from medicinal ones, and acknowledging many others kinds of plants with cultural uses. Knowledge of the diversity of plants, both domesticated and wild, was central to survival and those members of a community who had botanical wisdom commanded influence. However, this acquaintance and value of forest plants by traditional societies is today vanishing, along with the accompanying behaviors, perceptions, and appreciation of the forest. These days, logging and widespread removal of forests takes place on a regular basis and this loss results in changes in human behavior. A shift in interpretation of the value of the forest has occurred in many island communities, so that instead of perceiving it as a "garden" to utilize, it is now often perceived as "valueless" unless tamed.

SHIFT IN RESOURCE USE PURPOSE AND LAND OWNERSHIP

Although the total population of Oceania at present is only 15 million people, it is projected to rise to 24 million by mid-century (Population Reference Bureau, 2012). When coupled with the loss of traditional knowledge, human population growth in fragile areas such as New Guinea and several other Melanesian islands (Cincotta, Wisnewski and Engelman, 2000; Anonymous, 2006) has resulted in escalated land clearing. Mining activities in New Guinea and other parts of Melanesia have emerged and are particularly destructive trends. The endless demand for products such as wood, fish, shrimp, and sea cucumbers from the area by rich countries located elsewhere often leads to local destruction. In the Fiji Islands, the human population is low and increasing relatively slowly, but the redistribution of the population reflects significant increases in urbanization (Anonymous, 2008). This shift has contributed to a substantial demand for the commercial production of food and associated unsustainable land-use practices, which in turn have made land degradation a major problem in many parts of the region (see Lees and Siwatibau, 2007).

Much land clearing and overexploitation of resources such as mangroves and fish have occurred because of a substantial shift from subsistence to commercial resource use. In the past, for example, the majority of non-urban community members fished to feed their own families, with any excess routinely distributed to the rest of the community. Now, fishing or other resource use, outside special events, occurs primarily

with the aim of making a profit. Profit is desired to pay for basic items such as school fees, church contributions, cultural obligations (e.g., funerals), and "luxuries" such as television.

In most developed countries, the government controls and enforces regulations designed to prevent overuse. In contrast, community members directly own a very large proportion of the resources in many Pacific Island countries. Here, although regulations may exist, strong governance and enforcement capacity are often lacking.

In the Melanesian countries of Papua New Guinea, Fiji, and the Solomon Islands, the land tenure system plays a major role in achieving conservation outcomes. More than 80% of land is traditionally owned; such rights have evolved into a form of private landownership with a long history of human occupation, customary use, and ancestral relevance in terms of burial and customary sites.

SUSTAINABILITY AND CLIMATE CHANGE

Despite the obvious connection between sustainable development and the quality of Pacific Island human livelihoods, the clearly unsustainable exploitation of our natural resources continues unabated. The important question here is why do the majority of people or businesses continue to overexploit or trash our natural resources such as forests and coastlines, even in the face of obvious negative economic impacts and lifestyle loss? A clear example of this is provided by the increasing amount of human-made solid waste building up on our once pristine shorelines. These precious coastal resources are not only utilized by the public for recreational activities but perhaps more importantly form the basis for our high-value, income-producing, tourism activities.

In light of expected challenges related to climate change (Huang, 1997; Barnett, 2001) another good example of the environmental sustainability issues is the continued removal of mangrove habitat (Figure 21.1). Mangroves occur across the Pacific Islands with the world's center of species diversity in Papua New Guinea. Healthy mangroves are a vital first line of defense in many areas against natural disasters, such as cyclones and tsunamis (Barbier, 2006). The largest mangrove forests are found in the Melanesia region. In particular, Papua New Guinea supports over 70% of the region's mangrove area with at least 34 associated

Figure 21.1 Mangrove habitat removal in the Solomon Islands.
Photo courtesy of © Patrick Pikacha

Figure 21.2 Mangrove firewood for sale in the Fiji Islands.
Photo courtesy of © Gilianne Brodie

obligate species found only in mangroves (IUCN, 2009b).

Despite the obvious benefits mangrove ecosystems provide, they are one of the most threatened ecosystems in the Pacific. They are not only logged commercially but are also susceptible to overexploitation by local communities. In several Pacific Island areas, mangroves are routinely harvested and sold as firewood (Figure 21.2). It is hard to stop this behavior because many families are on or below the poverty line and may not be able to afford alternative fuels for cooking.

Pacific Islanders were, and still are, traditionally coastal resource based, with many early settlements close to mangrove areas. Mangroves continue to provide significant social, economic, and cultural benefits; however, because of a lack of adequate policy and legislation to guide and or constrain their use, the mangroves are rapidly disappearing. It is therefore not surprising that implementing governance arrangements over mangrove ecosystems, to sustain the benefits derived by humans, is now a key priority for the Pacific region (IUCN, 2009b). However, successful mangrove conservation still depends overall on changing human behavior.

VALUES, BELIEFS, AND RELIGION

From childhood, we grow to develop values and beliefs that determine our future behavior and priorities. In the past, for many Pacific Islanders this has included an intimate linkage to land and sea via cultural protocols such as totems. Sadly, with increasing urbanization, indigenous people are losing their bond with biodiversity. This can be seen today by asking young indigenous people in cities if they know what their totems are or if they can provide the name of 10 different plants in their backyards in their own native tongue.

There has been much recent discussion in the Pacific Islands on the topic of the use of religion in achieving conservation outcomes. This is a popular methodology with young conservationists who are keen to pass on their new ideas and education to their own personal communities. However, not all agree with such a use of religious principles, and some even blame elements of religion for a lack of individual social responsibility in the first place. So in the future we need to think hard about what the role of religion may be in conservation; perhaps it could become a key factor in changing human behavior with respect to sustainable natural resource use.

Religion has many assets for youth because it teaches principles of stewardship or being good proponents of the environment. In some Pacific Island areas, environmental sustainability is a common theme of youth programs, teaching young people how to manage and care for the environment. This has readily occurred because the basic principles of stewardship are slowly eroding with time, impacted by increasing commercialization, a desire for "better" education and the demand for wealth.

INCOME GENERATION

As in the majority of societies in the world today, money has come to be seen in most Pacific Island countries and territories as the highest necessity or indicator of quality of life. Thus, for communities without a strong focus on long-term benefits or long-term consequences, the immediate or promised cash rewards from activities such as commercial logging are hard to resist. How can we help land-owning communities find alternate environmentally sustainable sources of income generation?

Numerous papers have touched on this topic (e.g., Getz *et al.*, 1999; Isaacs 2000; Anonymous, 2003; Veitayaki *et al.*, 2005); however, terrestrially at least, the Sovi Basin conservation area (reserve) on the island of Viti Levu, Fiji, stands out. Here 13 different community groups (mataqali), with help from the Fiji government, Conservation International, and the University of the South Pacific, have united to conserve approximately 20,000 hectares of uninhabited forest of national significance. These organizations and the multiple land-owning communities have come together and set up a compensatory trust fund to generate income from leaving their forest intact and unlogged (Masibalavu and Dutson, 2006). Thus, a long-term benefit system has been put in place to reward landowners for keeping their native forest intact, demonstrating that successful cooperation is possible with sustained effort.

The focus of many communities over recent years has been on making changes in their behavior that improve only their own resources. This is a good start but what is really needed in addition is thinking more broadly and holistically about large-scale, long-term ecosystem function. This is a challenging obstacle that will require good governance and sound leadership at all levels if it is to be attained in the future.

LEADERSHIP, CORRUPTION, AND RESPONSIBILITY

Many will tell you that the grass roots people don't care much about biodiversity loss and the conservation of threatened species, but this is not the case in many areas once the right information is provided to the community in an understandable form. The Sovi Basin project discussed previously provides an excellent example of this relationship. However, there are many more successful initiatives, such as those linked by the widespread locally managed marine areas (LMMA) network (e.g., Veitayaki *et al.*, 2005).

The quality of governance and ethics of the leaders we have at all levels in society in the Pacific Islands is critical to current and future successful conservation outcomes and for developing a new generation of sound decision makers. Leaders with integrity, a willingness to listen, and the ability to think beyond current practices are needed.

Unfortunately, in some Pacific Island countries and territories, and in many sectors of society, it is not unusual to find leaders who take on pivotal leadership roles for their own gain and not because they wish to look after their people and their natural resources in the long term. A failure or unwillingness to recognize conflicts of interest is also apparent – for example, thinking it is acceptable to hold public office in the environment sector while being a director or major shareholder for an overseas logging company recognized for its unethical and unsustainable environmental practices.

We need to continue to support our youth and develop their leadership potential while encouraging them also to remain ethical and connected with their traditional indigenous roots and foundations.

YOUTH EDUCATION AND IMPLEMENTATION CAPACITY

A very large number of conservation programs in the Pacific Islands now actively involve previously marginalized groups (e.g., women and youth) when addressing environmental problem solving. The education of a community's children is an effective way of communicating information to adults and decision makers, particularly when several different communities are involved and landownership may be disputed. An excellent example of this is currently being implemented in Fiji's only official RAMSAR wetlands site, where the local non-governmental organization (NatureFiji/MareqetiViti) is investing considerable effort into educating the children and youth living in the area surrounding the site about the wealth of unique natural resources they own. Since this site is one of only six RAMSAR wetlands sites registered in the Pacific Islands (several more are in progress), and one of the 33 sites on the Wetland directory list for Oceania (Scott, 1993), it is a very important role model for inland water protected area management.

The bottom line is biodiversity surveys of known wilderness areas, and measures of extinction and biodiversity loss, cannot occur in the Pacific Islands without scientifically trained people to undertake the often difficult work required. Taxonomic and identification training and expertise are essential at a local level, as is local knowledge of the areas to be surveyed.

The need for such taxonomic and identification expertise, particularly at a national level is often not well acknowledged. Human resource capacity building needs to be funded to occur at a local level because to date Pacific Island governments are severely underresourced in the environment sector both in terms of funding and human resources (Watling, 2007). Although many international agencies contribute very effectively to overall conservation efforts, activities need to be nationally driven and directed (Keppel *et al.*, 2012).

CONCLUSION

It is clear that the key to addressing the biodiversity crisis and substantially improving our overall poor conservation outcomes to date lies in our ability to understand and produce changes in human behavior. Despite much good science, we are not succeeding in addressing biodiversity loss and conservation because we are not succeeding, at a large scale, in changing human behavior. The changes required need to be large scale and long term. Never before has any single animal species on the planet had the ability to manipulate and change the environment the way that humans can. The changes that are currently happening in the Pacific Islands are not reversible. Our fate lies in our own human hands.

REFERENCES

Anonymous (2003) Non-timber forest products: a sustainable income. *Appropriate Technology*, **30**, 44.

Anonymous (2006) Solomon Islands National Statistics Office. http://www.spc.int/prism/country/sb/stats/Publication/Public-Index-new.htm (accessed March 19, 2013).

Anonymous (2008) CENSUS 2007 results: population size, growth, structure and distribution. *Statistical News*, **45**. Fiji Island Bureau of Statistics. Report Release 1.

Barbier, E. B. (2006) Natural barriers to natural disasters: replanting mangroves after the tsunami. *Frontiers in Ecology and the Environment*, **4**, 124–131.

Barnett, J. (2001) Adapting to climate change in Pacific Island countries: the problem of uncertainty. *World Development*, **29**, 977–993.

Berkes, F., Colding, J. and Folke, C. (2000) Rediscovery of traditional ecological knowledge as adaptive management. *Ecological Applications*, **10**, 1251–1262.

Cincotta, R. P., Wisnewski, J. and Engelman, R. (2000) Human population in the biodiversity hotspots. *Nature*, **404**, 990–992.

COS (Centre for Ocean Solutions) (2008) *Ecosystems and People of the Pacific Ocean – Threats and Opportunities for Action: A Scientific Consensus Statement*. http://www.iucn.org/about/union/secretariat/offices/oceania/oro_programmes/oro_initiatives_pac2020 (accessed March 19, 2013).

Daily, G. C. (1997) *Nature's Services: Societal Dependence on Natural Ecosystems*. Island Press, Washington, D.C.

Getz, W. M., Fortmann, L., Cumming, D., du Toit, J., Hitty, J., Martin, R., Murphree, M., Owen-Smith, N., Starfield, A. M., Westphal, M. I. (1999) Sustaining natural and human capital: villagers and scientists. *Science*, **283**, 1855–1856.

Huang, J. C. K. (1997) Climate change and integrated coastal management: a challenge for small island nations. *Ocean and Coastal Management*, **37**, 95–107.

Isaacs, J. C. (2000) The limited potential of ecotourism to contribute to wildlife conservation. *Wildlife Society Bulletin*, **28**, 61–69.

IUCN (2009a) *Pacific Ocean 2020 Challenge*. http://www.sprep.org/att/irc/ecopies/pacific_region/532.pdf (accessed March 19, 2013).

IUCN (2009b) The Pacific Mangroves Initiative, in IUCN Oceania: Update on the Regional Programme 2009. http://cmsdata.iucn.org/downloads/iucn_oceania_mid_term_document_2009.pdf (accessed March 19, 2013).

Johannes, R. E. (1978) Traditional marine conservation methods in Oceania and their demise. *Annual Review of Ecology and Systematics*, **9**, 349–364.

Laird, S. A. (2002) *Biodiversity and Traditional Knowledge: Equitable Partnerships in Practice*. People and Plants Conservation Series, WWF/UNESCO.

Lees, A. and Siwatibau, S. (2007) *Review and Analysis of Fiji's Conservation Sector*. Report to Austral Foundation. http://www.australfoundation.org/uploads/9/8/3/5/9835787/final_report_pdf_lees.pdf (accessed March 19, 2013).

Keppel, G., Morrison, C. Watling, D., Tuiwawa, M. V. and Rounds, I. A. (2012). Conservation in tropical Pacific Island countries: why most current approaches are failing. *Conservation Letters*, **5**, 256–265.

Kingsford, R. T., Watson, J. E. M., Lundquist, J. C. Venter, O., Hughes, L. Johnson, E. L., Atherton, J., Gawel, M., Keith, D. A., Mackey, B. G., Morely, C, Possingham, H. P., Raynor, B., Recher, H. F. and Wilson, K. A. (2009) Major conservation policy issues for biodiversity in Oceania. *Conservation Biology*, **23**, 834–840.

Masibalavu, V. T. and Dutson, G. (2006) *Important Bird Areas in Fiji: Conserving Fiji's Natural Heritage*. Birdlife International Pacific Partnership Secretariat, Suva, Fiji.

Mittermeier, R. A., Robles-Gil, P., Hoffman, M., Pilgrim, J., Brooks, T., Goettsch-Mittermeier, C., Lamoreux, J. and da Fonseca, G. A. B. (2005) *Hotspots Revisited: Earth's Biologically Richest and Most Endangered Terrestrial Ecoregions*. University of Chicago Press, Chicago, IL.

Morauta, L., Pernetta, J. and Heaney, W. (1980) *Traditional Conservation in Papua New Guinea: Implications for Today*. Monograph 16, Institute of Applied Social and Economic Research, Boroko, Papua New Guinea.

Myers, N., Mittermeier, R. A, Mittermeier, C. G., da Fonseca G. A. B., Kent, J. (2000) Biodiversity hotspots for conservation priorities. *Nature*, **403**, 853–858.

O'Dowd, D. J., Green, P. T. and Lake, P. S. (2003) Invasional "meltdown" on an oceanic island. *Ecology Letters*, **6**, 812–817.

Population Reference Bureau (2012) *World Population Data Sheet 2012*. http://www.prb.org/pdf12/2012-population-data-sheet_eng.pdf (accessed March 19, 2013.

Rodda, G. H., Fritts, T. H. and Conry, P. J. (1992) Origin and population growth of the brown tree snake, *Boiga irregularis*, on Guam. *Pacific Science*, **46**, 46–57.

Scott, D. A. (ed.) (1993) *A Directory of Wetlands in Oceania*. IWRB (International Waterfowl and Wetlands Research Bureau), Slimbridge, UK, and AWB (Asian Wetland Bureau), Kuala Lumpur, Malaysia.

Thaman, R. (2010) Name it, record it, or lose it: tackling the island biodiversity crisis. *MaiLife*, **42**, 60–61.

Veitayaki, J., Tawake, A., Bogiva, A., Meo, S., Ravula, N., Vave, R., Radikedike, P. and Fong Sakiusa, P. (2005) Partnerships and quest for effective community-based resource management. *The Journal of Pacific Studies*, **28**, 328–349.

Veitch, C. R., Clout, M. N., and Towns, D. R. (2011) *Island Invasives: Eradication and Management*. IUCN, Gland, Switzerland.

Watling, D. (2007) Conservation crusading or Neocolonialism – the role of international NGOs in the Island Pacific? Invited plenary presentation at "The Biodiversity Extinction Crisis – An Australasian and Pacific Response," Society for Conservation Biology, University of New South Wales, Sydney, July.

Wiles, G. J., Bart, J. Beck, R. E. and Aguon, C. F. (2003) Impacts of the brown tree snake: patterns of decline and species persistence in Guam's avifauna. *Conservation Biology*, **17**, 1350–1360.

CHAPTER 22

WHEN WORLDS COLLIDE: CHALLENGES AND OPPORTUNITIES FOR CONSERVATION OF BIODIVERSITY IN THE HAWAIIAN ISLANDS

Carter T. Atkinson, Thane K. Pratt, Paul C. Banko, James D. Jacobi and Bethany L. Woodworth*

US Geological Survey, Pacific Island Ecosystems Research Center, Hawaii National Park, Hawaii
*Department of Environmental Studies, University of New England, Biddeford, Maine, ME, USA

INTRODUCTION

The Hawaiian Islands are the most isolated archipelago in the world, separated from the nearest continental landmass by more than 3200 km of open ocean (Juvik, 1998). A string of eight large islands and associated coral atolls and rocky islets, they extend for almost 2000 km in a broad arc from Hawaii Island in the southeast to Kure Atoll in the northwest. The islands range in elevation from the high volcanoes of Hawaii Island with peaks that exceed 4000 m to low coral atolls in the far northwest that emerge just above sea level. The interaction of the broad topographic relief with trade winds and local climatic patterns creates a remarkable diversity of habitats ranging from alpine deserts to montane rain forests with precipitation exceeding 7 m per year (Giambelluca and Schroeder, 1998). Many of the species of plants and animals able to reach the islands through natural dispersion underwent explosive radiation, eventually resulting in one the highest rates of endemism among terrestrial organisms in the world. Prior to human contact, 90% of

flowering plants, 98% of insects, and 100% of the forest birds recorded from the Hawaiian Islands were endemic, with many restricted to comparable habitats on different islands (Eldredge and Evenhuis, 2003; Pratt, 2009).

While Hawaii has always been a showcase of evolutionary processes, it is equally well known for high rates of extinction. As one of the last places to be reached by mankind, the islands experienced their first wave of extinctions soon after Polynesians arrived and began clearing lowland habitats for agriculture (Burney et al., 2001; Wilmshurst et al., 2011). Virtually every taxonomic group with endemic species has faced significant losses in the islands since then. A total of 53% of 1159 taxa of endemic and other indigenous flowering plants are extinct, endangered, vulnerable, or rare (Sakai, Wagner and Mehrhoff, 2002), and as many as 90% of the more than 750 endemic species of terrestrial land snails are extinct (Cowie, 2001). Today, only one-third of the estimated 142 species and subspecies of the archipelago's endemic birds still remain, and more than half of those are threatened (Banko et al.,

2001; Boyer, 2008; Banko and Banko, 2009a). Overall, more than 380 species of invertebrates, plants, and birds are currently listed as threatened or endangered by the US Fish and Wildlife Service (http://ecos.fws.gov/tess_public/pub/stateListingAndOccurrenceIndividual.jsp?state=HI&s8fid=112761032792&s8fid=112762573902), and loss of this rich, endemic biodiversity is likely to continue if degradation and fragmentation of remaining native ecosystems are not slowed.

In a recent review of the future of Hawaii's forest birds (Pratt *et al.*, 2009), we identified four key challenges that need to be resolved for remaining species of native forest birds to survive into the next century: alien species, landscape processes, social factors, and climate change. These challenges are also relevant to other threatened terrestrial taxonomic groups (i.e., plants and invertebrates) in the Hawaiian Islands. Such threats are familiar to conservation biologists the world over, but rarely do they act as synergistically as they do in the Hawaiian Islands. In the pages that follow, we expand the ideas of Pratt *et al.* (2009) to identify key challenges and opportunities for long-term conservation of Hawaii's unique biota. In reviewing conservation successes and failures in Hawaii, we provide an example of the possible future course of conservation in other island communities.

CONTROLLING THE IMPACTS OF INVASIVE SPECIES

Invasive species include intentionally and unintentionally introduced plants, vertebrates, invertebrates, and microorganisms that have the potential to destabilize native ecosystems both rapidly and radically through a wide variety of processes ranging from competitive exclusion and disruption of natural processes of succession (Zimmerman *et al.*, 2008), loss of pollinators and dispersers (Sakai, Wagner and Mehrhoff, 2002), alterations in nitrogen cycling (Vitousek and Walker, 1989), to disease (Warner, 1968). In Hawaii, the distribution and impacts of invasive species are generally greatest in the warmer, more disturbed lowlands, and decline as elevation increases, because many invasive species that have become established are from the tropics or have spread from ports of entry. As a result, most relatively intact native ecosystems are found at higher elevations, although individual native species can persist over a broad range of elevations and, in some cases, even in severely degraded non-native forests.

Hawaii faces two major challenges from invasive species: management and regulation. First, while potential damaging populations of new invaders can often be eliminated quickly and at relatively low cost, many are not recognized as problematic until they have become established. By then, control and eradication might be difficult, expensive, and, in most cases, impossible. Examples include the expensive efforts to contain the spread of miconia (*Miconia calvescens*) (Loope and Kraus, 2009) and the explosion of coqui frogs (*Eleutherodactylus coqui*) across the island of Hawaii after early opportunities to eradicate small, localized populations were missed (Kraus and Campbell, 2002). Many of Hawaii's high-impact alien species have been established in the islands for a long time, beginning with the Polynesian introduction of the Pacific rat (*Rattus exulans*) and the subsequent arrival of ungulates, additional rat species, cats, and a host of insect pests and weeds that came later during western colonization (Stone, 1985). These difficult-to-control species have become permanent components of Hawaii's biota, and today managers are left struggling to minimize their range and negative impacts.

The second major challenge is that state and federal regulatory agencies have not been able to stem the flow of new alien species into the islands because funding is limited and regulatory jurisdictions frequently do not fully overlap, allowing unregulated species to slip through. Of greatest concern are high-impact invasive species that have not yet been introduced, particularly those that can affect entire ecosystems (Loope *et al.*, 2001; Loope and Kraus, 2009). Until this flood of unwanted introductions is slowed, Hawaii will continue to face serious risk of additional catastrophic ecological damage from invasive species. Avoiding future introductions will require an expansion in focus to also include more proactive rather than just reactive solutions to the problem. Strengthening risk assessments would provide the justification to keep species presently not in Hawaii from being introduced. Development of strong public support for invasive species prevention and control could be further expanded through outreach and education, on par to what New Zealand and Australia have done (Loope and Kraus, 2009). There is also the need to promote existing collaborative efforts through agency partnerships such as the Coordinating Group on Alien Pest Species (CGAPS, http://www.hawaiiinvasivespecies.org/cgaps), invasive species action committees based on individual islands (http://www.hawaiiinvasivespecies.org/iscs), and watershed partnerships (www.hawp.org) that have been established on most of the main Hawaiian Islands. These partnerships are already striving to improve existing quarantine efforts and to identify and control both incipient introductions and larger scale invasions. It is clear, though, that invasive species prevention and control by themselves will not be effective at a scale that is sustainable over the long haul. An additional approach, albeit a controversial one, is biological control of plant and insect pests. This has been used successfully for some invasive weeds such as Banana Poka (*Passiflora tarminiana*) (Trujillo *et al.*, 2001) and may be the only feasible method to rein in other seemingly intractable invasive species (Denslow and Johnson, 2006; Messing and Wright, 2006; Price *et al.*, 2009).

IMPLEMENTING CONSERVATION AT AN EFFECTIVE SCALE FOR RECOVERY

When confronted with limited resources for protecting native species and ecosystems in Hawaii, natural resource managers in Hawaii have often been forced to make difficult choices about managing individual species or specific ecosystems, particularly when they occur in many different locations. While the greatest diversity of native species is currently found in upper elevation habitats of relatively intact wet and moderately wet native forest, significant areas of native-dominated plant communities that still contain much of their natural biodiversity of plant and invertebrate species are also found at lower elevations.

The importance of managing at the landscape scale is increasingly recognized since there are just too many declining species to focus on individually (Price *et al.*, 2009). Many Hawaiian forest birds need large areas of habitat across broad environmental gradients to track seasonal changes in food resources. For example, the Iiwi (*Vestiaria coccinea*) and Apapane (*Himatione sanguinea*) forage for nectar resources across vast expanses of wet forest habitat (Hart *et al.*, 2011), the critically endangered Palila (*Loxioides bailleui*) tracks ripening mamane (*Sophora chrysophylla*) seeds along an elevational gradient on Mauna Kea Volcano (Banko *et al.*, 2002), and the critically endangered Alala (*Corvus hawaiensis*) follows seasonal changes in fruit production in leeward forests on the island of Hawaii (Banko, 2009). Additionally, when only small areas of habitat are managed, the rates of ingress of invasive species may be extremely high (Nelson *et al.*, 2002). Habitat

fragmentation provides corridors for movement of invasive plants (Mortensen *et al.*, 2009) and facilitates movement of disease-carrying mosquitoes into the forest interior (LaPointe, 2008). Consequently, connectivity among habitat patches, either in the form of physical corridors or the maintenance of suitable matrix habitat among patches, is crucial for maintaining genetic diversity and long-term persistence of the birds listed earlier as well as less mobile plants and animals (Crooks and Sanjayan, 2006; Neel, 2008).

Researchers and managers are currently working to identify effective spatial scales for management actions, for prioritizing areas where these can be applied, and for coordinating these actions among diverse landowners and regulatory jurisdictions. The recent use of structured decision-making tools to resolve some of these problems shows some promise (Reynolds *et al.*, 2010; Paxton *et al.*, 2011). Even so, while alien species removal in some large protected areas (e.g., fencing and goat removal from Hawaii Volcanoes and Haleakala National Parks) has been remarkably successful in fostering recovery of native vegetation (Price *et al.*, 2009), some birds, arthropods, and plants are not responding. Lack of recovery for these species is because of other limiting factors that are currently difficult or impossible to control at large spatial scales, particularly rats, introduced social hymenoptera (wasps and ants), parasitoids, and some invasive weeds (Gambino, 1992; Banko *et al.*, 2001, Banko and Banko, 2009a, b; Wilson, Mullen and Holway, 2009).

Major advances in landscape management in Hawaii over the past two decades have included partnerships at various levels: on-the-ground watershed partnerships, island-specific invasive species control committees, and higher level partnerships such as CGAPS and the Hawaii Conservation Alliance (HCA, www. hawaiiconservation.org) that foster better linkage between research, management, and administrative efforts. Approximately 80% of the native-dominated conservation landscape falls within the ownership or jurisdiction of the HCA partners. These partnerships aim to effectively manage landscapes across the Hawaiian archipelago and thereby spur the recovery of Hawaii's native animal and plant communities.

SOCIAL FACTORS

Most people in Hawaii live and spend their time in lowland coastal areas that are highly altered and dominated by alien species, out-of-sight and out-of-mind of the spectacular endemic biodiversity of the islands. Landownership in most upland areas is restricted primarily to the state and federal government and a handful of large private landowners (Price *et al.*, 2009). While this restricted land tenure has protected montane areas from some forms of development, it has also isolated the public from native plants and animals; consequently, few people have developed a personal stake in protecting them. Because they are wholly distinct, Hawaii's unique flora and fauna are not included in most North American field guides, essentially divorcing the islands from a significant source of mainland public support for conservation actions (Leonard, 2009). Hawaii also receives relatively limited conservation funding compared with larger states, in spite of the scale of management problems in the Islands (Leonard, 2008). With fewer than 2 million people, the state has only four congressional representatives and much less clout at the national level in finding funding for natural resource conservation than do states with large populations (Leonard, 2009).

Without strong public support, conservation of biodiversity in Hawaii is often compromised by other interest groups or cultural values. For example, hunting groups, representing a small but vocal minority of citizens, lobby to maintain and promote high numbers of game mammals, especially pigs and mouflon sheep in native habitats, in spite of the environmental damage they cause (Stone, 1985; Stone and Loope, 1987). More recently, there have been well-publicized conflicts among county, state, and federal agencies tasked with protecting native wildlife and vocal groups that favor protection of some high-impact non-native species (Tummons, 2008). New conflicts between cultural values (both Hawaiian and modern) and conservation of native biodiversity often arise. For example, there was a recent outcry among local citizens on Kauai when the Hawaii Department of Land and Natural Resources enacted restrictions on popular nighttime high-school football games to protect night-flying Newell's Shearwaters (*Puffinus newelli*) from being injured after disorientation by bright stadium lights (http://www.newsvine.com/_news/2010/10/22/5333589-hawaii-birds-confuse-friday-night-lights-with-moon). Similarly, conservation is often thwarted by conflicting legal or regulatory mandates that may pit cultural resources, or recreational or commercial use of state lands, against natural resource management (Juvik and Juvik, 1984).

Public support for conservation among island residents and visitors could be built by increasing the exposure of residents and visitors to Hawaii's native wildlife and ecosystems, and by improving and expanding environmental education in Hawaii's school curricula. Significant educational opportunities already exist at Hawaii Volcanoes and Haelakala National Parks, but such opportunities could be expanded greatly on lands managed by the State of Hawaii, particularly state parks and forest reserves. While the State Department of Education (DOE) has made efforts to include environmental education in public schools, the only two centers operated by the DOE on Kauai and Hawaii islands were recently closed because of state budget problems (Anonymous, 2010). Other non-profit organizations, such as Kamehameha Schools and the Nature Conservancy of Hawaii can help fill this gap, but are not as likely to reach all of Hawaii's school children.

Many conservation organizations in Hawaii already have active volunteer and service programs and these can be promoted and expanded to connect more people with native ecosystems. Despite the success of forest restoration efforts on Maui (Medeiros, 2006), there is evidence that public support for ecosystem restoration and conservation is still largely untapped. At the national level, collaboration with mainland organizations will be critical for overcoming the state's political isolation from the rest of the US. The American Bird Conservancy and the National Fish and Wildlife Foundation have recognized the plight of Hawaii's forest birds and actively support increased funding for conservation activities in the islands (North American Bird Conservation Initiative, 2009). While these are largely targeted to conservation of forest birds, other components of native ecosystems should also benefit from this effort.

CLIMATE CHANGE

Climate change is perhaps the most formidable challenge facing conservation in Hawaii, because it is the result of global processes and is not under local control. It acts synergistically with other limiting factors, exacerbating impacts of habitat loss, invasive species, and disease, and elevating the importance of managing at the landscape scale. Conservation managers must anticipate future conditions and adapt management plans so as to maximize opportunities for long-term persistence and resilience of species and ecosystems. More detailed study is clearly needed to make the best possible projections of the fate of Hawaiian ecosystems and to take appropriate steps to conserve their biota. While current climate models predict an overall temperature increase in the tropics of about 2–3° C by the year 2100 (Neelin *et al.*, 2006), increases in mean temperatures of as much as 1°C have already occurred in Hawaii (Giambelluca, Diaz and Luke, 2008). At higher elevations, there has been a documented increase in mean nighttime low temperatures over the past decade and a steady rise in the mean freezing level isotherm (Giambelluca, Diaz and Luke, 2008; Diaz, Giambelluca and Eischeid, 2011). Associated with rising temperatures are changes in mean precipitation, increased occurrence of drought cycles, and increases in the intensity and number of hurricanes (Loope and Giambelluca, 1998), but their localized effects on the archipelago are less well understood. This is particularly true at higher elevations on the islands where impacts from rising temperatures and increased drought on plant distribution may be difficult to predict (Juvik *et al.*, 2011). Nevertheless, the probable effects of climate change are evident in the subalpine forest of Mauna Kea Volcano, where the Palila population has been declining sharply since 2003 mainly due to severe, prolonged drought that is curtailing the production of mamane seeds (Banko *et al*, 2013).

Both the height and stability of the tropical inversion layer in Hawaii are critical under various climate change scenarios because they control the upper limit of montane rain forests on the high Hawaiian Islands where the largest areas of high-quality native forest still remain. Forests will likely move upslope in response to global warming if mean height of the inversion layer increases as temperatures increase. However, if rates of climate change exceed the regenerative capacity of native forest to move upslope or if this movement is capped by an inversion layer that remains stable in altitude (Cao *et al.*, 2007; Diaz, Giambelluca and Eischeid, 2011), these changes could lead to loss and increasing fragmentation of existing montane habitats. Paleoecological studies of past rapid climate changes have shown that species migrate at different rates, leading to community disaggregation (Hunter, Jacobson and Webb, 1988); thus long-standing ecological relationships are likely to unravel as coevolved species migrate at different rates. Moreover, climate change will allow upslope movement of established invasive species that favor warmer temperatures, including mosquitoes and

bird disease, invasive plants, and parasitoids (Loope and Giambelluca, 1998; Benning *et al.*, 2002; Peck *et al.*, 2008). There is some evidence that upslope movement of avian disease is already underway, with corresponding declines in the Akikiki (*Oreomystis bairdi*) and Akekee (*Loxops caeruleirostris*) on the island of Kauai (Freed *et al.*, 2005; Atkinson and Utzurrum, 2010).

Options for responding to climate change fall into several categories. First, we must strive to make Hawaii's native ecosystems and their components resilient to change using multiple approaches. Specifically, we must use all the tools at our disposal to increase the carrying capacity of existing habitats for native species by reducing alien threats and restoring native vegetation over large areas. This will both increase the probability of persistence and dispersal of native species, as well as prevent further loss of existing genetic and phenotypic variability that could be crucial to adaptation. A prime example of the sort of action needed to increase habitat-carrying capacity is to remove introduced browsing ungulates permanently from Palila Critical Habitat, which is the first step in restoring forest structure and function (Banko *et al.*, 2013).

Second, we must continue to manage at the landscape scale, *particularly across environmental gradients*. To do this, we should foster the recent expansion of watershed partnerships and design or expand existing conservation units along the lines of the Ahupuaa concept of native Hawaiian culture. The Ahupuaa model views land tenure units as pie-like slices of habitat arranged around an island, extending from sea level to mountain top. This approach for reserve design incorporates a wide range of elevations to allow altitudinal movements of plants and animals as existing habitats become unsuitable. Establishing corridors between reserves, increasing reserve sizes, establishing buffer areas and expansions on the upper elevation boundaries of forest and species ranges, and decreasing the distances between reserves, are important management strategies to facilitate movement in a rapidly changing climate (Heller and Zavaleta, 2009).

Third, restoration of high-elevation habitats, such as reforestation efforts ongoing at Hakalau National Wildlife Refuge (Horiuchi and Jeffrey, 2002) and the leeward side of East Maui (Medeiros, 2006), are paramount. These high-elevation habitats will be a stronghold for many species, particularly forest birds that are susceptible to disease, and we must make sure they are ready to harbor immigrants moving in from lower elevation over the coming decades.

At the same time, we can seek to protect habitats and species that are likely to be resistant and/or resilient to changing climate (Dawson *et al.*, 2011), and many of these are found at low elevations. We can seek to identify climate refugia, areas that are buffered from changes in temperature and precipitation, such as forested ravines likely to remain cooler than surrounding hillsides. We may prioritize protection of habitats or populations that are likely to persist under the new conditions, such as dry, drought-tolerant leeward forests, and populations that are likely to be resilient, such as low-elevation populations of Hawaii Amakihi with some tolerance to disease (Woodworth *et al.*, 2005).

Finally, if climate change approaches worst-case scenarios, we might need to accept that native ecosystems may not be sustainable in their present composition and distribution. Many alien species may never be removed entirely from remnant native forests, and it might be necessary to recreate biological communities without the alien species in highly altered landscapes, such as leeward East Maui and on the island of Kahoolawe, which was used as a military bombing and weapons testing range for decades. Assisted migration of some elements of native ecosystems and translocations among islands may become necessary to assure their continued survival. In addition to uncertainties in climate change scenarios and imperfect knowledge about habitat requirements for each species, such actions can have negative consequences such as hybridization of closely related taxa, impacts on resident biota, and spread of disease (Ricciardi and Simberloff, 2009). However, "desperate times call for desperate measures," and research into these options today may provide us with a tool that we are thankful to have in the future.

FINAL THOUGHTS

The geographic isolation that fostered evolution of such a highly endemic flora and fauna has been shattered by the consequences of globalization, and we expect to face additional ecological challenges to Hawaii's native species and communities far into the future. Nevertheless, in the aftermath of the collision of Hawaii's unique biota with many new and potentially damaging invasive species from the rest of the world, we see opportunities for understanding how to save what remains, how to maintain or reconstruct critical interactions and functions in ecosystems, and how to develop adaptive strategies for coping with new threats. Hawaii's steep environmental gradients and diversity of habitat types also provide a natural laboratory for understanding and applying principles of ecological restoration in the face of global climate change, and the results of this work could also greatly benefit the rest of the world as it strives to achieve comparable results. The smaller scale of the Hawaiian Islands allows for more intensive management as well as experimentation compared with continents. Real progress in conserving Hawaii's natural heritage, however, hinges on how effectively we manage the substantial areas of remaining native habitat, encourage even more public awareness and support of conservation issues, and expand partnerships to bring more land under active management for native species. Coordinating strategic actions to prevent further losses of Hawaii's unique and spectacular native biodiversity and to foster the recovery of species and communities at landscape scales will require dedicated, long-term funding, strong leadership, and broad public education and participation.

ACKNOWLEDGMENTS

We thank the US Geological Survey Ecosystems Program for programmatic and financial support.

REFERENCES

Anonymous (2010) *DOE Operating Budget Request Fiscal Biennium 2011–2013.* http://doe.k12.hi.us/reports/budget/BienniumOperatingBudgetRequest_1113.pdf (accessed March 19, 2013).

Atkinson, C. T and Utzurrum, R. B. (2010) Changes in prevalence of avian malaria on the Alala'i Plateau, Kaua'i, 1997–2007. Hawaii Cooperative Studies Unit Techical Report HCSU-017, University of Hawaii at Hilo. http://www.uhh.hawaii.edu/hcsu/documents/TRHCSU017AtkinsonChangesinPrevalenceofAvianMalariaFINAL.pdf (accessed March 19, 2013).

Banko, P. C. (2009) 'Alalā, in *Conservation Biology of Hawaiian Forest Birds: Implications for Island Avifauna* (eds T. K. Pratt, C. T. Atkinson, P. C. Banko, J. D. Jacobi, and B. L. Woodworth), Yale University Press, New Haven, CT, pp. 473–486.

Banko, W. E. and Banko, P. C. (2009a) Historic decline and extinction, in *Conservation Biology of Hawaiian Forest Birds:*

Implications for Island Avifauna (eds T. K. Pratt, C. T. Atkinson, P. C. Banko, J. D. Jacobi, and B. L. Woodworth), Yale University Press, New Haven, CT, pp. 25–58.

Banko, P. C. and Banko, W. E. (2009b) Evolution and ecology of food exploitation, in *Conservation Biology of Hawaiian Forest Birds: Implications for Island Avifauna* (eds T. K. Pratt, C. T. Atkinson, P. C. Banko, J. D. Jacobi, and B. L. Woodworth),Yale University Press, New Haven, CT, pp. 159–193.

Banko, P. C., David, R. E., Jacobi J. D., and Banko, W. E. (2001) Conservation status and recovery strategies for endemic Hawaiian birds. *Studies in Avian Biology*, **22**, 359–376.

Banko, P. C., Camp, R. J., Farmer, C., Brinck, K. W., Leonard, D. L. and Stephens, R. M. (2013) Response of palila and other subalpine Hawaiian forest bird species to prolonged drought and habitat degradation by feral ungulates. *Biological Conservation*, **157**, 70–77.

Banko, P. C., Oboyski, P. T., Slotterback, J. W., Dougill, S. J., Goltz, D. M., Johnson, L., Laut, M. E., and Murray, T. C. (2002) Availability of food resources, distribution of invasive species, and conservation of a Hawaiian bird along a gradient of elevation. *Journal of Biogeography*, **29**, 789–808.

Benning, T. L., LaPointe, D. A., Atkinson, C. T. and Vitousek, P. M. (2002) Interactions of climate change with land use and biological invasions in the Hawaiian Islands: Modeling the fate of endemic birds using GIS. *Proceedings of the National Academy of Sciences of the USA*, **99**, 14246–14249.

Boyer, A. G. (2008) Extinction patterns in the avifauna of the Hawaiian Islands. *Diversity and Distributions*, **14**, 509–517.

Burney, D. A., James, H. F., Burney, L. P., Olson, S. L., Kikuchi, W., Wagner, W. L., Burney, M., McCloskey, D., Kikuchi, D., Grady, F. V., Gage, R. II, and Nishek, R. (2001) Fossil evidence for a diverse biota from Kaua'i and its transformation since human arrival. *Ecological Monographs*, **71**, 615–641.

Cao, G., Giambelluca, T. W., Stevens, D. and Schroeder, T. (2007) Inversion variability in the Hawaiian trade wind regime. *Journal of Climate*, **20**, 1145–1160.

Cowie, R. H. (2001) Invertebrate invasions on Pacific islands and the replacement of unique native faunas: a synthesis of the land and freshwater snails. *Biological Invasions* **3**, 119–136.

Crooks, K. R. and Sanjayan, M. (2006) *Connectivity Conservation*. Cambridge University Press, Cambridge.

Dawson, T., Jackson, S. T., House, J. I., Prentice, I. C., and G. M. Mace (2011) Beyond predictions: biodiversity conservation in a changing climate. *Science*, **332**, 53–58.

Denslow, J. S. and Johnson, M. T. (2006) Biological control of tropical weeds: research opportunities in plant-herbivore interactions. *Biotropica*, **38**, 139–142.

Diaz, H. F., Giambelluca, T. W. and Eischeid, J. K. (2011) Changes in the vertical profiles of mean temperature and humidity in the Hawaiian Islands. *Global and Planetary Change*, **77**, 21–25.

Eldredge, L. G. and Evenhuis, N. L. (2003) Hawaii's biodiversity: a detailed assessment of the numbers of species in the Hawaiian Islands. Records of the Hawaii Biological Survey for 2001–2002. *Bishop Museum Occasional Papers*, **76**, 1–30.

Freed, L. A., Cann, R. L., Goff, M. L., Kuntz, W. A. and Bodner, G. R. (2005) Increase in avian malaria at upper elevation in Hawai'i. *The Condor*, **107**, 753–764.

Gambino, P. (1992) Yellowjacket (*Vespula pensylvanica*) predation at Hawaii Volcanoes and Haleakala National Parks: identity of prey items. *Proceedings of the Hawaiian Entomological Society*, **31**, 157–164.

Giambelluca, T. W. and Schroeder, T. A. (1998) Climate, in *Atlas of Hawaii*, 3rd edn (eds J. O. Juvik, and S. P. Juvik), University of Hawaii Press, Honolulu, HI, pp. 49–59.

Giambelluca, T. W., Diaz, H. F. and Luke, M. S. A. (2008) Secular temperature changes in Hawai'i. *Geophysical Research Letters*, **35**, L12702.

Hart, P. J., Woodworth, B. L., Camp, R., Turner, K., McClure, K., Goodall, K., Henneman, C., Spiegl, C., LeBrun, J. Tweed, E. and Samuel, M. (2011) Temporal variation in bird and resource abundance across an elevational gradient in Hawaii. *The Auk*, **128**, 113–126.

Heller, N. E. and Zavaleta, E. S. (2009) Biodiversity management in the fact of climate change: a review of 22 years of recommendations. *Biological Conservation*, **142**, 14–32.

Horiuchi, B. and Jeffrey, J. (2002) Native plant propagation and habitat restoration at Hakalau Forest NWR, Hawaii, in *National Proceedings: Forest and Conservation Nursery Associations – 1999, 2000, and 2001* (eds R. K. Dumroese, L. E. Riley and T. D. Landis). Proceedings RMRS-P-24, Ogden, UT. US Department of AgricultureD Forest Service, Rocky Mountain Research Station. http://www.treesearch.fs.fed.us/pubs/29802 (accessed March 19, 2013).

Hunter, M.L., Jacobson, G. and Webb, T. (1988) Paleoecology and the coarse-filter approach to maintaining biological diversity. *Conservation Biology*, **2**, 375–385.

Juvik, J. O. (1998) Biogeography, in *Atlas of Hawaii*, 3rd edn (eds J. O. Juvik and S. P. Juvik), University of Hawaii Press, Honolulu, HI, pp. 103–106.

Juvik, J. O. and Juvik, S. P. (1984) Mauna Kea and the myth of multiple use: endangered species and mountain management in Hawaii. *Mountain Research and Development*, **4**, 191–202.

Juvik, J. O., Radomsky, B. T., Price, J. P., Hansen, E. W. and Kueffer, C. (2011) "The upper limits of vegetation on Mauna Loa, Hawaii": a 50th-anniversary reassessment. *Ecology*, **92**, 518–525.

Kraus, F. and Campbell, E. W. III (2002) Human-mediated escalation of a formerly eradicable problem: the invasion of Caribbean frogs in the Hawaiian Islands. *Biological Invasions*, **4**, 327–332.

LaPointe, D. A. (2008) Dispersal of *Culex quinquefasciatus* (Diptera: Culicidae) in a Hawaiian rain forest. *Journal of Medical Entomology*, **45**, 600–609.

Leonard, D. L. Jr. (2008) Recovery expenditures for birds listed under the US Endangered Species Act: the disparity between mainland and Hawaiian taxa. *Biological Conservation*, **141**, 2054–2061.

Leonard, D. L. Jr. (2009) Social and political obstacles to saving Hawaiian birds: realities and remedies, in *Conservation Biology of Hawaiian Forest Birds: Implications for Island Avifauna* (eds T. K. Pratt, C. T. Atkinson, P. C. Banko, J. D. Jacobi and B. L. Woodworth), Yale University Press, New Haven, CT, pp. 533–551.

Loope, L. L. and Giambelluca, T. W. (1998) Vulnerability of island tropical montane cloud forests to climate change, with special reference to East Maui, Hawaii. *Climatic Change* **39**, 503–517.

Loope, L. L. and Kraus, F. (2009) Preventing establishment and spread of invasive species: current status and needs, in *Conservation Biology of Hawaiian Forest Birds: Implications for Island Avifauna* (eds T. K. Pratt, C. T. Atkinson, P. C. Banko, J. D. Jacobi and B. L. Woodworth), Yale University Press, New Haven, CT, pp. 359–380.

Loope, L. L., Howarth, F. G., Kraus, F. and Pratt, T. K. (2001) Newly emergent and future threats of alien species to Pacific birds and ecosystems. *Studies in Avian Biology*, **22**, 291–304.

Medeiros, A.C. (2006) *Restoration of native Hawaiian dryland forest at Auwahi, Maui*. USGS Fact Sheet 2006–3035, US Geological Survey, Reston, VA.

Messing, R. H. and Wright, M. G. (2006) Biological control of invasive species: solution or pollution? *Frontiers of Ecology and the Environment*, **4**, 132–140.

Mortensen, D. A., Rauschert, E. S. J., Nord, A. N. and Jones, B. P. (2009) Forest roads facilitate the spread of invasive plants. *Invasive Plant Science and Management*, **2**, 191–199.

Neel, M. C. (2008) Patch connectivity and genetic diversity conservation in the federally endangered and narrowly endemic plant species *Astragalus albens* (Fabaceae). *Biological Conservation*, **141**, 938–955.

Neelin, J. D., Münnich, M, Su, H., Meyerson, J. E., and Holloway, C. E. (2006) Tropical drying trends in global warming models and observations. *Proceedings of the National Academy of Sciences of the USA*, **103**, 6110–6115.

Nelson, J. T., Woodworth, B. L., Fancy, S. G., Lindsey, G. D. and Tweed, E. J. (2002) Effectiveness of rodent control and monitoring techniques for a montane rainforest. *Wildlife Society Bulletin*, **30**, 82–92.

North American Bird Conservation Initiative, 2009. *The State of the Birds, United States of America, 2009*. US Department of Interior, Washington, D.C.

Paxton, E. H., Burgett, J., McDonald-Fadden, E., Bean, E., Atkinson, C. T., Ball, D., Cole, C., Crampton, L. H., Kraus, J., LaPointe, D. A., Mehrhoff, L., Samuel, M. D., Brewer, D. C., Converse, S. J., and Morey, S. (2011) *Keeping Hawai'i's Forest Birds One Step Ahead of Avian Diseases in a Warming World: A Focus on Hakalau Forest National Wildlife Refuge*. A case study from the Structured Decision Making Workshop, February 28–March 4, 2011, Hawaii Volcanoes National Park, Hawaii.

Peck, R. W., Banko, P. C., Schwarzfeld, M., Euaparadorn, M. and Brinck, K. W. (2008) Alien dominance of the parasitoid wasp community along an elevation gradient on Hawai'i Island. *Biological Invasions*, **10**, 1441–1455.

Pratt, T. K. (2009) Origins and evolution, in *Conservation Biology of Hawaiian Forest Birds: Implications for Island Avifauna* (eds T. K. Pratt, C. T. Atkinson, P. C. Banko, J. D. Jacobi and B. L. Woodworth), Yale University Press, New Haven, CT, pp. 3–24.

Pratt, T. K., Atkinson, C. T., Banko, P. C., Jacobi, J. D., Woodworth, B. L. and Mehrhoff, L. A. (2009) Can Hawaiian forest birds be saved? In *Conservation Biology of Hawaiian Forest Birds: Implications for Island Avifauna* (eds T. K. Pratt, C. T. Atkinson, P. C. Banko, J. D. Jacobi and B. L. Woodworth), Yale University Press, New Haven, CT, pp. 552–580.

Price, J. P., Jacobi, J. D., Pratt, L. W., Warshauer, F. R. and Smith, C. W. (2009) Protecting forest bird populations across landscapes, in *Conservation Biology of Hawaiian Forest Birds: Implications for Island Avifauna* (eds T. K. Pratt, C. T. Atkinson, P. C. Banko, J. D. Jacobi and B. L. Woodworth), Yale University Press, New Haven, CT, pp. 381–404.

Reynolds, M., McGowan, C., Converse, S. J., Mattsson, B., Hatfield, J. S., McClung, A., Mehrhoff, L., Walters, J. R. and Uyehara, K. (2010) Trading off short-term and long-term risk: minimizing the threat of Laysan Duck extinction from catastrophes and sea-level rise. A case study from the Structured Decision Making Workshop, January 25–29, 2010, National Conservation Training Center, Shepherdstown, WV.

Ricciardi, A. and D. S. Simberloff (2009) Assisted colonization: good intentions and dubious risk assessment. *Trends in Ecology and Evolution*, **24**, 476–477.

Sakai, A. K., Wagner, W. L. and Mehrhoff, L. A. (2002) Patterns of endangerment in the Hawaiian flora. *Systematic Biology*, **51**, 276–302.

Stone, C. P. (1985) Alien animals in Hawai'i's native ecosystems: toward controlling the adverse effects of introduced vertebrates, in *Hawai'i's Terrestrial Ecosystems: Preservation and Management* (eds C. P. Stone and J. M. Scott), Cooperative National Park Resources Studies Unit, University of Hawaii, Honolulu, HI, pp. 251–297.

Stone, C. P. and Loope, L. L. (1987) Reducing negative effects of introduced animals on native biotas in Hawaii: what is being done, what needs doing, and the role of national parks. *Environmental Conservation*, **14**, 245–258.

Trujillo, E.E., Kadooka, C., Tanimoto, V., Bergfeld, S., Shishido, G. and Kawakami, G. (2001) Effective biomass reduction of the invasive weed species banana poka by Septoria leaf spot. *Plant Disease*, **85**, 357–361.

Tummons, P. (2008) Controversy flares over proposal to control Waiawi with scale insect. *Environment Hawai'i*, **19**, 8–9.

Vitousek, P. M. and Walker, L. R. (1989) Biological invasion by *Myrica faya* in Hawai'i: plant demography, nitrogen fixation, ecosystem effects. *Ecological Monographs*, **59**, 247–266.

Warner, R. E. (1968) The role of introduced diseases in the extinction of the endemic Hawaiian avifauna. *The Condor*, **70**, 101–120.

Wilmshurst, J. M., Hunt, T. L., Lipo, C. P. and Anderson, A. J. (2011). High-precision radiocarbon dating shows recent and rapid initial human colonization of East Polynesia. *Proceedings of the National Academy of Sciences of the USA*, **108**, 1815–1820.

Wilson, E. E., Mullen, L. M. and Holway, D. A. (2009) Life history plasticity magnifies the ecological effects of a social wasp invasion. *Proceedings of the National Academy of Science of the USA*, **106**, 12809–12813.

Woodworth, B. L., Atkinson, C. T., LaPointe, D. A., Hart, P. J., Spiegel, C. S., Tweed, E. J., Henneman, C., LeBrun, J., Denette, T., DeMots, R., Kozar, K. L., Triglia, D., Lease, D., Gregor, A., Smith, T. and Duffy, D. (2005) Host population persistence in the face of introduced vector-borne diseases: Hawaii amakihi and avian malaria. *Proceedings of the National Academy of Sciences of the USA*, **102**, 1531–1536.

Zimmerman, N., Hughes, R. F., Cordell, S., Hart, P., Chang, H. K., Perez, D., Like, R. K. and Ostertag, R. (2008) Patterns of primary succession of native and introduced plants in lowland wet forests in eastern Hawai'i. *Biotropica*, **40**, 277–284.

CHAPTER 23

THE CHIMERA OF CONSERVATION IN PAPUA NEW GUINEA AND THE CHALLENGE OF CHANGING TRAJECTORIES

Phil Shearman

University of Papua New Guinea, Papua New Guinea

Papua New Guinea (PNG) is an exception from the many stories of Southeast Asian conservation challenges for the simple reason that it is perhaps 20 years behind other countries in the process of liquidating its environmental capital – its forests, clean rivers, and healthy seas – and transforming them into portable financial capital. It is in a comparatively marvelous condition; approximately 70% of the country is forested, it has little in the way of a polluting manufacturing sector, and its population density is overall quite low. Currently, the population is 7 million people, but it is projected to nearly double to 13.3 million by midcentury (Population Reference Bureau, 2012); efforts to alleviate poverty and a desire for increased standards of living, coupled with high overseas demand for such products as wood and minerals are likely to accelerate the destruction of natural communities in the coming decades. PNG has a massive biotic diversity that has originated as a product of a series of mountain-building events and island arcs speciations sandwiched sequentially onto the main islands. While some of the largest river systems have been polluted by the spoils of the mining industry, most of the smaller rivers remain undammed and clean. However, things are now changing fast. The logging industry has been through much of the accessible lowlands and is now grasping for re-entry permits, or, worse still, rights to clear-cut forested lands. The population is growing rapidly, as we have seen, putting stresses on what are partly collapsed health, transport, and education sectors. And there is barely a region of the country that has not been allocated to mining exploration with a slew of new mines under development. There are challenges aplenty on the conservation horizon. For a long while, financial capital was afraid of the risks of PNG, but now a chink in the levee has opened and money-seeking "growth" is flooding in.

PNG is a country that is becoming wealthier but at the same time becoming substantially more corrupt. This has far-reaching implications for the practicalities and potential for conservation or for responsible environmental management. In the future, PNG could well become a textbook example of the "Dutch Disease," the condition under which economies that are based principally on the extraction and sale of raw resources are prone to corruption, mismanagement, and inflation. In the recent Transparency International index of corruption (Transparency International, 2012), PNG was

Conservation Biology: Voices from the Tropics, First Edition. Navjot S. Sodhi, Luke Gibson, and Peter H. Raven.
© 2013 John Wiley & Sons, Ltd. Published 2013 by John Wiley & Sons, Ltd.

ranked 150 out of 176, and the situation is rapidly deteriorating still further. PNG as a nation "State" in the common sense of the word does not really exist. It is governed under a parliamentary democracy, with a Prime Minister as the elected head of government, and a Governor General as the head of State. In late 2010, the country had no head of government after the Prime Minister was suspended on account of a corruption investigation. Additionally, it had no Deputy Prime Minister or Governor General on account of breaches of the constitution. Indeed, there was really no government in the normal sense of the word, yet for the majority of people its absence will have gone unnoticed. Government interaction with most people is scarce or even non-existent. Most people describe themselves by their language group, of which there are somewhere in the order of 850, and within those groups are further divisions of clans and families. Clan and tribal ties are far more important in most societal roles than any other factor. The "wantok" ("belonging to my language") system is the network through which resources and patronage flow most freely. It is hard to say no to a "wantok" and if you are in a position of influence, power, or finance it is hard to avoid being persuaded into using that influence for the benefit of your kin. While this system is certainly one of the key factors that has dragged the country to the bottom quintile of all the corruption indices, it is also the system whereby many people are fed, clothed, and schooled. If the wantok system were to disappear one morning, there would be many more hungry mouths by nightfall.

The need for conservation in most of its interpretations is based on the notion of scarcity, at present or in the future. But for most PNG people this idea is largely an alien one. If one asks a villager about the population of an animal that he has rarely if ever seen, the usual answer is *"planti I stap!"* – there are plenty of them out there – yet on more careful questioning it might be revealed that the species has not been seen or caught for rather a long time. This might be partly based on the restricted range of most human groups over much of their 60,000-year presence on these islands. Rugged topography reduces the spaciousness of people's outlook; it creates an insularity of sorts, and most rural villages are surrounded by a sea of mountains rather than water. I have visited many remote villages where most people have never been out of their valley system. When you are surrounded by steep mountains cloaked in forest, it must be easy to imagine that the

forest goes on forever, even though people live in and have cleared the neighboring valleys. The wonder in a villager's eyes when he can see his terrain in a satellite photo, or in some cases from the window of a small plane, is a sight to behold. It is only then that he realizes that the forest does not go on forever. Conservation in most Western interpretations is not overly relevant to PNG people, because most human-environment interactions at the village or valley scale are not limited by scarcity. Therefore, the challenge for Western conservation paradigms in PNG is that most people are not concerned with environmental issues because the ecosystems in which they live are still intact.

Many Papua New Guineans are brilliant bushmen, and their ability to travel through and survive in the jungle is awe-inspiring. But it has to be. In much of the mountainous interior, survival does not come easily. Protein has been and continues to be in common shortage, with few big animals and low densities of game. Nonetheless, most rural Papua New Guineans consume much of their protein from bushmeat; an estimated 4 to 8 million vertebrate animals, or 10–20 thousand tonnes of biomass, are captured to support the rural population each year, with much of it consisting of small animals (Mack and West, 2005). Nothing is too small to eat. Many populations of animals are likely to be under significant pressure in all but the most remote places. About two-thirds of PNGs forest is within a day's walking distance from a village. But it also suggests that some notions of conservation, in this case voluntarily relinquishing the option to eat meat, is not a luxury that most Papua New Guineans have. They generally do not see their environment as something that needs to be protected from their actions – for most, the opposite would ring true. The environment is something from which food is won, that is fought back to clear areas for gardens or that is the domain of snakes, spirits, and disease.

Historically, the boundaries between different tribal groups were always contested places, and outbreaks of hostilities were and still are common. Especially in the highlands, payback and compensation are perennial parts of life. As an aside, I have often suspected that some forested but contested grounds, commonly running along the spines of mountain ranges, would be the simplest areas to "protect" in a concept akin to the peace parks that have been promoted around the world to soothe such borders. But it also means that the notion of expansive "environmental protection" being something of a common good is an alien concept –

what counts is what is good for the clan or tribe. In the early 1990s, I was shocked to see the relish with which landowners from the environs of the Porgera gold mine learned about the problems the discharge of tailings into the river system was causing people living downstream. The affected communities were of a different tribe with whom the Porgerans had long-held enmity, so hearing of their plight was not unwelcome news. In a perhaps more humorous example, in the 1980s the Amazonian weed *Eichornia crassipes* spread *up* the Sepik River far faster than had been expected. Groups hostile to each other were deliberately seeding their neighbors' waterways so that they too could suffer from the same problems – a form of low-tech biological warfare. The conservation ethic is one foreign to the rural people of PNG.

Although local people are still clearing substantial areas of forest for gardening purposes or via deliberate use of fire, there is little prospect of limiting this behavior. Nor is there perhaps, from a conservation perspective, a convincing rationale for attempting to do so. Local people in PNG place one of the most minimal *per capita* impacts on the global environment of most people on the planet. But more critically, pretty much everything that can be done that is within the inventory of reasonable external intervention to ameliorate the impact of subsistence activities can only, at best case, have no impact – but in most cases will actually make such impacts greater. Anything that increases potential future rents from deforestation, such as realizing increased yields through intensification, is likely to increase, not decrease, deforestation. In regard to recent discussions on the UN's Reducing Emissions from Deforestation and Forest Degradation (REDD) programme, the very act of having a carbon market competing with the international wood and agricultural markets, in the absence of concerted policies aimed at reducing global consumption, and in the absence of a functional tropical government willing and able to enforce compliance, will translate into more agricultural and logging related deforestation – because they become more profitable (Angelsen, 1999).

In PNG, environmental despoliation tends to follow a trend of punctuated equilibria – long periods of apparent stasis, or incremental loss, followed by brief periods of major change. An example is the El Niño fires of 1997–1998 and 2002 that destroyed 15% of the country's sub-alpine forests and impacted large areas of lowland forest (Shearman and Bryan, 2011). Forest loss in the next El Niño is likely to be very signifi-

cant indeed, especially now that huge areas have been primed for conversion by logging activity. What is accelerating, however, is the number and intensity of industrial resource exploitation projects throughout the country. Old growth logging, agroforestry, and mining, in part facilitated by corruption within the government, all demand consideration as key drivers of the changes now occurring across the country.

The logging industry deserves the gold medal in this group for despoiling so much forest – simplifying forest structure, reducing its biomass and diversity while predisposing it to conversion – but also for its role in giving the country its first decent push on the roller coaster of corruption.[1] It was in this sector that corruption gained its foothold, and from there it spread out until it became a systemic infection. The jockeying for government positions that could intercept the favor of the industry still continues, although now has become so institutionalized that it barely gets discussed. This is also perhaps because the other daily scandals appear to be occurring in increasing orders of magnitude relative to those of the loggers. In short, the forest industry has led the way in dropping the trajectory of development away from maximizing benefit across PNG society and firmly towards short-term outlooks, local self-interest, and ingrained corrupt practices.

As the unlogged areas become fewer and are restricted to more remote and inaccessible terrain, the industry is increasingly sourcing lumber by re-entering previously logged areas, a practice that is intensely damaging to regeneration – as well as illegal. While the logging code of practice stipulates that an area can be logged only once in a 35-year rotation, some areas have now been logged three times within 20 years. The pressure to cut corners is not just coming from the industry or from the revenue demands of government: it is also coming from forest owners themselves. With many rural communities dependent on logging companies for most of their revenue, employment, and basic services (and so very basic they are), the evaporation of timber landowners' royalties as the big trees are finished comes as a shock, and this is a catalyzing call for re-logging or, more catastrophically, for conversion into agroforestry projects.

Agroforestry in PNG includes many new projects, which to date have little to do with agriculture and lots

---

[1] The loggers have been in at least half the accessible forests, and almost all of those possessing high timber volumes.

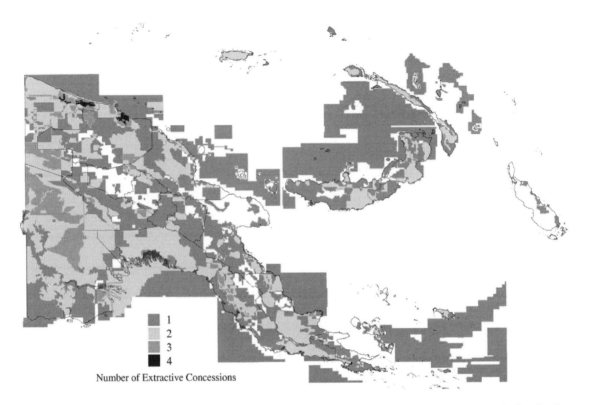

Figure 23.1 There is barely a square kilometer of PNG that is not covered by one or more extractive leases, whether they be for oil or mineral exploration, timber extraction, or, more recently, agroforestry projects. This map shows the coverage of these resource concessions. Most areas have at least one concession, but huge swathes have two or more, now including the sea-floor where an enterprising firm plans to start strip-mining hydrothermal vents. What will be left to apply to human development in 50 years when the revenue from these projects has been squandered or misappropriated?.

to do with circumventing the Forestry Act to gain cheap access to hardwoods, and, in the medium term, land. Collusion between the Department of Agriculture and Livestock, and the Department of Environment and Conservation (DEC) has seen the massive use of loopholes to obtain agriculture leases and, in turn, forest clearance licenses. As much as 15% of the country now sits within 99-year agricultural leases that legally strip the local traditional owners' use rights for three generations. While most of these leases, mostly given to Asian firms with little if any capital or experience, would never stand up to legal scrutiny, competent lawyers are prohibitively expensive for villagers. It is likely that these leases will remain in place and will be on-sold repeatedly until they do end up in the hands of those with sufficient capital to develop agroforestry projects. It is land alienation by stealth,

and, given its extent, has the distinct possibility of resulting in huge clearances of forests.

Then there are the miners being granted exploration licenses over pretty much every square inch of the country with the apparent intent by the Government of mining every deposit at once (see Figure 23.1). Five major rivers are now seriously polluted with mining wastes, and many more are about to follow suit.[2] Yet no one in Government circles has said:

[2] The Panguna Copper Mine (Bougainville Copper Ltd) polluted the Jaba River, the Ok Tedi Mine the Ok Tedi and Fly Rivers, the Porgera Gold Mine the Lagaip and Strickland Rivers, the Tolokuma Gold Mine the Angabanga River, and the recent Hidden Valley mine in Morobe has dumped large amounts of potentially acid-forming overburden into the Watut/Markham river system.

[O]ur economy cannot handle all of these projects at once, we don't have near enough skilled people as a result of dutifully ignoring the education sector, and we will kick off frightening inflation as our economy has no depth – so why don't we just have a couple of massive projects, rather than permitting all at once?

The most recent example of the collusion between Government leaders and industrial developers at the expense of the national good is the allocation of the vast Ramu nickel deposit in Madang province to a largely state-owned Chinese company MCC (China Metallurgical Group Corporation). The agreement between the Somare Government and the Chinese exempts the company from paying taxes or royalties and allows them to bring in all their workers from Guangdong and Fujian. With virtually no revenue being generated for PNG, and no employment for its citizens, it is difficult to understand why anyone would agree to the project at all.

Over the past decade, the failure of DEC to mitigate, control, or regulate the extractive sectors, far less to examine holistic land management regimes, has been tragic. This is despite a massive injection of international funds on account of PNG's charismatic fauna and intact ecosystems. Donors and their consultants have congregated around DEC like polychaetes on an abyssal hydrothermal vent. In most developing countries, one can expect mining and agriculture departments to be somewhat laissez-faire with environmental issues, but then look towards environment departments to restore some balance. Not so in PNG. Here in the late 1990s, DEC ceased supporting all protected areas in the country. National parks were "handed back" to the local people with no accompanying finance. This was done citing insufficient revenue in spite of the fact that large sums were spent on international travel. This trend continued throughout the 2000s with DEC staff being directed not to process new applications for Wildlife Management Areas (WMAs) in areas that were likely to be of interest to the logging industry.[3] It is serious when senior staff from mining or forestry departments write to DEC appealing to them to use their legislative power to rein in resource projects, or to take up conservation set asides – but to no avail. No, senior DEC staff consider that their job is to facilitate development – making sure the environmental laws are not an impediment to any project – even going to the length of changing the laws if necessary, with a helping hand from industry.

Emasculating the water quality schedules in the early 2000s was a severely retrograde step, but this was minor in comparison to the recent "veto-on-demand" amendments to the Environment Act, written by MCC's lawyers and illegally run through Parliament in 2010 to allow for millions of tons of tailings to be pumped into the sea. This occurred in response to a legal challenge initiated by landowners disputing the government's issuance of permits to the Ramu Nickel (MCC) operation to undertake Submarine Tailings Discharge (STD) off the coast near Madang. It was the miner's own lawyers who wrote the laws and handed them to the government to pass through Parliament. These laws gave one man, the Secretary of DEC, the power to exempt any project from the environmental protection laws and any need to compensate local people (or the State) for "unexpected" impacts, should the development be deemed by the same man as in the national interest.[4] Surely this should get an international award for prostituting one's country to international capital? Thankfully, in early 2012, the O'Neill-Namah Government overturned this amendment, thus removing these powers.

On this particular issue, with its massive ramifications for all aspects of environmental protection, the big three international conservation groups were noticeably silent. Not a squeak was heard from them about this. Apparently this is because they did not wish to jeopardize their relationship with the PNG government, but, if true, this could only be for two reasons – either they are doing so much good that on balance it would be foolish to put their achievements at risk, or

---

[3] While the designation of an area as a Wildlife Management Area (WMA) does not exclude logging or mining from taking place within its borders, there is the perception in corners of industry that it could make their activities more difficult.

[4] The Environment Act Amendment Bill (2010) removes the rights of landowners to mount any legal challenge against any mining or development application approved by the government; it infers that environmental damage will happen in the course of doing business as an inevitable consequence of business, and explicitly excuses corporations from damage, removing any responsibility or obligation for clean-up and restoration or recompense.

alternatively because they are raising so much money on behalf of their projects in PNG that they have become afraid of the tap being turned off if matters get rough.

It is with regret to come to the conclusion that, in PNG at least, these conservation groups have become little more than a business model, a model predicated on the periodic need for "new" exotic areas, the protection of which they can sell to amnesic donors. "New" areas are needed, not because the "old" ones have been protected, but because the old areas are now no-go zones for conservation organizations because of project-related disputes with local people, provincial governments, or the donors themselves, or just the need for new projects to sell.[5] Very few of the areas these groups were involved with a decade ago still have their presence; far fewer have been conserved. Nor has the expenditure of massive amounts of cash transformed these regions into State-backed conservation initiatives, partly perhaps because there are none of them. PNG has little if anything to show for the rivers of cash that have been poured into these large conservation "firms." In this reconciliation, some of the analyses that have been undertaken in the field of "prioritization" need to be included.

In most countries, what is commonly "conserved," and by this I mean put into some form of protected area and managed by the state, is what is left at the end of the steep bit of a development trajectory – the mountainous bits, the infertile regions, the deserts. Conservation in PNG has been haunted by the spurious goal that the end point of conservation effort is the protection of the "right" bits. Make no mistake – if at some time in the future this country has managed to transition from 70% to 10–20% forest cover, even if this is in protected areas, it has failed in some really intrinsic ways. What are the "right bits" anyway? Those that are "representative," as if we are trying to turn a hierarchical web of systems into a set of living museum specimens, or those that enclose as much biodiversity in as small an area as possible, or contain some charismatic, flagship, rare, sexy species? A common theme is their

presence in relatively small discrete locations, out of the way of development.[6]

Over the years, a suite of unimplementable, map-based studies have proposed various smatterings of protected areas (PAs) as a worthy future target. I have difficulty with the paradigm that underpins this work, and that is not because we just do not have, nor will we have, sufficient biological knowledge to make these calls at any time in the relevant future (see Figure 23.2). Nor is my concern that, if only 10% of the country remains ecologically intact in a hundred years, these final locales will quickly be destroyed or degraded.[7] My reticence at new rounds of target setting is also not principally because the ecological processes that maintain this country in a fundamental sense cannot survive such a reduction without being massively compromised and cease to support these PAs. My disquiet occurs because these exercises display such an impoverished vision of the potential that PNG still holds to not end up like other countries. Is it not possible for PNG to develop its human potential, to increase living standards without following the same path of ecological destruction as other countries?

A major problem with map-based prioritization exercises aimed at protecting "the best bits" is that, in a political environment in which implementation is not and could not be entertained, these plans get used by industrial players to justify the negative environmental consequences of their activity. Many a time, I have had an argument with one of the proponents of despoilation whose defense goes something like, "What is the problem if we trash this place, it is not a 'high conservation area,' a 'priority resource management unit,' or 'hotspot,' or 'ecoregion'" . . . and on it goes. As they are unlikely to ever be implemented, the only function such prioritization exercises serve is to justify environmental

---

[5]The chapter titled "Fiasco" in the book *Conservation Refugees, The Hundred-Year Conflict between Global Conservation and Native Peoples* by Mark Dowie (The MIT Press, Cambridge, MA, 2009) describes a text-book example of such a problem.

[6]It is worth noting how some prioritization exercises have deliberately skewed the selection of the "right bits," away from those sites that have high timber volumes, or productive soils, so as to not constrain development – yet by doing so guaranteeing that the endpoint, if these plans were ever followed, would be akin to that of the West protecting the bits no one with money wants.

[7]Why would the forces that destroyed the rest stop at the PA boundaries? Loggers have access to around 20% of the "protected forests" and the miners to most of them (Shearman and Bryan, 2011).

Figure 23.2 *Bulbophyllum polyblepharon*, an orchid species known only from three localities now within a logging concession in Madang Province. The flower is 12 mm high and has a metallic iridescence. Prior to 2010, the species was last recorded or examined with scientific intent by Rudolph Schlecter in 1913. There are probably about 3500–4000 species of orchids recorded in PNG, with an estimated 700–1200 of them yet to be described. Between 90 and 95% of these species are endemic. Of those that have been described, more than 60% are known from only one or two collections. For the majority, we don't know if they are common and poorly collected, or rare and restricted. This paucity of knowledge is not going to be rectified within the time period in which decisions regarding which regions to keep and which to destroy are made.

degradation of areas outside those identified as worthy of protection. But maybe one should not be worried because there is little chance of any of these plans being implemented anyway, even despite recent DEC fantasies as to financing compulsory land acquisitions with carbon money.

In thinking about PNG's environmental future, I keep coming back to the "enclave" model of development discussed widely in the 1970s and 1980s, but now all but forgotten. In this model, it was recognized that some areas would need to be sacrificed to generate the money necessary to invest in the well-being of people, the only true application of the word "development". Back in the 1970s and 1980s, it was sensible and realistic. But if the money is not applied to such ends, instead siphoned off, wasted, or stolen, while turning ever greater areas into "enclaves" in the hope

that more will be used for the right purposes, either insanity, corruption, or both have prevailed.

The majority of people in PNG are not benefiting from the flood of resource cash now lapping at its shores. Tragically, they are going to receive a lesser share in the future as the efficiency of theft improves. As there is more money around, paradoxically the number of people involved in its interception increases, and those already involved become more serious in their efforts. Papua New Guineans are a resilient and resourceful people – the lot of the majority can be improved dramatically by the provision of basic services, allowing them to retain a rural existence while not watching their children die from preventable diseases or not having the educational opportunities much of the world now takes for granted. But their future is dependent on a decent environment: this is perhaps more important in PNG than for many of the world's other peoples. The current trajectory of ever more mines, logging concessions, and speculative opportunism coupled with greater efficiency in stealing is a tragedy that could have been avoided, and maybe still can. PNG is rich enough to not have to make the same choices that other less fortunate countries have had to make – it could keep most of its environments intact and have a decent quality of life for its people, but it desperately needs new leaders who can set a course for a different horizon without personal enrichment as their primary objective. Only then it seems can the state reengage with local people through initiatives to manage the environment, both directly and in the honest application of PNG's environmental laws.

REFERENCES

Angelsen, A. (1999) Agricultural expansion and deforestation: modelling the impact of population, market forces and property rights. *Journal of Development Economics*, **58**, 185–218.

Dowie, M. (2009) *Conservation Refugees, The Hundred-Year Conflict between Global Conservation and Native Peoples.* The MIT Press, Cambridge, MA.

Mack, L. and West, P. (2005) *Ten thousand tonnes of small animals: wildlife consumption in Papua New Guinea, a vital resource in need of management.* Resource Management in Asia-Pacific Working Paper no. 61, Resource Management in Asia-Pacific Program, Research School of Pacific and Asian Studies, Australian National University, Canberra, Australia.

Population Reference Bureau (2012) *World Population Data Sheet 2012*. http://www.prb.org/pdf12/2012-population-data-sheet_eng.pdf (accessed March 19, 2013.

Shearman, P. L. and Bryan, J. (2011) A bioregional analysis of the distribution of rainforest cover, deforestation and degradation in Papua New Guinea. *Austral Ecology*, **36**, 9–24.

Transparency International (2012) *Corruption Perceptions Index 2012*. http://www.transparency.org/research/cpi/overview (accessed 27 March 2013).

PART 2

THOUGHTS FROM DIASPORA

CHAPTER 24

COMPLEX FORCES AFFECT CHINA'S BIODIVERSITY

Jianguo Liu

Center for Systems Integration and Sustainability, Michigan State University, MI, USA

SUMMARY

Enormous global efforts have been put into biodiversity conservation, but biodiversity loss continues rapidly in many parts of the world, including China. While China is one of the most biodiversity-rich countries, its biodiversity has been affected by numerous socioeconomic, demographic, technological, political, and biophysical forces both inside the country and elsewhere. The forces behind biodiversity maintenance or loss are complex – they occur at different times and with different strengths, have non-linear relationships and time-lag effects, and interact with each other directly and indirectly in many ways (e.g., enhance or offset). Over the past few decades, negative forces have become much stronger than positive ones. To fundamentally alter the trajectory of biodiversity loss, it is essential to explicitly analyze various forces as well as their interactive effects, recognize that biodiversity is part of coupled human and natural systems, and integrate natural and social sciences. China should also elevate its leadership role in biodiversity conservation to the global level and help protect biodiversity in other developing countries. The future of biodiversity in China (and elsewhere) will depend on the relative strengths of existing and emerging positive and negative forces.

INTRODUCTION

Global biodiversity continues along a trajectory of decline despite enormous conservation efforts (Rands *et al.*, 2010). This is also true in China, one of the most biodiversity-rich countries in the world (Liu and Raven, 2010; Ministry of Environmental Protection of China, 2010a). To fundamentally change this trajectory, it is essential to understand complex interactions among various forces affecting biodiversity, develop effective strategies to maintain it, and take bolder actions. This chapter outlines the overall status of China's biodiversity, highlights major forces behind biodiversity since the establishment of the People's Republic of China in 1949, and discusses strategies and actions that could lead to a brighter future for biodiversity.

OVERALL STATUS OF CHINA'S BIODIVERSITY

China's diverse ecosystems include forests, wetlands, lakes, oceans, meadows, shrublands, grasslands, and deserts. Although the tropical region, located in southern China, accounts for only 3.2% of China's territory, it includes one-quarter of the ecosystem types and

Conservation Biology: Voices from the Tropics, First Edition. Navjot S. Sodhi, Luke Gibson, and Peter H. Raven.
© 2013 John Wiley & Sons, Ltd. Published 2013 by John Wiley & Sons, Ltd.

one-third to half of the species found in China (Wang *et al.*, 2008). China is one of the richest countries in number of species and genetic resources, with about 32,500 vascular land plant species (third highest in the world; We, Raven and Hong, 2011), over 10,000 fungal species, and 6445 vertebrate species (13.7% of the global total) (Ministry of Environmental Protection of China, 2010a). China apparently has as much as 10% of the world's eukaryotic diversity, which would then amount to some 1.2 million species, the great majority of them unknown (Raven, personal communication).

Unfortunately, China's biodiversity has been declining at an alarming rate (Ministry of Environmental Protection of China, 2010a). Virtually all ecosystems are affected by human activities. For example, more than 90% of grasslands have been degraded to various extents (Ministry of Environmental Protection of China, 2010a), surface water has suffered quite severe pollution (Ministry of Environmental Protection of China, 2010b), wetlands have declined by 11.5% from 360,000 km^2 in 1990 to 324,000 km^2 in 2008 (Chen, 2010), and many species are disappearing in marine and coastal areas along with their habitats (Ministry of Environmental Protection of China, 2010a). Although forest cover has increased from 12% in 1981 to more than 20% in 2008 (Xinhua News Agency, 2009a), a large proportion of forests are single-species plantations, consequently with limited capacity to resist pests and diseases (Ministry of Environmental Protection of China, 2010a). China had the largest tree plantation area in the world, with a total of 62 million hectares (ha) as of 2008 (Xinhua News Agency, 2009a). This is also true in tropical regions. Plantations account for 48.3% (4.7 million ha) of woodland in China's tropical regions. In southern Yunnan Province, tropical forests declined by two-thirds over a period of 27 years (1976–2003), mainly because of the conversion of native forests to rubber plantations, a process that is still continuing (Li *et al.*, 2007).

At the species level, 15–20% of vascular plants (including more than 40% of gymnosperms and orchid species) are endangered, 233 vertebrate species are on the brink of extinction, and 44% of all wildlife species are declining. Many of these species are endemic, rare, and extremely local in distribution. Genetic resources for crops, trees, flowers, livestock, poultry, and fish are also being lost rapidly as their habitats are being destroyed (e.g., 60–70% of original habitat of wild rice has been lost) (Ministry of Environmental Protection of China, 2010a).

COMPLEXITY OF INTERACTING FORCES AFFECTING BIODIVERSITY

With the largest human population and second-largest gross domestic product in the world (Liu and Raven, 2010), China has experienced numerous positive and negative forces affecting biodiversity directly and indirectly (Figure 24.1). Positive forces are those whose impacts on biodiversity are positive or mostly positive, while negative forces have opposite impacts (e.g., Bearer *et al.*, 2008; Lepczyk *et al.*, 2008; Rutledge *et al.*, 2001; He *et al.*, 2008; Liu, Cubbage and Pulliam, 1994). Although many of these forces were created without biodiversity conservation or destruction in mind, their impacts have been enormous. Forces influencing China's biodiversity are complicated and have complex interactions.

Positive forces are largely remedies to earlier negative actions or inactions. For instance, the one-child policy was started in 1979 in response to inaction that had led to a population of 975 million, 80% higher than in 1949 (~541 million) (China Population and Development Research Center, 2011). The decline in birth rate resulting from the implementation of the one-child policy reduced human pressures on biodiversity. Other positive examples include the Natural Forest Conservation Program and the Grain-to-Green Program (Figure 24.2a), two of the world's largest conservation programs, both providing payments for ecosystem services (Liu *et al.*, 2008). The former aims to protect and restore natural forests through logging bans and plantations, and had an investment of 85.3 billion Yuan (US$13.3 billion) from the central government as of 2008 (Figure 24.2b). The latter project converts cropland on steep slopes to forests or grasslands, and had an investment of 189.4 billion Yuan ($US29.6 billion) as of 2009 (Figure 24.2c). These two programs have produced important ecological and socioeconomic outcomes nationwide (Liu *et al.*, 2008). They began in 1998 and 1999, respectively, in response to the devastating floods of 1998 that were widely believed to be the result of soil erosion related to the extensive deforestation nationwide that was greatly accelerated by certain national movements. During the "Great Leap Forward" in the 1950s, the clearing of forests to fuel backyard furnaces for steel production led to large-scale deforestation of at least 10% of China's forests (Liu, 2010a). Later, during the "Learn from Dazhai in Agriculture" movement in the 1960s and 1970s, the government promoted the agricultural model of ter-

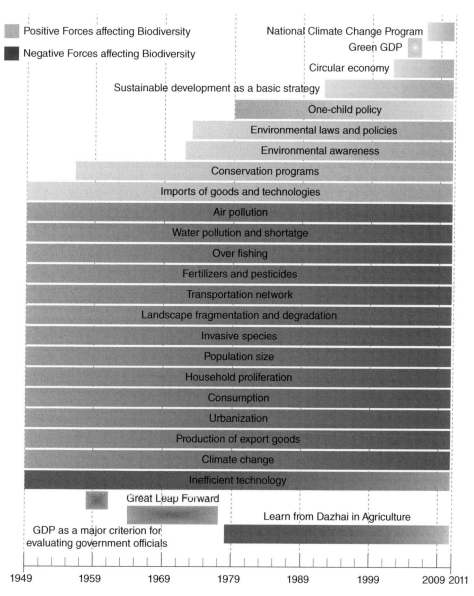

Figure 24.1 Examples of driving forces affecting the status of China's biodiversity. These particular forces were chosen for the following reasons: they are important at the national level; they represent different types of factors (socioeconomic, demographic, political, policy, technological, biophysical, international); they have different beginning and ending years; some of them become stronger while others become weaker over time; and they are useful to illustrate various interactions among them. Shaded gradients illustrate approximate general trends between starting and ending years. Lighter/heavier shades represent worse/better conditions and weaker/stronger forces, but the degrees of difference are not scaled. For the sake of simplicity, interactions among different forces are not shown.

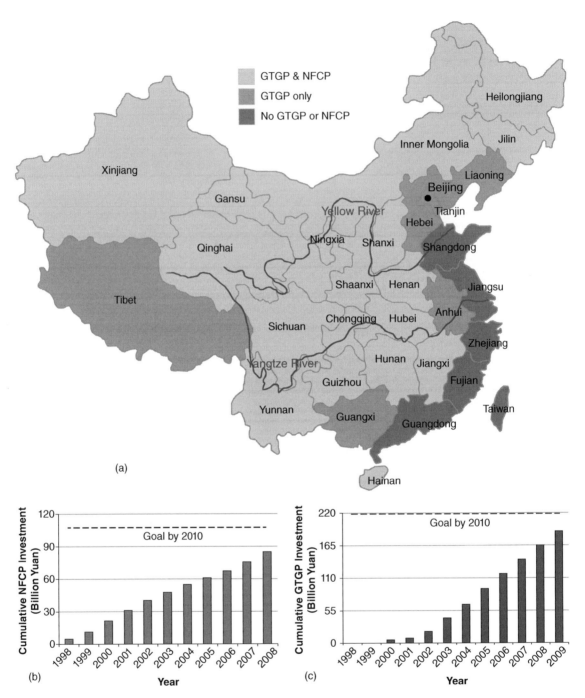

Figure 24.2 (a) Distribution of the Natural Forest Conservation Program (NFCP) and Grain-to-Green Program (GTGP) in China. Shown are names of provinces, autonomous regions, municipalities, and two major rivers (from Liu *et al.*, 2008. Proceedings of the National Academy of Sciences 105, 28). (b) Cumulative amount of investment in NFCP from 1998 to 2008. (c) Cumulative amount of investment in the GTGP from 1999 to 2009. Because data on separate investment in the GTGP for 1999 and 2000 are not available, the total amount of investment during 1999 and 2000 is shown in 2000 due to the lack of data on separate investment during these two years. In both (b) and (c), the dashed line indicates the goal for 2010 and data are from the State Forestry Administration. Reproduced from Annual Report of China's Forestry Development. State Forestry Administration, China.

racing mountains to grow grain in Dazhai Brigade of Shanxi Province, a policy that led to the transformation of numerous landscapes and filling of many lakes, wetlands, and coastal areas, disregarding local topographic, climatic, and socioeconomic conditions (Shapiro, 2001).

Some negative forces are weakening as positive ones emerge. Helping to stem biodiversity loss, the nature reserve system now includes 2541 reserves and occupies 1.47 million km[2], or 15.3% of China's territory (higher than the world average) (Liu and Raven, 2010). These figures are impressive given that China's first nature reserve was established in 1956; only 34 reserves existed in 1978 when China's economic reform started. Most of the reserves were established in the past decade to bring China's natural areas under increasing protection (Editorial Committee of Environmental Protection in China, 2000) (Liu and Raven 2010).

Unfortunately, many positive forces for conservation are offset by negative ones. For instance, while the one-child policy averted more than 300 million births by 2005 (National Development and Reform Commission of China, 2007), the number of households has increased much more rapidly than population since 1979. The rapid increase in household number has been related to factors including increasing numbers of divorces (Yu and Liu, 2007) and fewer traditional multigeneration households functioning (Liu *et al.*, 2003a). Reduction in *household size* alone added 80 million more households from 1985 to 2000 (Liu *et al.*, 2003a). More households consume more resources, release more waste, and destroy and fragment more habitat; smaller households lower the efficiency of resource use (Liu *et al.*, 2003a).

While environmental protection has been a basic national policy since 1983, GDP growth has been the major criterion for evaluating government officials. Many laws and policies have been enacted to protect biodiversity, but most of them have limited or ineffective implementation and enforcement (Ministry of Environmental Protection of China, 2010a). Although sustainable development has been a national strategy since 1994 (Zhu, 2007), short-term economic gain still receives priority. Despite increasing awareness about biodiversity, most biodiversity-friendly words and attitudes have not been translated into actions. Some positive actions, such as the calculation of green GDP (Chinese Academy for Environmental Planning 2006) (Figure 24.1), were short-lived due to much more powerful negative forces such as rapid economic development. In response to the Rio conference in 1992, China developed and released its biodiversity action plan

(General Report Writing Group of the "China Biodiversity Conservation Action Plan," 1994) with a series of biodiversity conservation goals.

As part of economic globalization, production of many products for other countries has come at the expense of China's biodiversity, with natural resources consumed and pollution left behind, resulting in biodiversity loss and degradation. Although tropical areas cover only a small fraction of China's territory, the rising imports of resources (e.g., tropical timber, minerals, and soybeans grown on land converted from tropical forests) are compromising biodiversity in many other countries, especially tropical ones (Liu and Diamond, 2005). In this way, rapid development has put China in the position of an industrialized country, one that gathers resources worldwide to support its economy. After the implementation of the Natural Forest Conservation Program, which prohibits timber logging in natural forests, in the late 1990s and early 2000s (Liu *et al.*, 2008), China has become the world's largest timber importer (Yang, Nie and Ji, 2010). Many tropical countries, such as Malaysia, Papua New Guinea, New Zealand, and Gabon are among China's major timber suppliers (Yang, Nie and Ji, 2010). To minimize or avoid cutting down tropical forests, it is important that China increases its timber supply. It is encouraging that China has been planting more trees for timber production, but it may take several decades for these new plantations to become useful timber materials. Meanwhile the resources of other countries are being drained to support China's economic growth on a one-time basis, the supply being renewable neither in the source countries nor for China.

Interactions and consequences of various forces are often non-linear, yielding drastic changes when they reach thresholds. Also, their effects usually have time lags – effects of various forces might not show up until years or even decades later. In many instances, the government's intention is good, but surprises may then occur and upset calculations. For instance, in the early 1980s, the government and the World Food Program built a large apartment complex in Wolong Nature Reserve to relocate farmers away from habitat areas of the endangered giant panda, but no one moved because there was no land near the apartment complex for the farmers to grow crops (Liu *et al.*, 2003b). When the Natural Forest Conservation Program was introduced in Wolong in 2001, it led to a sudden large increase in the number of households, because subsidies were allocated at the household level and splitting one family into two could double the subsidies (Liu *et al.*, 2007).

STRATEGIES AND ACTIONS FOR BIODIVERSITY CONSERVATION

A wide range of strategies and actions for biodiversity conservation has been proposed (Liu *et al.*, 2003b; Xu *et al.*, 2009; Ministry of Environmental Protection of China, 2010a). The list is long and includes:

(a) improving biodiversity conservation law, policy, and regulations

(b) coordinating conservation efforts among government agencies and with non-governmental organizations

(c) incorporating biodiversity conservation into socio-economic development planning

(d) enhancing the capacity of biodiversity conservation (e.g., obtaining baseline biodiversity information, monitoring biodiversity dynamics, and training research and conservation staff)

(e) strengthening conservation in priority areas and nature reserves

(f) establishing and improving preservation systems for genetic resources

(g) promoting sustainable use of biological resources

(h) restoring endangered ecosystems and species

(i) developing early-warning systems to prevent invasive species

(j) increasing the ability of addressing new threats (e.g., climate change and biofuel production) to biodiversity

(k) raising investment in conservation

(l) enhancing the public's awareness and participation in biodiversity conservation.

While these and other strategies and actions are necessary, they might not be sufficient to stop biodiversity loss. In other words, it is not clear whether biodiversity loss can be halted even if all strategies are implemented. Below are several complementary yet essential strategies to achieve biodiversity sustainability.

1. It is imperative to consider and explicitly analyze various forces as well as their interactive effects (e.g., enhancing or offsetting). China released a new strategic action plan in 2010 that aims to fully protect biodiversity by 2030 (Ministry of Environmental Protection of China, 2010a). However, without analyzing driving forces behind biodiversity, it is unknown whether this aim can actually be achieved. Furthermore, it is important to model, simulate, and track the long-term consequences of human activities to biodiversity and humans. Such analyses and simulations would help detect possible, important gaps in the 2010 action plan and avoid the outcome of the 1994 action plan

(General Report Writing Group of the "China Biodiversity Conservation Action Plan," 1994), whose major strategies and planned actions were largely implemented but failed to maintain or improve biodiversity. For positive forces to overtake negative ones, the former must be stronger than the latter and be introduced before or right after the latter forces emerge. The relative strength needed will depend on the damages caused by negative forces and time lags between occurrences of positive and negative forces. More damages and longer time lags will require stronger positive forces. Unfortunately, positive forces usually do not occur until negative forces have already produced severe damages. Even if negative forces are eventually eliminated, their legacy effects are often enormous and long-lasting. For example, the absence of the one-child policy before 1979 increased China's population base by several hundred million, which will continue to have amplifying effects on population growth for a long time and create problems that will be increasingly difficult to solve in the future (Liu, 2010a).

Bolder actions are needed to weaken the negative forces and strengthen the positive ones (Liu and Diamond, 2008). Here are three examples. First, the success of their efforts to manage biodiversity sustainability should be a major criterion for evaluating government officials because strong leadership is critical to implement biodiversity actions effectively. This requires a fundamental shift from economic performances as the sole or major evaluation criterion. Second, because climate change is a major and rapidly increasing threat to biodiversity, China should reduce the absolute amount of CO_2 emissions by going beyond reducing CO_2 emissions per unit of GDP by 40–45% from the 2005 levels by 2020 (Xinhua News Agency, 2009b; Liu and Raven, 2010). This goal could be achieved by transforming the economic development model, using renewable energy, and reducing waste. Since climate change is a global problem, China should assume a leadership role in mitigating future increases of greenhouse gases. Third, biodiversity actions need also to take place at the household level. While global pressures for biodiversity conservation have been mainly on industry, households are basic socioeconomic units of consumption and production (Linderman *et al.*, 2005). Because most industrial products are manufactured to meet household needs, energy consumption is largely related directly or indirectly to the functioning of households. As households proliferate (Liu, 2010b), building more housing units converts more natural land to residential use, and resource consumption, emissions of CO_2, and waste at the household level rise rapidly. There are many effective ways in which households can increase resource-use efficiency and reduce emissions (Dietz et al., 2009), and these should be implemented to the extent possible in China and throughout the world.

2. It is important to explicitly recognize that biodiversity is a part of coupled human and natural systems (Liu et al., 2007). A key characteristic of coupled human and natural systems is reciprocal interactions between nature and humans, such as how humans affect biodiversity and how changes in biodiversity in turn affect human well-being and behavior. While many biodiversity strategies and actions are proposed in the 2010 action plan (Ministry of Environmental Protection of China, 2010a), nothing is mentioned about reciprocal interactions – how these strategies and actions will affect biodiversity, how changes in biodiversity will affect people, how people will react to these changes in biodiversity, how people's reactions will affect biodiversity, and how impacts of people's reactions on biodiversity will and should change biodiversity strategies and actions in the future.

3. To untangle the complexity of positive and negative forces as well as the effects of their interactions on biodiversity, the integration of natural and social sciences is an important key. As in other countries, however, social sciences in China are largely isolated from natural sciences (Liu, 2008), even though their integration could provide crucial insights for promoting more effective policies for conservation. In the apartment complex case mentioned earlier, understanding farmers' needs for cropland would have avoided the construction of the complex and led to designing other solutions to the fundamental problem. In the case of distributing subsidies from the Natural Forest Conservation Program, understanding human behavior and basing the subsidies on amounts of land rather than households would have prevented the sudden formation of many households. Social norms also affect farmers' enrollment in conservation programs such as the Grain-to-Green Program (Chen et al., 2009), thus influencing the efficiency and effectiveness of conservation investment. As a general principle, improving the efficiency and effectiveness of conservation investment requires not only new technology but also changes in human attitudes and behaviors.

4. As China continues to grow as an economic superpower and becomes increasingly interconnected with the rest of the world, it should increase its leadership role in biodiversity conservation at the global level, learn successful experience from other nations, and help protect biodiversity in developing countries. China's biodiversity has profound implications for the entire planet. It is not only a treasure to Chinese people but also crucial to human well-being in other parts of the world (Liu and Diamond, 2005; Millennium Ecosystem Assessment, 2005).

Since the early 1990s, China has begun participating in many international treaties (e.g., Convention on Biological Diversity) and has collaborated with a number of international organizations on biodiversity conservation. Sustaining large conservation programs such as the Natural Forest Conservation Program and Grain-to-Green Program (Liu et al., 2008) would provide valuable lessons (e.g., large investment from the central government and transparency in distributing the investment) for other developing countries. Moreover, many forces affecting China's biodiversity also influence biodiversity elsewhere (Liu and Diamond, 2005). Protecting biodiversity at home should not be at the expense of biodiversity in other countries, especially in tropical countries that export timber, agricultural products, and other materials to China. Helping protect biodiversity in other countries will also benefit China in the long run, politically, economically, and environmentally.

THE FUTURE OF CHINA'S BIODIVERSITY

The future of China's biodiversity will depend on the relative strengths of existing and emerging positive and negative forces. On one hand, major existing positive forces (Figure 24.1) will continue at least for the foreseeable future. Some forces, such as many conservation programs and the public's awareness of them, will grow stronger. Other forces, although still supported, are getting weaker. The one-child policy seems to be an example of this trend, with increasing numbers of couples allowed to have more than one child and growing resistance to the limitations imposed. On the other hand, many negative forces (Figure 24.1) will also persist or even become stronger. There are also emerging negative forces. For instance, prompted by the global financial crisis, China is shifting its focus on exporting manufactured goods and raw materials to increasing domestic consumption. China will face more daunting challenges in achieving biodiversity sustainability with rapid expansions in its economy and urbanization, continued population growth, even faster household proliferation and consumption, global climate change, and shrinking essential resources such as water (Figure 24.1).

Over the short term, the interactions between current positive and negative forces indicate that China's overall biodiversity conditions will continue to worsen. Over the long term, however, positive forces might eventually outweigh the negative ones and lead to biodiversity sustainability. If the strategies outlined in this chapter are adopted, the biodiversity in China and other tropical countries will be more likely to persist and flourish.

ACKNOWLEDGMENTS

I thank Joanna Broderick, Luke Gibson, Shuxin Li, and Peter Raven for constructive comments on an earlier manuscript and helpful assistance in preparing the manuscript, and the National Science Foundation and National Aeronautics and Space Administration for financial support. This article is dedicated to the memory of Navjot S. Sodhi.

REFERENCES

Bearer, S. L., Linderman, M., Huang, J., An, L., He, G., and Liu, J. (2008) Effects of Fuelwood Collection and Timber Harvesting on Giant Panda Habitat Use. *Biological Conservation* **141**, 385–393.

Chen, X., Lupi, F., He, G. and Liu, J. (2009) Linking social norms to efficient conservation investment in payments for ecosystem services. *Proceedings of the National Academy of Sciences of the United States of America*, **106**, 11812–11817.

Chen, Y. (2010) Remote sensing distribution map of wetlands shows the total area of wetland decreased nationwide. *Wetland Science and Management*, **6**, 12.

China Population and Development Research Center (Total Population of China (1949–1998) National Bureau of Statistics of China 2011. China Statistical Yearbook 2011. China Statistics Press. Beijing, China. (China POPIN, Beijing).

Chinese Academy for Environmental Planning (2006) *China Green National Accounting Study Report 2004*. Beijing, China.

Dietz ,T., Gardner, G., Gilligan, J., Stern, P. and Vandenbergh, M. (2009) Household actions can provide a behavioral wedge to rapidly reduce US carbon emissions. *Proceedings of the National Academy of Sciences of the United States of America*, **106**, 18452–18456.

Editorial Committee of Environmental Protection in China (2000) *Environmental Protection in China*. China's Environmental Science Press, Beijing, China.

General Report Writing Group of the "China Biodiversity Conservation Action Plan," (1994) *China Biodiversity Conservation Action Plan*. China Environmental Science Press, Beijing, China.

He, G., Chen, X., Liu, W., Bearer, S., Zhou, S., Cheng, L., Zhang, H., Ouyang, Z., and Liu, J. (2008) Distribution of Economic Benefits from Ecotourism. *Environmental Management* **42**, 1017–1025.

Lepczyk, C. A., Flather, C. H., Radeloff, V. C., Pidgeon, A. M., Hammer, R. B. and Liu, J. (2008) Human Impacts on Regional Avian Diversity and Abundance. *Conservation Biology* **22**, 405–416.

Li, H., Aide, T. M., Ma, Y., Liu, W. and Cao, M. (2007) Demand for rubber is causing the loss of high diversity rain forest in SW China. *Biodiversity and Conservation*, **16**, 1731–1745.

Linderman, M. A., An, L., Bearer, S., He, G., Ouyang, Z. and Liu, J. (2005) Modeling the spatio-temporal dynamics and interactions of households, landscape, and giant panda habitat. *Ecological Modelling* **183** (1): 47–65.

Liu, J., Cubbage, F. and Pulliam, H. R. (1994) Ecological and economic effects of forest structure and rotation lengths: Simulation studies using ECOLECON. *Ecological Economics* **10**, 249–265.

Liu, J. and Diamond, J. (2005) China's environment in a globalizing world. *Nature*, **435**, 1179–1186.

Liu, J. and Diamond, J. (2008) Revolutionizing China's environmental protection. *Science* **319**, 46–47.

Liu, J. (2008) Integrate disciplines. *Nature*, **454**, 401.

Liu, J. and Raven, P.H. (2010) China's environmental challenges and implications for the world. *Critical Reviews in Environmental Science and Technology*, **40**, 823–851.

Liu, J., Daily, G. C., Ehrlich, P. R. and Luck, G. W. (2003a) Effects of household dynamics on resource consumption and biodiversity. *Nature*, **421**, 530–533.

Liu, J., Li, S., Ouyang, Z., Tam, C. and Chen, X. (2008) Ecological and socioeconomic effects of China's policies for ecosystem services. *Proceedings of the National Academy of Sciences*, **105**, 9477–9482.

Liu, J., Ouyang, Z., Pimm, S., Raven, P., Wang, X., Miao, H. and Han, N. (2003b) Protecting China's Biodiversity. *Science*, **300**, 1240–1241.

Liu, J. G. (2010a) China's road to sustainability. *Science*, **328**, 50.

Liu, J. G. (2010b) Sustainability: a household word. *Science*, **329**, 512.

Liu, J. G., Dietz, T., Carpenter, S. R., Alberti, M., Folke, C., Moran, E., Pell, A. N., Deadman, P., Kratz, T., Lubchenco, J., Ostrom, E., Ouyang, Z., Provencher, W., Redman, C. L., Schneider, S. H. and Taylor, W. W. (2007) Complexity of coupled human and natural systems. *Science*, **317**, 1513–1516.

Millennium Ecosystem Assessment (2005) *Synthesis*. Island Press, Washington, D.C.

Ministry of Environmental Protection of China (2010a) *China National Biodiversity Conservation Strategy and Action Plan*

(2011–2030). China Environmental Science Press, Beijing, 95 pages.

Ministry of Environmental Protection of China (2010b) *The 2009 Report on the State of the Environment in China* (in Chinese). http://cn.chinagate.cn/infocus/2011-06/14/content_22778913.htm (accessed on April 2, 2013)

National Development and Reform Commission of China (2007) *China's National Climate Change Program* (in Chinese). Beijing, China.

Rands, M. R., Adams, W. M., Bennun, L., Butchart, S. H., Clements, A., Coomes, D., Entwistle, A., Hodge, I., Kapos, V., Scharlemann, J. P., Sutherland, W. J. and Vira, B. (2010) Biodiversity conservation: challenges beyond 2010. *Science,* **329**, 1298–1303.

Rutledge, D., Lepcyzk, C. Xie, J. and Liu, J. (2001) Spatial and temporal dynamics of endangered species hotspots in the United States. *Conservation Biology* **15**, 475–487.

Shapiro, J. (2001) *Mao's War against Nature.* Cambridge University Press, Cambridge.

Wang, H., He, B., Zeng, L., Zhou, Q. and Zhong, W. (2008) Studies on the distribution, types and area of the tropical secondary forests in China. *Guangdong Forestry Science and Technology* (in Chinese), **24**, 65–73.

Wu, Z., P. H. Raven, D. Hong. 2011. *Flora of China.* Science Press, Beijing.

Xinhua News Agency (2009a) *Forest Cover in China has Increased to 20.36% and Forest Resources Enter the Phase of Fast Development* (in Chinese). http://news.xinhuanet.com/fortune/2009-11/17/content_12476936.htm (accessed on April 2, 2013)

Xinhua News Agency (2009b) *China Announces Targets on Carbon Emission Cuts.* http://news.xinhuanet.com/english/2009-11/26/content_12544181.htm (accessed March 19, 2013).

Xu, H., Tang, X., Liu, J., Ding, H., Wu, J., Zhang, M., Yang, Q., Cai, L., Zhao, H. and Liu, Y. (2009) China's progress toward the significant reduction of the rate of biodiversity loss. *BioScience,* **59**, 843–852.

Yang, H., Nie Y., and Ji, C.(2010) Study on China's timber resource shortage and import structure: natural forest protection program outlook, 1998 to 2008. *Forest Products Journal,* **60**, 408–414.

Yu, E. and Liu, J. (2007) Environmental impacts of divorce. *Proceedings of the National Academy of Sciences of the United States of America* **104**, 20629–20634.

Zhu, T. (ed.) (2007) *China's Environmental Protection and Sustainable Development* (in Chinese), Science Press, Beijing. China.

CHAPTER 25

GOVERNANCE AND CONSERVATION IN THE TROPICAL DEVELOPING WORLD

Kelvin S.-H. Peh

Conservation Science Group, Department of Zoology, University of Cambridge, Cambridge, UK

Policy makers must come to grasp the importance of the linkages between biodiversity conservation and the broader challenge of national governance, as well as to appreciate the linkages between different countries that suffer from the same governmental problems. Corruption levels differ between developed and developing nations and these differences help to explain the scale of the environmental meltdown in many tropical biodiversity hotspots. The challenge is to develop government-sponsored anti-corruption programs in which governments, communities, businesses, civil society organizations, and charity sectors coordinate their efforts to uphold good governance principles. One shortcut between the broader need to eliminate or contain corruption (which is by nature a slow process) and the immediate requirements of environmental protection might be the establishment of somewhat independent statutory bodies to protect biodiversity hotspots in developing countries. If handled correctly and if supported by international donors, these statutory bodies could become oases of corruption-free practices, and more efficient vehicles for the administration of aid designed to protect the environment than those that now exist.

Governments made significant progress at the UN climate conference held in Cancun, Mexico, during the final weeks of 2010; the likelihood of reaching a landmark deal that would protect tropical forests because of their role in sequestering carbon has noticeably

improved. And, with this, there has been a marked improvement in the chances of slowing deforestation, one of the major causes of species extinction in the tropics, where biological diversity is extraordinarily high. The mood among the government delegates attending the Cancun gathering was triumphant: they cheered what arguably proved to be the most fruitful UN climate change conference. Among the concrete achievements at the summit was the development of plans to channel money from developed nations to developing countries. For example, the Green Climate Fund (first touted at the failed Copenhagen summit the year before) became a reality; it is intended to raise and disburse US$100 billion a year by 2020 to help poor nations deal with the impact of climate change. Accordingly, measures to compensate developing countries in return for not cutting down their forests were also outlined. In essence, a global framework is now in place to assist and engage with developing countries in the grander effort of low-carbon economic development while implementing measures to reduce deforestation.

But are we rejoicing too soon? Although the Cancun summit, for the first time, allowed leaders from 190 countries to reach an agreement that proclaims that they have a "shared task to keep the planet healthy and keep it safe," the hurdles that lurk ahead remain huge. While we have a better idea from the subsequent climate summit – held in Durban in late 2011 – on how

---

Conservation Biology: Voices from the Tropics, First Edition. Navjot S. Sodhi, Luke Gibson, and Peter H. Raven.
© 2013 John Wiley & Sons, Ltd. Published 2013 by John Wiley & Sons, Ltd.

the Green Climate Fund might be run and who would control it, there remains the much more sensitive issue of how the money will be used. For the moment, we have every reason to expect that the Fund's resources would not be misappropriated, as there are a number of safeguards built into the agreement regarding the transparency of the resources used. First, the Green Fund will initially use the World Bank as a trustee – as the US, EU and Japan had demanded all along – while the task of oversight over the new body is balanced between developed and developing countries. Second, developing countries will have their emission-curbing measures subjected to international verification when these are funded by Western money, a formulation that seems to satisfy even China – which bridled at more "intrusive" verification procedures – as well as developed nations led by the US, which insisted on strict verification measures.

No doubt, these agreements constitute significant steps towards implementing the Reducing Emissions from Deforestation and Forest Degradation (REDD+) programme – a scheme backed by the United Nations offering financial incentives to developing countries to stop destroying their forests. And some funds have already started flowing for this purpose. Even at the Copenhagen climate change conference in 2009, which ended in disarray, developed nations committed US$4.5 billion to finance REDD-related investments over the next few years. Some of this money is already being distributed. Indonesia, for instance, has already received a US$30 million payment, part of a US$1.4 billion package pledged by Norway to fund projects that reduce carbon emissions by preserving tropical forests. Nevertheless, these plans remain vulnerable, because the funds could still be mismanaged or simply lost to corrupt practices.

The gap between commitments made by both developed and developing countries to reduce deforestation remains wide, and the necessary trust is often absent. Consequently, most of the recent big environmental talks have become platforms for haggling between those interested in obtaining funds without many pre-conditions, and those interested in imposing conditions before offering cash. For example, at the 10th Conference of Parties to the Convention on Biological Diversity held in Nagoya, Japan, in 2010, many developing countries threatened to boycott any negotiations unless new substantial financial commitments were made by the developed world governments. In addition, at the climate talk in Cancun, some of the tropical forest

countries argued that they should receive money with no strings attached. Their argument is baffling: that there should not be any safeguards to prevent them from using the money earned from REDD+ for actions that, if not subjected to peer review, might even destroy more tropical forests. While the sensitivities of all developing countries to any measures that may restrict their sovereignty should be acknowledged, it is also clear that attempts to improve their national budget transparency and increase budget accountability remain vital for the success of such incentive schemes. These checks do not need to involve an intrusive supervision by outside governments, but they must be transparent and amenable to constant checks. The governments of the recipient countries should ensure that the global funds are used to pay local communities that choose to preserve their forested lands instead of clearing them for commercial gain; the greatest challenge in this multi-party process is ensuring that the money paid reaches the indigenous and rural communities. For, ultimately, the present biodiversity crisis – and especially its pervasive and enduring severity – has its roots in poor governance in the developing world, a problem that remains unresolved and one that will continue to foil many conservation efforts.

It is by now an uncontroversial proposition that good governance spurs economic growth and social progress. However, what often remains less documented is just how much the ability to contain corruption – arguably one of the most important dimensions of good governance – impacts the environment and wildlife. Conversely, a good environmental record is often a primary indicator of good governance. A recent study found a strong correlation between environmental performance and the credibility of governments in the developing world, even after accounting for the compounding effect of income poverty (Peh, 2009; Figure 25.1). The latest empirical analysis also demonstrates that, although the effects of corruption are complex and are worsened by other factors such as poverty and the lack of social capital, corruption clearly exerts a negative impact upon environmental performance in developing countries (Peh and Drori, 2010; Figure 25.2). According to findings by Transparency International (2012) and a recent World Bank report (Kaufmann, Kraay and Mastruzzi, 2010), many biodiversity-rich countries in the tropics are riddled with corruption. There are 74 countries with territories containing tropical biodiversity hotspots where there are significant reservoirs of biodiversity (Conservation International,

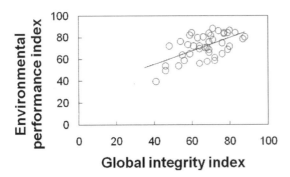

Figure 25.1 Positive correlation between global integrity index (2008) and environmental performance index (2008) among 42 developing countries, using Pearson coefficient (r). The correlation is highly significant ($n = 42$, $r = 0.593$, $p < 0.001$).

Adapted with permission from Peh (2009) The environment and corruption in developing countries. *RUSI Newsbrief* 29, 24–26.

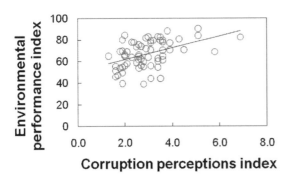

Figure 25.2 Positive correlation between environmental performance index (2008) and corruption perceptions index (2008) among 66 tropical developing countries using Pearson coefficient (r). The correlation is highly significant ($n = 66$, $r = 0.446$, $p < 0.001$).

Reproduced with permission from Peh and Drori (2010) Fighting corruption to save the environment. *Ambio* 39, 336–339. Royal Swedish Academy of Sciences.

2007); almost all these countries are considered as developing economies and most are perceived as afflicted by corruption, to varying degrees. The average corruption perceptions index (CPI) score among these countries was 3.1, while the least corrupt country in the world scored 9.3 (Table 25.1). On this standard indicator, the average corruption scores for hotspot countries in sub-Saharan Africa (2.8) and Asia Pacific (2.8) remain well below the world's average CPI, which

is 4.0. The other governance indicators that provide a good source of information and a reasonable basis for comparison are the World Bank's "government effectiveness," "rule of law," and "control of corruption" measurements (Kaufmann, Kraay and Mastruzzi, 2010; for definitions, see Table 25.1). For all countries under analysis for the purpose of conservation, only two – Bahamas and Chile – are ranked in the top quarter of the world's nations for the effectiveness of their governments and the control of corruption; the rest of the nations perform badly in these measurements, suggesting that they suffer significantly from poor governance (Table 25.1).

Critics of national-level indicators often argue that these measurements have limited utility, because "corruption is a highly localized and inconsistent phenomenon" (Clynes, 2010). Accounting for local variables is a problem with all indices, and the difficulties are clearly more acute with any measurement technique that seeks to quantify an inherently flexible – if not hazy – notion such as corruption. Nevertheless, while acknowledging Clynes' criticism, it is worth pointing out that, for the purposes of this chapter, his observation is not as significant as it might seem. The corruption indices we use usually measure government performance at the highest level – i.e., observance of existing legislation, transparency of financial disbursements, the correct collection of taxes, and so forth – all the ingredients that contribute to good governance. A failure to respect any of these yardsticks is acknowledged to contribute to a poorer record of conservation, whether directly or indirectly. And the fact that many of the current international efforts to promote conservation entail relatively large transfers of cash to central governments in the expectation that these would be disbursed correctly means that the relevance of the indices we use remains highly pertinent.

Poor performance in the wildlife and forestry sectors of many tropical developing countries is part of the wider problem of governance of these nations. Over the past few decades, protected areas – the cornerstone of global conservation efforts – enjoyed a remarkable period of increase in their number and geographical extent across the tropics (IUCN-UNEP, 2010), a trend that has lulled many into a false sense of security about the international community's ability to manage natural resources and deal with biodiversity crises. However, a recent study by Craigie *et al.* (2010) should awaken most nations from their complacency. Their work revealed that African protected areas have generally

Table 25.1 Assessments and perceptions of corruption in the countries that have territories falling within tropical biodiversity hotspots. Countries ranked in the top 25% for the control of corruption are in bold

Country in biodiversity hotspots	CPI 2010[a]	Worldwide governance indicators 2009[b]		
		Government effectiveness[c]	Rule of law[d]	Control of corruption[e]
		Percentile rank (1–100)[f]		
Developed				
Singapore	9.3	100.0	92.5	99.0
Brunei	5.5	75.2	72.2	79.0
Developing				
Argentina	2.9	39.5	29.7	38.1
Bahamas	**n/a**	**81.0**	**72.6**	**90.0**
Bangladesh	2.4	16.7	27.8	16.7
Belize	n/a	41.4	44.3	56.7
Benin	2.8	38.1	28.8	31.4
Bolivia	2.8	27.6	9.9	27.6
Brazil	3.7	57.6	49.5	56.2
Burma	1.4	1.0	3.8	1.0
Burundi	1.8	12.9	11.8	10.5
Cambodia	2.1	25.7	16.0	8.6
Cameroon	2.2	23.3	15.6	18.6
Chile	**7.2**	**85.7**	**87.8**	**89.5**
China	3.5	58.1	45.3	36.2
Colombia	3.5	56.2	39.6	48.1
Comoros	2.1	1.4	14.6	24.8
Costa Rica	5.3	65.7	65.6	72.9
Côte d'Ivoire	2.2	9.5	6.1	9.0
Cuba	3.7	37.6	31.1	64.8
Democratic Republic of Congo	2.0	1.9	1.9	2.9
Dominican Republic	3.0	38.6	28.3	30.0
Ecuador	2.5	21.9	7.5	17.6
El Salvador	3.6	52.9	22.6	53.3
Equatorial Guinea	1.9	2.9	8.0	1.0
Eritrea	2.6	5.7	9.4	45.7
Ethiopia	2.7	40.5	23.1	26.7
Ghana	4.1	56.7	51.9	59.5
Guatemala	3.2	29.0	13.7	32.4
Guinea	2.0	7.6	2.4	6.7
Haiti	2.2	4.8	5.7	11.0
Honduras	2.4	28.1	20.8	20.5
India	3.3	54.3	55.7	46.7
Indonesia	2.8	46.7	34.4	28.1
Jamaica	3.3	58.6	36.8	40.0
Kenya	2.1	31.0	15.1	11.9
Laos	2.1	14.8	18.4	9.5
Liberia	3.3	10.5	14.2	34.3
Madagascar	2.6	32.9	25.9	50.5
Malawi	3.4	36.7	48.6	38.6
Malaysia	4.4	79.5	65.1	58.1

(Continued)

Table 25.1 (*Continued*)

Country in biodiversity hotspots	CPI 2010[a]	Worldwide governance indicators 2009[b]		
		Government effectiveness[c]	Rule of law[d]	Control of corruption[e]
		Percentile rank (1–100)[f]		
Mauritius	5.4	72.9	80.2	73.8
Mexico	3.1	60.5	34.0	49.0
Mozambique	2.7	43.8	33.5	41.4
Namibia	4.4	61.0	61.3	63.3
Nicaragua	2.5	14.3	21.7	24.3
Nigeria	2.4	8.6	10.4	14.8
Oman	5.3	71.4	69.3	70.5
Panama	3.6	62.4	52.4	49.5
Papua New Guinea	2.1	22.4	17.0	3.8
Paraguay	2.2	19.5	16.5	21.0
Peru	3.5	43.3	30.2	45.2
Philippines	2.4	50.0	35.4	27.1
Puerto Rico	5.8	66.2	73.1	71.9
Rwanda	4.0	48.6	36.3	61.9
Sao Tome and Principe	3.0	29.5	27.4	42.4
Saudi Arabia	4.7	51.9	58.5	62.9
Seychelles	4.8	60.0	56.6	66.7
Sierra Leone	2.4	10.0	17.5	16.2
Solomon Islands	2.8	16.2	29.2	43.8
Somalia	1.1	0.0	0.0	0.5
South Africa	4.5	67.6	56.1	60.5
Sri Lanka	3.2	49.0	53.3	44.8
Sudan	1.6	7.1	5.2	6.2
Tanzania	2.7	39.0	40.1	40.5
Thailand	3.5	59.5	50.9	51.0
Timor Leste	2.5	11.0	11.0	15.7
Togo	2.4	6.2	20.3	14.3
Uruguay	6.9	33.8	40.6	21.4
Vanuatu	3.6	45.2	63.2	68.6
Venezuela	2.0	18.6	2.8	8.1
Vietnam	2.7	46.2	41.5	36.7
Yemen	2.2	11.4	13.2	15.2
Zimbabwe	2.4	2.4	0.9	1.9
Average (developing countries)	3.1			
Regional average				
Americas	3.6			
Asia Pacific	2.8			
Middle east	4.1			
Sub-saharan Africa	2.8			

[a] Corruption Perceptions Index 2010; countries with higher scores are perceived as less corrupt. Source: Transparency International (2012)
[b] World Bank worldwide governance indicators 2009 dataset. Source: Kaufmann, Kraay and Mastruzzi (2010)
[c] Measures the quality of public and civil services and the degree of its independence from political pressures
[d] Measures the extent to which agents have confidence in and abide by the rules of society
[e] Measures the extent to which public power is exercised for private gain
[f] Indicates rank of country among all countries in the world. 0 corresponds to lowest rank and 100 to highest

failed to mitigate human-induced threats to African large mammal populations. Many of these protected areas have performed badly: on average, population abundance of their key species suffered a 57% decline between 1970 and 2005. Further, it has been reported that the annual illegal bushmeat consumption in the Congo Basin may now exceed 1 million metric tonnes, the equivalent in weight of about 4 million cattle. As these protected areas were transformed into veritable "killing fields", fatal flaws in conventional thinking came to light. One of the most notable realizations is just how poorly we have grasped the linkages between conservation and broader governance – as well as the linkages between different countries that suffer from the similar governmental problems.

Not everything is gloomy, however. Singapore, one of the world's least corrupt countries according to the last Transparency International study, was once in a similar governance "slump"; it can offer lessons for other tropical developing countries seeking a cure from the same affliction. As recent as half a century ago – when Singapore attained self-governance from British rule – corruption was rife, permeating all sectors of society. There were also problems with weak anti-corruption legislation and the habits of a less-educated population who often preferred to get things done through bribery. Today, Singapore is one of the few nations to have made great strides in eliminating corruption, and to have restored the population's confidence in the role of government as an efficient and often impartial distributor of resources. All sectors of life are scrutinized to keep the systems clean; the Corrupt Practices Investigation Bureau (CPIB) concentrates on both public and private sectors. Despite its remarkably low levels of corruption, Singapore was ranked as having the highest proportional (relative to resource availability per country) environmental impact of any country in the world (Bradshaw, Giam and Sodhi, 2010). Thus, although better governance does not always translate into better natural resource management, *poor* governance indisputably represents a chief obstacle to environmental protection. And other countries may be well advised to follow Singapore's efficient governance model, to help combat egregious environmental transgressions.

Singapore's example should interest other tropical developing nations because the Asian city-state has employed very distinct ideas and policies in eradicating corruption, sometimes in contradiction to what has – or is – being practiced in parts of the developing world.

In short, the example of Singapore is of an energetic anti-corruption effort that does not entail a slavish adoption of Western-invented methods; that is, precisely what other developing nations are seeking to achieve. Singapore has a focused approach, involving scrutinizing with special care the upper echelons of political office-holders or civil servants. No ministers or officials are spared punishment if they are found to have obtained extra sources of unexplained or unexpected income; even the Prime Minister and the Director of CPIB are scrutinized. A broad political will and a strong, specific commitment to fighting corruption remain the most important pillars upon which Singapore's national governance structure is founded.

Anti-corruption legislation in Singapore is also distinctive. Besides applying a stern punishment policy against offenders designed to serve as deterrence, Singapore's laws also put the burden of proof on the defendant or accused if he/she is in possession of more assets or income than could be derived from regular employment or plausible business activities. The suspect, therefore, has to disprove the presumption of guilt, a far more onerous legal burden than simply inspecting records. This practice adds more bite to graft-busting, and comes in addition to other requirements for mandatory declarations of property ownership by Singapore's state officials, similar to measures that are already in place in some developing countries, as well as most developed countries, including the US (Peh, 2008).

What could other developing countries learn from Singapore in this regard? First, that a key pillar in this anti-corruption effort is an overhaul of the judicial system, the key driver for safeguarding natural resources by creating an effective backbone for investigating corrupt practices. The importance of this measure cannot be overestimated. For it is not enough to identify corruption if the culprits are either subsequently released without proof or punishment, or "bent" courts and lawyers transform the legal system into a mockery by dragging out corruption cases forever, as is all too often the case in the developing world. To rid the judiciary of corruption and to ensure its integrity, a government agency devoted to fighting corruption needs to be established, overseen by a panel that includes scholars, experts, and community leaders, thereby giving this body a legitimacy distinct from that of other state institutions. The judiciary should also form a team to monitor its own personnel. In reality, many of the countries that urgently need such measures do not lack institutions to investigate corruption, but their systems currently rely on the political willpower of people in charge to launch investigations; the results are, therefore, unpredictable and haphazard.

A second requirement is the improvement of a nation's resistance to fraud by creating a preventive environment that eliminates the temptation to engage in corrupt practices. Ministers and government officials must be paid reasonably, so that they can live according to their expected or aspired station in society without being tempted to accept bribes or gifts. As long as the salaries are derisory – whether for policemen, custom officers, or national park wardens – corruption will continue as the necessary means for survival. Singaporean ministers, judges, and civil servants are among the highest paid officials in the industrialized world. This attracts the occasional criticism, but the fact remains that an investment in their salaries – which is never great in terms of the overall national budget and can, therefore, be afforded by many other developing nations – has amply paid off by providing civil servants with both a respected social status and a disincentive to corruption. This also makes the civil service an attractive employer, and thereby creates a "virtuous circle," with new recruits reinforcing the manpower quality of existing ones.

Third, efforts should be made to raise public awareness about the effects of corruption on environmental management and conservation projects through education and the support of the media. It hardly needs explaining that, if people are better informed, they are more likely to make sound decisions and might also actively seek to improve the quality of governance. The media could also have a contributory role in actively educating the public about anti-corruption legislation. This might indirectly encourage the public to "blow the whistle" on any suspected corrupt practices. What the media could provide is an informative platform to instill in the population the belief that they have a stake in natural resource management, and broader, participatory governance.

Many might doubt that Singapore, with its small and by now highly educated population, has much relevance to the developing world. Admittedly, Singapore's anti-corruption model might not be adequate in many developing countries where both needs and challenges are greater and financial resources sparser. Nevertheless, it is important for the developing countries to study and emulate nations where corruption is brought under control. An anti-corruption practices identification and facilitation framework could be useful for developing countries, for it will increase the probability of success. This framework requires policy makers to identify the effective anti-corruption legislation and policies that are already in operation in other developing countries with similar natural resources, and with a per capita income higher than their own. If the legislation against corruption is already present, these countries should ensure that law enforcement is effective, and should further refine their code of governance for the charity sector. An example of useful action is to tighten the key guidelines related to staff pay and the distinction between the board and operational roles to uphold good governance principles and the greater accountability of charities that are in receipt of public funds. In countries where anti-corruption laws are still not in place, policy makers could resort to emulating existing legislation in other nations, thereby reducing the time necessary to adopt such measures. After decades of fighting corruption around the world, there is a solid body of evidence and experience that can help developing nations in managing what, admittedly, will always be a daunting challenge, and an area in which progress will always be incremental and relative rather than absolute.

Furthermore, while the fight against corruption often requires new legislation and regulatory frameworks, in some cases corruption can also be held in check by actually reducing existing regulations. India's experience clearly illustrates the problems of extensive regulations at a national level, but also the way by which this can be reversed. The root of corruption in India's environmental sectors – as in much of the rest of the country's economy – can be traced to the all-embracing so-called "Permit Raj" concept introduced in India's administration from the late 1940s, with its licensing requirements and accompanying red tape covering most economic activities, such as development, exports, and investments. It did not take long for civil servants to work out that the extensive system of licensing could provide the means to supplement their income, while politicians began to see the licenses as a commodity that could be bartered for favors. India's economy really took off only when much of this cobweb of regulation was swept aside. So, a simplification of cumbersome work methods and governmental procedures might be an additional essential element in good governance, as well as eliminating the opportunities for bribery.

The complex nature of the currently proposed international environmental aid schemes – which involve massive bureaucratic requirements from both donors and the recipient governments – makes them an easy target for corruption. Consequently, the need to decentralize forest governance from the direct control of national government by devolving powers to local statutory bodies, with safeguards in place for the nature-based economic activities and cultural safeguards for the indigenous and rural communities, becomes even more urgent. This will ensure that international financial support could be directly offered to these statutory bodies without any administrative bottlenecks from national governments. In turn, this opens up the opportunity for international donors to directly monitor areas such as sustainable forest management and certification, management of protected areas, community management of forest and fauna resources, capacity building, training and research at a local level. Another advantage to this new framework is that it could save the national governments from embarrassment if global schemes to halt deforestation fail. And the verification procedures will be less intrusive than the ones currently being demanded by Western governments, since this proposed verification framework will concentrate more on the agencies that manage the distribution of cash, rather than the entire performance of governments in developing countries. The outcome could be the creation of "corruption-free oases" in what otherwise will remain nations with seemingly intractable corruption challenges.

Citizen-led movements are another important facet in dealing with corruption. In countries that are heavily centralized, where the government dominates most economic and social activities, it is easy enough to ignore civic monitoring groups. But when national institutions are inadequate, an empowered civil society could step up to the plate and help to condemn errant officials, as recently happened in Tunisia, a North African state whose people rebelled against an administration which was, by African standards, quite efficient, but also deeply corrupt. Clearly, the bedrock of a successful civil society is a well-informed and well-connected citizenry. And poor infrastructure, as well as the lack of education opportunities, remains a major hindrance for civil society activism in many underdeveloped countries. However, the robustness of citizenry could be improved through technology such as the use of networking tools. For example, microblogs (Twitter-like blogs) already give citizens the means by which to access and disseminate accurate information and to rally ordinary people to stand up against any improper decision made in the public sector that benefits a particular group or a person who made the payment. As microblogging services can be accessed on mobile phones, information spreads much faster and on a larger scale than on more "traditional" media. To a large extent, this is already happening in China and many other Asian countries, and also in the Middle East and Eastern Europe, areas where corruption has been rife and governance standards leave much to be desired. Thus, microblogs could potentially intensify the pressure on corruption-prone wildlife and forestry departments, prodding the authorities into action. This was the goal of DeforLeaks, an online platform "for any concerned individual anywhere in the world to report on deforestation events and/or their impacts on wildlife and humans" (www.deforleaks.org). In addition, with people spreading news through decentralized networks, it is easy for information to be disseminated quickly without identifying the original whistle-blower.

Unfortunately, in developing countries with authoritarian governments, censorship remains strict and environmental activists could be black-listed and have their microblogging services shut down. For these countries, therefore, access to knowledge through other means will remain vital, and technological advances might not be the first instrument in improving governance. Instead, civil society organizations could play a vital role by empowering ordinary citizens to take action against unlawful practices. It is wrong to assume that such bodies do not exist, or that civic initiative is lacking: Anti-Corruption Cameroon – a law-enforcement non-governmental organization – is proof that, even in poor countries, people are ready to mobilize members of community forest groups and lead them in confronting local forestry department officials who extort bribes from them.

Turning back to the broader international effort dimension, a better understanding of how the big-spending nations influence the poor but resource-rich countries remains essential for knowing how inter-state initiatives can play a role in improving governance. Illegal cross-border commerce often enjoys the complicity of government officials, from all sides. For example, China has created a special economic zone near the border with Laos; it would be naïve to think that economic cooperation is the only motive in establishing this zone between countries where corruption remains endemic. Within this special zone, body parts of endangered and protected wildlife from Laos are traded openly across the border, in violation of the Convention on International Trade in Endangered Species to which both Laos and China are signatories (Ghosh, 2010a). This is clearly a case of poor governance amplified and made worse by the encouragement of two countries where the institutional rules of procedures and decision-making processes of administrative bodies in implementing government policies simply flout the law. As a result, Laos' wildlife is bearing the brunt of illegal wildlife trade – an "unprecedented plunder" of biodiversity driven by new cash-rich Chinese companies (Ghosh, 2010b).

This type of cross-border crime was a contributor to the US$10–20 billion revenue generated from the illegal wildlife commerce, according to the United Nations Congress on Crime Prevention and Criminal Justice (Wilkins, 2010). It now seems obvious that anti-corruption agencies in developing countries cannot work alone. In today's globalized economy, fraudulent activities that affect more than one country frequently use cutting-edge technology. Developing countries that lack the capacity to create oversight institutions should join various international and regional initiatives – akin to the Asian Development Bank/Organisation for Economic Co-operation and Development (ADB/OECD) Anti-Corruption Initiative – that promote dialogue and cooperation. Exchanging knowledge and ideas on anti-corruption is a key step toward establishing a global norm on governance and accountability.

The potential for conservation gains when national governance improves is illustrated by the Indian Environment Ministry's recent law enforcement efforts. The constructions of dams, power and steel plants have all been delayed in order to balance the needs of economic development with the preservation of the environment. India's Environment Ministry's assertive actions remain the largest counterbalance to that country's big business interests; major industrial development projects were subjected to an assessment of their environmental impact, something that is becoming more common in countries with emerging economies. Similarly, in Brazil, the largest enforcement action ever launched against illegal logging has enabled that country to reduce forest loss across the Amazon basin by three-quarters, from 2.7 million hectares (ha) in 2004 to 650,000 ha in 2009 (Nepstad, 2009). Such shake-ups in both India's and Brazil's environmental management should serve as an inspiration for other countries. And they are a reminder that, despite the political and economic problems facing developing nations, substantial progress can be made.

Poor governance is a persistent issue in the developing world and policy makers should admit that a great deal can be done to stop undermining environmental equity. The much-needed improvement requires stronger regulation as well as constant scrutiny from a citizenry that takes more ownership of its environment. The latter process may take a generation to come about. Nevertheless, as both large, well-organized rackets that involve the highest echelons of governments combine with small-scale corruption of local bureaucrats to destroy the respect and trust that characterizes good governance, the need for developing countries to shift course has never been greater. Analysts often assumed that the population in developing countries is apathetic and cowed into submission. But, as recent events in North Africa and the Middle East indicate, this is not the case. The people are no longer apathetic, and they do organize themselves, even outside traditional politics. Governments that ignore the sheer scale of corruption in their countries are ultimately devoured by it.

REFERENCES

Bradshaw, C. J. A., Giam, X. L. and Sodhi, N. S. (2010) Evaluating the relative environmental impact of countries. *PLoS One*, **5**, e10440.

Clynes, T. (2010) Confronting corruption. *Conservation Magazine*, **11**, 4.

Conservation International (2007) *Biodiversity hotspots*. http://www.biodiversityhotspots.org/Pages/default.aspx (accessed March 19, 2013).

Craigie, I. D., Baillie, J. E. M., Balmford, A., Carbone, C., Collen, B., Green, R. E. and Hutton, J. M. (2010) Large mammal population declines in Africa protected areas. *Biological Conservation*, **143**, 2221–2228.

Ghosh, N. (2010a) Little China in Laos. *Straits Times*. 2 October, p. C6.

Ghosh, N. (2010b) Bearing the brunt of illegal wildlife trade. *Straits Times*. 2 October, p. C6.

IUCN-UNEP (International Union for Conservation of Nature and United Nations Environment Programme (2010) The World Database on Protected Areas Annual Release. IUCN-WCMC (World Conservation Monitoring Centre) www.wdpa.org (accessed March 19, 2013).

Kaufmann, D., Kraay, A. and Mastruzzi, M. (2010) *Worldwide Governance Indicators*. The World Bank. http://info.worldbank.org/governance/wgi/index.asp (accessed March 19, 2013).

Nepstad, D., Soares-Filho, B. S., Merry, F., Lima, A., Moutinho, P., Carter, J., Bowman, M., Cattaneo, A., Rodrigues, H., Schwartzman, S., McGrath, D. G., Stickler, C. M., Lubowski, R., Piris-Cabezas, P., Rivero, S., Alencar, A., Almeida, O. and Stella, O. (2009) The end of deforestation in the Brazilian Amazon. *Science*, **326**, 1350–1351.

Peh, K. S.-H. (2008) Cameroon's lessons in conservation for sub-Saharan Africa. *BioScience*, **58**, 678–679.

Peh, K.S.-H. (2009) The environment and corruption in developing countries. *RUSI Newsbrief*, **29**, 24–26.

Peh, K. S.-H. and Drori, O. (2010) Fighting corruption to save the environment. *Ambio*, **39**, 336–339.

Transparency International (2012) *Corruption Perceptions Index 2012.* http://www.transparency.org/research/cpi/overview (accessed 27 March 2013).

Wilkins, C. (2010) Central Africa: four-nation "sting" operation busts wildlife smuggling ring. *The Observer.* 17 December, p. 17.

CHAPTER 26

KNOWLEDGE, INSTITUTIONS, AND HUMAN RESOURCES FOR CONSERVATION OF BIODIVERSITY

Kamaljit S. Bawa

Department of Biology, University of Massachusetts, Boston, MA, USA;
Sustainability Science Program, Harvard University, Cambridge, MA, USA;
Ashoka Trust for Research in Ecology and the Environment, Bangalore, India

INTRODUCTION

I am glad to have this opportunity to present a personal perspective on constraints to conservation in India. Conservation constraints often are context and place specific, and in this chapter I focus on examples from India. Of course we need generalizations, but, as my colleagues and I have argued elsewhere, general principles must emerge from local case studies (Bawa, Seidler and Raven, 2004). Indeed, India, a biologically and culturally complex subcontinent, can hardly be considered a "local" focus, but my perspectives are an amalgam of experiences at specific locations throughout the country. India has certain unique and contradictory features: great biocultural diversity combined with high population density; enormous human resources together with chronic and acute development needs; diverse knowledge-based organizations and expertise in information technologies combined with persistently high rates of illiteracy; democratic institutions within a fractious social fabric; and commitment of the state to decentralized governance competing with long-standing traditions of centralized

authority. All of these, as I will argue, are relevant to conservation not only in India but also elsewhere in the tropics.

To put the conservation issues in perspective, India is a mega-diversity country containing parts or wholes of four "global hotpots." In total, India harbors as much as 8–10% of global biodiversity (Bawa, 2010). Land cover is monitored every two years by a government agency, the Forest Survey of India (FSI), which places the reports of its surveys on its website. The FSI estimates that forest cover in India is currently increasing (FSI, 2005; FSI, 2009), apparently due to forest plantations, though the area under *native and dense natural forest* is decreasing (Puyravaud, Davidar and Laurance, 2010). These trends hint at ongoing processes of forest degradation coexisting with an increase in *moderately dense* and *open* forest cover (Figure 26.1).

On the other hand, estimates of land cover change by researchers *un*affiliated with FSI indicate relatively high rates of deforestation (Puyravaud, Davidar and Laurance, 2010, and references therein). Information about changes in other habitats – terrestrial, freshwater, and marine – is lacking, but it is widely believed

Conservation Biology: Voices from the Tropics, First Edition. Navjot S. Sodhi, Luke Gibson, and Peter H. Raven.
© 2013 John Wiley & Sons, Ltd. Published 2013 by John Wiley & Sons, Ltd.

Figure 26.1 Forest–agriculture interface in the Eastern Himalayas: limited knowledge and institutions for sustaining multifunctional landscapes.
Photo courtesy of © Kamal Bawa

that the biodiversity decline in such habitats is substantial.

One of India's major conservation challenges is clearly the pressure on biodiversity from its more than 1.26 billion people, of whom some 35% live below the poverty line (as defined by the government of India according to a complex formula incorporating 13 criteria for rural households, 7 for urban; 76% living on less than US$2 per day, Population Reference Bureau, 2011). The aspirations of (some of) these people for a better life are currently symbolized by economic growth rates that are among the highest in the world. However, current models of economic development also place acute stress on natural habitats, not only within the country but in neighboring countries as well (Bawa et al., 2010). Demographic and economic pressures are exacerbated by chronic deficits of usable knowledge, functional institutions, and good governance. Here I suggest ways to mitigate these pressures. Over the next 13 years, 200 million people are projected to be added to the population, a huge rate of growth (Population Reference Bureau, 2012), and a very large number of people to accommodate.

USABLE KNOWLEDGE

"Usable knowledge" might be defined for the realms of biodiversity science and sustainability science as any form of interpreted information that helps to bridge the gaps between theory and practice, or between science and policy. Cash et al. (2003) highlight some characteristics of usable knowledge: relevancy, credibility, and saliency. They also comment that our stocks of such knowledge for sustainability science are meager. More than 10 years ago, Ehrenfeld (2000) pointed out the shortage of usable knowledge in conservation science, and urged conservation biologists to reflect on the future of conservation biology. During the ensuing 10 years, conservation science has indeed become a good deal more trans-disciplinary and less parochial. Nevertheless, our ability to undertake action – or even "action research" – and to transform policy in the tropics remains painfully limited.

What constrains the production of useful knowledge, and why is there a deficit of usable knowledge despite the tremendous growth in the production of knowledge in biodiversity science in recent decades? One answer, argued by Ehrenfeld (2000), is that much of the research enterprise is not clearly oriented towards the production of usable knowledge. This remains too true for both natural and social sciences. Another answer, ironically, is that the more information available in a particular field, the greater the level of uncertainty, due to the proliferation of perspectives on any particular issue (Kropp and Wagner, 2010). Uncertainty is conducive neither to good policymaking nor to action on the ground.

For biodiversity science, as for other environmental sciences, a principal limitation in the production of usable knowledge may be the nature of the production process itself. The production of usable knowledge is often – quite correctly – motivated by the desire to link knowledge with policy and decision making. However, the concerns of policy makers may thus become paramount in determining research agendas. This can be a good thing when policy makers' interests coincide or overlap with those of communities, but in India, as in many other countries, this cannot be taken for granted.

In the Western democracies, (at least some) academic work has been able to exert considerable influence on policy. Academic work and "think-tanks" are recognized by (at least some) policy makers and government agencies as potential resources for knowledge and information. Moreover, the public – through its elected representatives and other institutions – has opportunities to express its views. Once formulated, policies are generally implemented. Although there are significant exceptions to these generalizations, many of them highly contested, it is clear that knowledge production is (at least) normatively linked with policy formulation.

However – especially in developing countries – policy makers may also be isolated from realities on the ground, and from civil society. In some cases, individual policy makers may not subscribe to certain policies. Take, for example, the Recognition of Forest Rights Act (RFRA) of 2006, enacted in India to correct historic injustices to forest dwellers (Government of India, 2006). Throughout the nineteenth and early twentieth centuries, forest dwellers' lands were regularly expropriated by the state and incorporated into state forests without settlement of community rights. Not until the RFRA of 2006 did the state formally recognize certain rights over land as well as over forest products. Nevertheless, disagreement continues among policy makers at many levels about the interpretation and application of the rules governing the Act (FRA Committee, 2010). As a result, there are serious shortfalls in the implementation of the Act (Bawa *et al.*, 2010).

Thus we have two dilemmas. One is the continuing deficit of usable knowledge; the other is the fact that even the limited amounts of such knowledge available are not put to full use because of ineffective governance for policy implementation. The characteristics of usable knowledge and its sources (epistemology) (Dilling and Lemos, 2011) have so far received more attention than the reasons behind the shortage of such knowledge (etiology) and the avenues for its application to pressing societal needs (pragmatics). In biodiversity science, one constraint to the production and application of usable knowledge may be the absence of an important stakeholder group: people.

Biodiversity conservation is ineluctably linked with specific places and people. Maintenance of biodiversity is dependent on social, cultural, and economic factors prevalent at particular places. People's social and political histories (not limited to their history of resource use) are important determinants of the types of knowledge they need and will be able to use. Communities have significant stores of knowledge about the structure and functioning of their local systems, as well as about the drivers of changes to them. Given the amount of new knowledge we need rather quickly, the involvement of society in knowledge production will be critical in addressing contemporary challenges.

Yet local people are rarely consulted either about the types of knowledge they need or about the role they and their accumulated knowledge can play in augmenting collective knowledge. Involvement of local constituents – communities, local policy makers, and elected representatives – from the planning stages of biodiversity management programs can partly overcome the isolation of policy makers while promoting wider acceptance and effectiveness of policies. Unfortunately, knowledge is often generated far from specific places, often continents away (Bawa, Seidler and Raven, 2004). More importantly, it is often generated in an exclusive manner, keeping out important stakeholders and other knowledge systems (the "creed of expertise," Adams, 2004). Even when included, other knowledge systems may not be equally valued. Locally embedded knowledge or experiences have thus rarely had an impact on setting the agenda for usable knowledge generation.

But making these assertions doesn't obviate the need for systematic ways of merging or integrating local knowledge with scientific observation. One might say that the modern environmental sciences are almost entirely about devising novel ways of extending the observational powers of the senses in space and time, in order to incorporate trends and effects at large and small scales. Such observations may often appear to contradict local, community, or traditional knowledge; we need to rapidly develop systematic processes for understanding and reconciling these apparent contradictions. Such processes are largely lacking at the moment.

Climate change typifies the difficulties in linking research with policy and in marshaling public opinion toward specific actions (Sarewitz and Pielke, 2007; Dilling and Lemos, 2011). Perhaps symptomatically, the foregoing sentence hints at an underlying ambivalence about public opinion that is typical of the science/policy divide. If public opinion needs to be marshaled in a particular direction, who decides on the direction and who does the marshaling? Jasanoff (2010) notes that climate change science has to overcome a problem of acute disconnect between *globalized* knowledge and *universal* action on the one hand, and *localized* human culture and experience on the other. A related point is made by Hulme (2010) when he cautions against the dangers of reducing a large, complex, multi-scalar climate problem to the single parameter of global mean temperature, in isolation from specific places and cultures.

Progress in achieving conservation goals is thus contingent upon unraveling some of the interactions between nature and society. Societal involvement –

particularly the engagement of people living in and around the ecosystems we wish to sustainably manage or protect – can increase the pace at which usable knowledge is generated, as long as there are systems in place for vetting and integrating the knowledge so generated. It could also slow the process down considerably, since in many cases competing interests are at stake. Much depends on the ways in which people are integrated into research programs.

Recent contributions (e.g., Sutherland *et al.*, 2009) setting the knowledge generation agenda for biodiversity science give long lists of critical questions, many of them quite basic. In the interest of generating knowledge that is useful for the management of numerous local biodiversity crises, it might be useful to ask fewer fundamental questions but pay more attention to the manner in which questions are addressed – and to the pathways over which the answers we get can be translated into conservation action.

The idea of involving society closely in the generation of scientific knowledge that is relevant to society's concerns is not novel. Participatory approaches in the field of public health are common (Seshadri, 2003; Pullin and Knight, 2009). Regarding biodiversity science, Ehrenfeld (2000) mentioned that conservation biologists could learn from the health sciences for addressing urgent problems. Agricultural researchers, too, have long advocated participatory approaches (Witcombe *et al.*, 1996; Altieri, 2002; Virk *et al.*, 2003; Johnson, Lilja and Ashby, 2003). In the field of the environment, research on water issues has usefully engaged local communities (Sengupta *et al.*, 2003). Even in biodiversity science, participatory decision making for managing parks has recently emerged as a research issue (Sordoni *et al.*, 2010). However, engagement of the public in all these cases has mostly been in pursuit of specific actions, rather than for transforming the way we produce usable knowledge.

INSTITUTIONS

Increase in usable knowledge will require institutions – both formal and informal – that can bring together the perspectives of researchers, policy makers, and society. For formal institutions, such as think-tanks and major research centers, the blending of knowledge and policy work is a relatively commonplace goal, but such centers are rare in the tropics. Furthermore, such institutions have limited incentives, mandates, or

opportunities for integrating local, site-specific perspectives and knowledge for addressing conservation issues.

The Ashoka Trust for Research in Ecology and the Environment (ATREE), a knowledge-generating institution in India (www.atree.org), is experimenting with an unusual approach. At several of its research sites, the organization has set up community conservation centers to facilitate two-way flow of knowledge between policy-oriented researchers and local resources. It is too early to cite concrete results from this approach, but it has the potential to make a major difference to our deficit of usable knowledge.

ATREE is not unique in its attempt to harness usable knowledge. Other centers such as Earth University in Costa Rica and Africa Conservation Center, to name two, are developing innovative models of integrating knowledge and making it more useful. However, given the challenges, the number of such centers remains low. International agricultural research centers under the Consultative Group on International Agricultural Research (CGIAR) do fill a gap to some extent, but few of these are focused on biodiversity conservation. The magnitude of the problem warrants more innovative indigenous centers, and more support for existing centers (Bawa, Balachander and Raven, 2008).

Informal institutions – community-based groups, resource-user groups, self-help groups, and other village-level bodies – have a critical role to play in advancing biodiversity science in partnership with formal institutions. Such informal institutions tend to be weak in most tropical countries with centralized state systems of governance. Democratization of usable-knowledge production will require the strengthening of informal institutions and of connections among various types of institutions, both formal and informal. Although social scientists have recognized the importance of informal institutions in natural resource management (Dietz, Ostrom and Stern, 2003), their links with formal knowledge-generating institutions remain largely unexplored.

HUMAN RESOURCES

"Usable knowledge' integrates concepts and tools from different disciplines, integrates knowledge systems, and meets pressing conservation challenges while advancing the frontiers of science. This kind of knowledge requires different types of institutions as well as human resources. Only strong institutions of the type described earlier can produce the type of human resources needed. Developing and strengthening appropriate human capacities is a major challenge for both national and international agencies, and for their increasingly outdated, ineffective, and inappropriate programs.

International foundations, donors, and national governments must pay more attention to the development of independent indigenous knowledge-generating institutions, and through them the development of human resources to meet contemporary challenges at all levels. However, national governments in many tropical countries remain wary of local civil society organizations, often creating a vacuum most easily filled by international non-government organizations. These tend to establish local units that often perpetuate the agendas and approaches of a globalized model developed far away (Bawa, Seidler and Raven, 2004). Perhaps surprisingly – despite decades of massive investments, and despite a few bracing exceptions – the record of international conservation organizations and donors in building appropriate local indigenous institutions and the associated human resources remains less than stellar.

EPILOGUE

In conclusion, it appears from my perspective that the biodiversity management crises presently facing developing nations including India require at least three fundamentally new approaches to the generation of knowledge that has some hope of usefully informing policy. First, research programs must become more sensitive to the needs of recipient and user groups, whether communities or management agencies. An example of a successful application of this principle in India has been the development of more reliable tiger censusing methods. However, the results of these censuses have only intermittently been able to affect the development and implementation of tiger-related policy, and that brings us to a second point: we need renewed attention to the interfaces of environmental science and environmental policy, to systematize the pathways over which knowledge can usefully enter the policy arena. Third, the institutions and human resources capable of enacting these approaches to research and knowledge generation need to be cultivated and supported on the ground. As regards the challenges of biodiversity conservation – and perhaps equally in related research areas such as climate change – this means developing indigenous institutions with the mission and the wherewithal to bridge chronic gaps between generalized global science and place-specific localized observation and knowledge.

ACKNOWLEDGMENTS

I thank Reinmar Seidler, with whom I have freely exchanged ideas and information over the least 15 years, for providing suggestions for improvement of the manuscript. Bill Clark stimulated my interest in usable knowledge. Pashupati Chaudhary and Uttam Shrestha have continued to act as sounding boards for my ideas. Sharad Lele prevents me from going astray in fields about which I know very little. This work was completed during my tenure as a visiting scholar in the Sustainability Science Program at Harvard University.

REFERENCES

Adams, D. (2004) Usable knowledge in public policy. *Australian Journal of Public Administration*, **63**, 29–42.

Altieri, M. (2002) Agroecology: the science of natural resource management for poor farmers in marginal environments. *Agriculture, Ecosystems, and Environment*, **93**, 1–24.

Bawa, K. S. (2010) Cataloguing life in India: the taxonomic imperative. *Current Science*, **98**, 151–153.

Bawa, K. S., Balachander, G. and Raven, P. (2008) A case for new institutions. *Science* (editorial), **319**, 136.

Bawa, K., Seidler, S. R. and Raven, P. H. (2004) Reconciling conservation paradigms. *Conservation Biology*, **18**, 859–860.

Bawa, K. S., Koh, L. P., Lee, T. M., Liu, J., Ramakrishnan, P. S., Yu, D. W., Zhang, Y. and Raven, P. H. (2010) China, India, and the environment. *Science*, **327**, 1457–1459.

Cash, D. W., Clark, W. C. Alcock, F., Dickson, N. M., Eckley, N., Guston, D., Jäger, J. and Mitchell. R. (2003) Knowledge systems for sustainable development. *Proceedings of the National Academy of Sciences of the USA*, **100**, 8086–8091.

Dietz, T., Ostrom, E. and Stern, P. C. (2003) The struggle to govern the commons. *Science*, **302**, 1908–1912.

Dilling, L. and Lemos, M. C. (2011) Creating usable science: opportunities and constraints for climate knowledge use and their implications for science policy. *Global Environmental Change*, **21**, 680–689.

Ehrenfeld, J. (2000) Industrial ecology: paradigm shift or normal science. *American Behavioral Scientist*, **44**, 229–244.

FRA Committee (2010) Report of the National Committee on Forest Rights Act, Dec. 2010; also compare Alternative

Summary and Conclusions; both available at https://sites.google.com/site/fracommittee/file-cabinet (accessed March 19, 2013).

FSI (2005) *State of Forest Report 2005*. Ministry of Environment and Forests, Government. of India, Dehra Dun, India.

FSI (2009) *State of Forest Report 2009*. Ministry of Environment and Forests, Government of India, Dehra Dun, India.

Hulme, M. (2010) Cosmopolitan climates: hybridity, foresight and meaning. *Theory, Culture and Society*, **27**, 267–276.

Jasanoff, S. (2010) A new climate for society. *Theory, Culture and Society*, **27**, 233–253.

Johnson, N. L., Lilja, N. and Ashby, J. A. (2003) Measuring the impact of user participation in agricultural and natural resource management research. *Agricultural Systems*, **78**, 287–306.

Kropp, C. and Wagner, J. (2010) Knowledge on stage: scientific policy advice. *Science, Technology and Human Values*, **35**, 812–838.

Population Reference Bureau (2011) *2011 World Population Data Sheet*. http://www.prb.org/pdf11/2011population-data-sheet_eng.pdf (accessed March 19, 2013).

Population Reference Bureau (2012) *World Population Data Sheet 2012*. http://www.prb.org/Publications/Datasheets/2012/world-population-data-sheet.aspx (accessed March 19, 2013).

Pullin, A. S. and Knight, T. M. (2009) Doing more good than harm: building an evidence-base for conservation and environmental management. *Biological Conservation*, **142**, 931–934.

Puyravaud, J., Davidar, P. and Laurance, W. F. (2010) Cryptic destruction of India's native forests. *Conservation Letters*, **3**, 390–394.

Government of India (2006) *The Scheduled Tribes and Other Traditional Forest Dwellers (Recognition of Forest Rights) Act 2006*. Ministry of Tribal Affairs, New Delhi, India. http://tribal.nic.in/writereaddata/mainlinkFile/File1033.pdf (accessed March 19, 2013).

Sarewitz, D. and Pielke Jr., R. A. (2007) The neglected heart of science policy: reconciling supply of and demand for science. *Environmental Science and Policy*, **10**, 6–10.

Sengupta, S., Mitra, K., Saigal, S., Gupta, R., Tiwari, S. and Peters, N. (2003) *Developing Markets for Watershed Protection Services and Improved Livelihoods in India*. Winrock International India, New Delhi, and International Institute for Environment and Development, London.

Seshadri, S. R. (2003) Constraints to scaling-up health programmes: a comparative study of two Indian states. *Journal of International Development*, **15**, 101–104.

Sordoni A., Briot, J., Alvarez, I., Vasconcelos, E., Irving, M. A., Melo, G. and Sebba-Patto, V. (2010) Design of a participatory decision making agent architecture based on argumentation and influence function: application to a serious game about biodiversity conservation. Proceedings of the Symposium on COGnitive systems with Interactive Sensors (COGIS'09), SEE, Paris, France.

Sutherland, W. J., Adams, W. M. Aronson R. B., Aveling, R., Blackburn, T. M., Broad, S., Ceballos, G., Côté, I. M., Cowling, R. M., Da Fonseca, G. A. B., Dinerstein, E., Ferraro, P. J., Fleishman, E., Gascon, C., Hunter Jr., M., Hutton, J., Kareiva, P., Kuria, A., MacDonald, D. W., MacKinnon, K., Madgwick, F. J., Mascia, M. B., McNeely, J., Milner-Gulland, E. J., Moon, S., Morley, C. G., Nelson, S., Osborn, D., Pai, M., Parsons, E. C. M., Peck, L. S., Possingham, H., Prior, S. V., Pullin, A. S., Rands, M. R. W., Ranganathan, J., Redford, K. H., Rodriguez, J. P., Seymour, F., Sobel, J., Sodhi, N. S., Stott, A., Vance-Borland, K., Watkinson, A. R. (2009). One hundred questions of importance to the conservation of global biological diversity. *Conservation Biology*, **23**, 557–567.

Virk, D. S., Singh, D. N., Kumar, R., Prasad, S. C., Gangwar, J. S. and Witcombe, J. R. (2003) Collaborative and consultative participatory plant breeding of rice for the rainfed uplands of Eastern India. *Euphytica*, **132**, 95–108.

Witcombe, J. R., Joshi, A., Joshi, K. D. and Sthapit, B. R. (1996) Farmer participatory crop improvement. *Experimental Agriculture*, **32**, 445–460.

CHAPTER 27

PEOPLE, PLANTS AND POLLINATORS: UNITING CONSERVATION, FOOD SECURITY, AND SUSTAINABLE AGRICULTURE IN EAST AFRICA

Dino J. Martins

Insect Committee of Nature Kenya, The East Africa Natural History Society, Nairobi, Kenya;
Museum of Comparative Zoology, Harvard University, Cambridge, MA, USA;
Turkana Basin Institute, Stony Brook University, Kenya

SUMMARY

A major challenge facing both social and development issues across the world today is that of meeting not just food security, but nutritional security for a rapidly growing human population. This is the reality against which many decisions around conservation will need to be made. An overlooked "free" ecosystem service, pollination, is essential to both crops and most terrestrial habitats with some 80% of angiosperms dependent on wild pollinators. In developing regions like Eastern Africa, pollinators are primarily wild insects that travel between farms and natural habitat, and are extremely vulnerable to habitat loss and destruction. Pollinators make a direct connection between wild species and food security. Conserving pollinators provides a platform for increasing nutritional security and connecting small-scale agriculture with conservation and management of natural habitats. This chapter highlights some case studies showing the links between wild pollinators, natural habitat, and rural farmers.

INTRODUCTION

Why we need pollinators

A major issue for all of the world's developing countries today, and especially in Africa, is that of increased and sustainable food production. With the world's population now at 7.1 billion, this is an issue of great concern to governments, one that also affects all aspects of conservation and environmental issues (UNEP, 2004). Hunger, primarily due to poverty, is one of the greatest challenges to conservation in the developing world. Despite the advances in technology and intensification of food production worldwide, over 1 billion people are malnourished (Global Hunger Index, 2010). Many of them are rural subsistence farmers living within or adjacent to high-biodiversity habitats. The future of many species, especially in tropical forest patches in Africa, lies in their hands. With the population of sub-Saharan Africa projected to grow from its present level of 902 million to 2.1 billion over the next 38 years,

Conservation Biology: Voices from the Tropics, First Edition. Navjot S. Sodhi, Luke Gibson, and Peter H. Raven.
© 2013 John Wiley & Sons, Ltd. Published 2013 by John Wiley & Sons, Ltd.

adding 55,000 people net per day, the problem becomes all the more urgent (Population Reference Bureau, 2012).

"One in three bites of food can be attributed to a pollinator" is a relationship widely touted by scientists, policy makers, and ecologists today (Buchmann and Nabhan, 1996). An overlooked ecosystem service, pollination is essential to humanity. Many crops depend on wild insect pollination to set fruit/produce seed. In total, around two-thirds of crop species in cultivation today rely on pollinators (Kevan, Clark and Thomas, 1990; Free, 1993; Kremen, Williams and Thorp, 2002).

In Africa, pollinators are primarily wild insects that travel between farms and natural habitats; they are extremely vulnerable to habitat loss and destruction. Pollinators intimately link wild species with basic human livelihoods (Roubik, 1995, 2002; Allen-Wardell *et al.*, 1998; Martins 2004, 2007a, 2009). Saving pollinators justifies the conservation of small species-rich habitats such as forest patches and contributes to food security and to the livelihoods of rural communities living close to nature, oftentimes alongside critically endangered species (Klein *et al.*, 2007).

The relationships between insects and flowers are at once ancient, beautifully intricate, and correspondingly fragile. Nowhere is this diversity of species and partnerships more evident than in and around tropical forests. Millions of species, most of them as yet undescribed and many of them rare or threatened, call these forests their home. In East Africa, many of these forests today persist only as isolated patches on mountains, following increasingly arid conditions on the African continent as well as recent forest clearing (Bennun and Njoroge, 1999). Some of these forest patches have been recognized as global biodiversity hotspots, such as the Eastern Arc Mountains of Kenya and Tanzania, which include the last stronghold of charismatic species like the African Violet (*Saintpaulia teitensis*), a plant that forms the basis for a huge global ornamental plant/horticultural industry today.

East Africa's precious forests today are islands in a sea of rural subsistence agriculture. For example, only 1.7% of Kenya's land remains forested (National Atlas of Kenya, 1970). For millennia people have farmed around these forests. However, increasing pressures are being placed on forests due to changing farming systems, charcoal production, and, more recently, population growth and climate change. Conservationists are pressed with finding an urgent, tangible way of linking human livelihoods with forest conservation. This is critical for the survival of thousands of species and also for the farmers who benefit from these forests.

One direct yet under-appreciated link between forest fragments and rural agriculture is through pollinators. Many crops in East Africa require pollination, and most pollinators in East Africa that visit crops are wild insects. Saving pollinators justifies the conservation of small species-rich habitats, such as forest patches, because of their contribution to food security and rural livelihoods of the communities around them.

However, given negative attitudes towards insects, many rural farmers and other members of communities adjacent to forests fail to appreciate the importance of the ones that pollinate their crops. Farmers often kill insects that visit flowers through sheer ignorance and the lack of information about the necessity of pollination. Poor habitat management and direct destruction through clearing of natural vegetation, burning of trees for charcoal, and overgrazing also impact pollinators and the resources on which they depend (Murren, 2002; Pauw, 2007). This directly reduces crop yields and threatens food security.

The challenge today is to produce food more intensively as well as more sustainably to feed a rapidly growing, often already-needy human population. Pollinators contribute an essential link in the interface of rural subsistence farming and biodiversity conservation (Kremen *et al.*, 2004; Ricketts, 2004) (Figure 27.1). Many of the crops pollinated by wild insects are rich in nutrients, including essential vitamins and micronutrients. Additionally, many of the pollinator-dependent

Figure 27.1 Carpenter bee (*Xylocopa* sp.) pollinating pigeonpea (*Cajanus cajan*), Nyegesi Bay, Mwanza, Tanzania.

crops are also varieties, land-races, and species that are unique to local farming systems, better adapted to local climatic conditions and interwoven with the cultures and traditions of the peoples that cultivate them. In this sense, pollinators are essential not just for food security, but equally important for nutritional security and sustainable development.

CONSERVATION ISSUES AROUND SUBSISTENCE AGRICULTURE

Pollinators can provide a very real and tangible connection between natural habitats, including national parks and forest reserves, and daily human life. Rural subsistence farmers in many parts of East Africa are located at the edges of forest patches and other biodiversity-rich habitats. It is in these rural communities that the daily battle against extinction is fought. It is here that a crucial constituency and awareness about pollinators and sustainable human livelihoods need to be built.

Conservation work on pollinators involves moving from discovery and documentation of links between agriculture and nature to building on these remarkable links between subsistence farms and wild pollinators to form a strong grassroots movement for the protection of habitats, better farming practices, and the restoration of pollination services, especially where they have suffered or been negatively impacted by environmental degradation. The interactions between flowers and their pollinators are also intricate and interesting in their own right (Martins and Johnson, 2007; Martins, 2010). Sharing the sense of wonder they inspire with future generations is also a powerful way of engaging with young people (through schools) who live around the forest patches to better manage and sustain both farming and habitat. This can also help build pollination research, which is largely lacking in the region (Martins, 2004; Rodgers, Balkwill and Gemmill, 2004).

The following are the major threats faced by pollinators and by much of biodiversity in general in rural East Africa, identified through working with rural subsistence farmers:

(a) Clearing of natural vegetation at forest edges and hedgerows, draining wetlands, and clearing within forest for farming.

(b) Clearing of forest, woodland, and trees for charcoal production reduces wild pollinator forage.

(c) Unsustainable overharvesting of wild bee colonies: overharvesting is the destruction of wild honeybee and other social bee colonies in a given geographical area. The colonies are destroyed for harvesting their honey (breaking open nests, use of fire, and burning of old trees with nesting cavities). In the past, the wild harvesting of colonies was patchy and limited by low human populations. In East Africa today, it typically involves the excavation and complete destruction of a wild colony from a natural cavity in a natural area. In a short time, this depletes the number of wild colonies in an area and can even lead to local extinctions of colonies of stingless bees (D. Martins, Insect Committee of Nature Kenya, Kenya unpublished results). This trend applies to both wild honeybees (*Apis mellifera scutellata*) and stingless bees (*Meliponula* spp., *Plebeina* spp., *Hypotrigina* spp.). Wild harvesting destroys the wild genetic variation and colonies that serve as a source for filling domestic/managed hives.

(d) Poisoning by pesticides used both in agriculture as well as for poisoning wildlife.

(e) Destruction of nesting sites through clearing, sand and murram harvesting, and other practices.

(f) Direct killing of flower-visiting insects through misidentification/ignorance about the importance of pollinators.

(g) Introduced diseases and the spread of parasites, such as the varroa mite (*Varroa destructor*) are posing a growing threat to wild and managed East African honeybee populations.

TILLING COMMON GROUND: POLLINATORS AS A FLAGSHIP FOR MITIGATING CONSERVATION CHALLENGES

Involving farmers in conservation through pollinator monitoring

Involving farmers in monitoring and awareness building can serve to develop an "army for conservation" on the very frontline of habitat destruction. In East Africa, these are the edges of subsistence farms surrounding forest patches. Through working with hundreds of rural farmers, some key practices and solutions that conservationists can engage in to promote pollinator conservation have been found to include the following:

(a) The identification of crops that are more susceptible to a "pollinator deficit" and clearly document their pollinators and pollination needs. This is best done through working with smaller core groups of farmers or women's groups.

(b) The development of a monitoring program, especially for groups of pollinators that are potentially more vulnerable, such as in the case of bees and pesticides, or seasonal, as in the case of pollinators such as hawkmoths. This can be done both with farmers and local schools, through wildlife clubs, young farmers' associations and the training of parataxonomists (local people trained about taxonomy) from the community. As farmers and other members of the local community spend more time closely watching their flowering crops, they become more and more aware of the diversity of wild species that directly contribute to their livelihoods.

(c) Identifying the key areas of habitat that pollinators use and highlight the necessity of protecting these as part of a wider process to maintain ecosystem services. This should ultimately be undertaken using GIS (geographic information system) over time to develop a local-scale landscape map of both pollinator distribution and the spatial patterning of their resource needs. This serves to highlight the links between natural habitats, wild species, and sustainable agricultural productivity.

(d) Promoting best practices for pollinator conservation. This is best done by identifying individual farmers who are currently protecting and encouraging pollinators through either their own innovations or traditional practices, and then inviting and supporting these farmers to present their work in their own words to a wider audience.

(e) Drawing clear links between protection of forest patches and agricultural productivity. This is best done by using the data collected by farmers and schools through presentations at farmer field days. A key component of this is the discussion of improved yields from better management of the local landscape and pollinators.

Wild pollinators provide pollination services for African crops

Studies in East Africa at several different sites have revealed that wild insect pollinators are an important component of subsistence agriculture by providing pol-

lination services to farms. Following are three examples from a wide range of habitats showing the links between pollinators and crops as well as the bigger picture of biodiversity conservation. Understanding the links between wild biodiversity and agricultural productivity can help bridge the gap between agricultural development and biodiversity conservation.

Hawkmoth pollination of papaya (*Carica papaya*) in Kenya

The Kerio Valley is an extension of the Great Rift Valley system. The valley runs from north to south and is connected to the Great Rift Valley with Lakes Baringo and Bogoria lying east of the valley. Altitude ranges from 900 m in the Lake Baringo and Lake Kamnarok basins to over 2300 m in the adjacent highlands.

Rainfall is highly variable and patchy, and strongly linked to elevation. The mean annual temperature ranges from 14°C in the highlands to 24°C in the semi-arid lowlands. Rainfall in the lower areas of the valley is rarely more than 400 mm annually; long-term cultivation therefore requires irrigation from permanent and semi-permanent streams originating in the highlands. The Kerio Vallye is inhabited by the Keiyo, Elgevo, Pokot, Tugen, and Marakwet of the Nilotic peoples who practice a mixture of pastoralism and small-scale subsistence agriculture using irrigation. Population density in the valley is about 15 people per square kilometer, which is very low in contrast to the highlands nearby where maximum densities are over 150 people per square kilometer. Crops grown include maize, beans, and millets seasonally as well as small orchards of mango, citrus, banana, and papaya, which are gathered locally for sale and transport to towns and cities.

Papaya is dioecious and requires pollination to set fruit. Hawkmoths were found to be the most abundant and reliable visitors to both "female" pistillate and "male" staminate papaya flowers, and accounted for over 95% of visits and the xenogamous pollination of papaya flowers. Among hawkmoths, *Hippotion celerio*, *Nephele comma*, and *Agrius convolvuli* were the main pollinator species (Martins, 2007c, Martins and Johnson, 2009).

Hawkmoth abundance and visitation rates declined sharply with increasing distance of natural habitat patches from the crop. Fruit set was also lower at sites with high levels of disturbance or poor agricultural practices. Natural habitat containing larval food-plants and other nectar resources for hawkmoths supports adjacent cultivated papaya with pollination services.

Farmers in the Kerio Valley showed a variety of perceptions on the question of pollination of papaya. Some farmers, who had been educated, or participated in farmer-training sessions at a local farmer's development society, were aware of the necessity of pollination. Other farmers were not aware of pollination as a process.

The main practices being carried out at this site in relation to maintaining papaya pollination are:
(a) protection and encouragement of alternative nectar sources for pollinators
(b) planting of larval food plants in hedgerows
(c) protection of trees in surrounding bush and woodland areas.

Wild bee pollination of Pigeonpea (*Cajanus cajan*) in Mwanza, Tanzania

The greater Mwanza area by Lake Victoria is inhabited by the Sukuma people, who practice subsistence agriculture. Along the lakeshore a number of communities rely on fishing for a livelihood, but the vast majority of people are small-scale mixed subsistence farmers. This area is also one of the most densely populated parts of Africa: the typical farm is a small family plot tilled by hand by members of the family.

The climate is characterized by distinct wet and dry seasons, with a peak of rainfall in March–May and again in November. The most important food crops grown in the Mwanza region are maize, paddy rice, sorghum/millet, cassava, sweet potatoes, bananas, beans, and pigeonpeas.

Because of poverty among the majority of peasant farmers, subsistence crops are also sold as cash crops whenever possible. Such dual-purpose crops include pigeonpeas and beans. This indicates that pigeonpea is both an important source of nutrition and income for small-scale farmers in Mwanza District.

Pigeonpea pollinators were found at this site to be several different native bee species. Of these, the large carpenter bees (*Xylocopa* spp.) and leaf-cutter bees (Megachilidae) were the most reliable and effective visitors. These bees landed on the flowers and moved constantly between different individual plants and between fields of the crop (Martins, 2007b).

Pollen loads on the bees were abundant and the ability of these bees to disperse swiftly across large areas suggests that they are able to cross-pollinate the pigeonpea flowers regularly and efficiently. Both kinds of bees were seen using areas of the farmstead and homestead to nest in. Nests were found both in wooden structures and earthen walls. Many other wildflowers growing in the vicinity were also visited by both the carpenter and leaf-cutter bees.

Wild pollinators overlap between crops and endangered species: African violets and subsistence farms in the Taita Hills

One of the world's most iconic and endangered plants, the African violet (*Saintpaulia teitensis*), occurs in the Eastern Arc global biodiversity hotspot. Forest fragments on the Taita Hills in southeastern Kenya hold the only known wild populations of this plant. The Taita Hills in Kenya are considered to be the northernmost outlier of the Eastern Arc biodiversity hotspot.

The hills are surrounded by flat plains covered in dry, arid bushland. This bushland extends up onto the flanks of the hills giving way to intensive small-scale agriculture. The farms are typically small and grow a wide range of subsistence crops alongside fruit trees, climbers such as passion fruit and tubers, primarily cassava and yams. As a result of many decades of intensive farming, the forests of the Taita Hills survive only as tiny patches on the hilltops and in general on slopes and areas inaccessible for cultivation.

Pollinators of the endangered African violet were found to be exclusively wild bee species of the genus *Amegilla*. The same individual bees visiting the African violets were also observed to feed from other forest-floor plant species and on crops in adjacent small-scale mixed agriculture farms. Dispersal of pollen between sub-populations of the plant and movement by bees across forest fragments and adjacent cultivation are key to the survival of both the plants and their pollinators (Martins, 2008).

The sharing of a pollinator between agricultural crops and an endangered flower (Martins, 2009) provides a clear link between biodiversity and rural livelihoods of small-scale farmers living adjacent to the forest, and highlights the importance of sustainable farming practices to preserve both rural livelihoods and intact native biodiversity.

CONCLUSION

Saving small patches of habitat and increasing the understanding and appreciation of pollinators represent direct ways of improving food security and alleviating poverty through increased crop yields. It is in the hands of rural farmers that the future of so many habitats sits. Subsistence farmers need to be engaged as partners in conservation. The beauty of pollination is that it really is where a strong and clear link can be drawn between human livelihoods, sustainable agriculture, and protection of natural areas, and the myriad species they contain.

The global die-offs of honeybees have recently shown the dangers of relying on just one species. Once people recognize that something as fundamental as food production is tied to biodiversity, and given the staggering diversity of just bees alone (circa 20,000 described species) (Roubik, 1989; Torchio, 1990; Canto-Aguilar and Parra-Tabla, 2000), there is great potential here to sustainably manage biodiversity in ways that directly reduce poverty and improve human health and nutrition.

ACKNOWLEDGMENTS

I thank the participating farmers in Kenya and Tanzania who have worked with us on pollinator research and conservation. Assistance and useful insights were provided by A. Powys, B. Gemmill (FAO, Food Agricultural Organization), C. Eardley (ARC-PPRI, Agricultural Research Council-Plant Protection Research Unit), C. Ngarachu, C. Odhiambo, N. Pierce (Harvard), S. Miller (Smithsonian Institution), W. Kinuthia, and the scientists and staff of the National Museums of Kenya and Nature Kenya. Research and conservation was supported by a fellowship to D. J. Martins from the Smithsonian Institution Women's Committee and the Mpala Research Centre, the FAO-GEF (Global Environment Facility) Global Pollinator Project, the Ashford Fellowship at Harvard-GSAS (Graduate School of Arts and Sciences), the Whitley Fund for Nature, the National Geographic Society and the Turkana Basin Institute – Stony Brook University.

REFERENCES

Allen-Wardell, G., Bernhardt, P., Bitner, R., Burquez, A., Buchmann, S., Cane, J., Cox, P. A., Dalton, V., Feinsinger, P., Ingram, M., Inouye, D., Jones, C. E., Kennedy, K., Kevan, P., Koopowitz, H., Medellin, R., Medellin-Morales, S., Nabhan, P., Pavlik, B., Tepedino, V., Torchio, P. and Walker, S. (1998) The potential consequences of pollinator declines on the conservation of biodiversity and stability of crop yields. *Conservation Biology*, **12**, 8–17.

Bennun, L. and Njoroge, P. (1999) *Important Bird Areas in Kenya*. East Africa Natural History Society, Nairobi, Kenya.

Buchmann, S. L. and Nabhan, G. P. (1996) *The Forgotten Pollinators*. Island Press, Washington, D.C.

Canto-Aguilar, A. and Parra-Tabla, V. (2000) Importance of conserving alternative pollinators: assessing the pollination efficiency of the squash bee, *Peponapis limitaris* in *Cucurbita moschata* (Cucurbitaceae). *Journal of Insect Conservation*, **4**, 203–210.

Global Hunger Index (2010) *IFPRI* (International Food Policy Research Institute) *Concern Worldwide and Welthungerhilfe*. Bonn, Washington D.C., Dublin.

Free, J. B. (1993) *Insect Pollination of Crops*. Academic Press, London.

Kevan, P. G., Clark, E. A. and Thomas, V. G. (1990) Insect pollinators and sustainable agriculture. *American Journal of Alternative Agriculture*, **5**, 12–22.

Klein, A. M., Vaissiere, B. E., Cane, J. H., Steffan-Dewenter, I., Cunningham, S. A., Kremen, C. and Tscharntke, T. (2007) Importance of pollinators in changing landscapes for world crops. *Proceedings of the Royal Society, B* **274**, 303–313.

Kremen, C., Williams, N. M. and Thorp, R. W. (2002) Crop pollination from native bees at risk from agricultural intensification. *Proceedings of the National Academies of Sciences of the United States of America*, **99**, 16812–16816.

Kremen, C., Williams, N. M., Bugg, R. L., Fay, J. P. and Thorp, R. W. (2004) The area requirements of an ecosystem service: crop pollination by native bee communities in California. *Ecology Letters*, **7**, 1109–1119.

Martins, D. J. (2004) Foraging patterns of managed honeybees and wild bee species in an arid African environment: ecology, biodiversity and competition. *International Journal of Tropical Insect Science*, **24**, 105–115.

Martins, D. J. (2007a) *Papaya in Kenya, In Crops, Browse and Pollinators in Africa: An Initial Stock-Taking*. Food and Agricultural Organization of the United Nations. Rome, Italy.

Martins, D. J. (2007b) *Pollination Profile of Pigeonpea (Cajanus cajan) in Mwanza, Tanzania. Global Survey of Good Pollination Practices*. Food and Agricultural Organization of the United Nations. Rome, Italy.

Martins, D. J. (2007c) *Papaya Pollination in Kenya, In Crops, Browse and Pollinators in Africa: An Initial Stock-Taking*. Food and Agricultural Organization of the United Nations. Rome, Italy.

Martins, D. J. (2008) Pollination observations of the African violet in the Taita Hills, Kenya. *Journal of East African Natural History*, **97**, 33–42.

Martins, D. J. (2009) Pollination and facultative ant-association in the African leopard orchid *Ansellia africana*. *Journal of East African Natural History*, **98**, 67–77.

Martins, D. J. (2010) Pollination and seed dispersal in the endangered succulent *Euphorbia brevitorta*. *Journal of East African Natural History*, **99**, 9–17.

Martins, D.J., and Johnson, S. D. (2007) Hawkmoth pollination of aerangoid orchids in Kenya with special reference to
nectar gradients in the floral spurs. *American Journal of
Botany*, **94**, 650–659.

Martins, D. J. and Johnson, S. D. (2009) Distance and quality
of natural vegetation influence hawkmoth pollination of
cultivated papaya. *International Journal of Tropical Insect
Science*, **29**, 114–123.

Murren, C.J. (2002) Effects of habitat fragmentation on pollination: pollinators, pollinia viability and reproductive
success. *Journal of Ecology*, **90**, 100–107.

National Atlas of Kenya (1970) Drawn and printed by Survey
of Kenya. Kenya Government, Nairobi, Kenya.

Pauw, A. (2007) Collapse of a pollination web in small conservation areas. *Ecology*, **88**, 1759–1769.

Population Reference Bureau (2012) *World Population Data
Sheet 2012*. http://www.prb.org/pdf12/2012-population-
data-sheet_eng.pdf (accessed March 25, 2012.

Ricketts, T. (2004) Tropical forest fragments enhance pollinator activity in nearby coffee crops. *Conservation Biology*, **18**,
1262–1271.

Rodgers, J. G., Balkwill, K. and Gemmill, B. (2004) African
pollination studies: where are the gaps? *International Journal
of Tropical Insect Science*, **24**, 5–28.

Roubik, D. W. (1989) *Ecology and Natural History of Tropical
Bees*. Cambridge University Press, Cambridge.

Roubik, D. W. (1995) *Pollination of Cultivated Plants in the
Tropics*. Food and agriculture organization of the United
Nations, Rome, Italy.

Roubik, D. W. (2002) The value of bees to the coffee harvest.
Nature, **417**, 708.

Torchio, P. F. (1990) Diversification of pollination strategies
for U.S. crops. *Environmental Entomology*, **19**, 1649–1656.

UNEP (2004) *Global Environment Outlook – GEO4, Environment for Development*. United Nations Environment Programme, Progress Press, Valletta, Malta.

CHAPTER 28

BALANCING SOCIETIES' PRIORITIES: A SCIENCE-BASED APPROACH TO SUSTAINABLE DEVELOPMENT IN THE TROPICS

Lian Pin Koh

Department of Environmental Sciences, Zurich, Switzerland

SUMMARY

Rising global demands for water, food, and energy are intensifying land-use conflicts, contributing to greenhouse gas emissions, and worsening threats to natural ecosystems and wildlife. It is imperative that we develop ways to balance our desires for increasing levels of consumption with environmental protection, particularly in the tropics, where population growth has been most rapid, the people are poorest, and biodiversity is richest and yet most threatened. Environmental and social scientists can help by developing decision-support tools that will enable decision makers to evaluate the consequences and trade-offs of pursuing alternative development options in relation to the biophysical, socioeconomic, and technical constraints and considerations within individual societies and landscapes. Ultimately, scientists play a crucial role in helping decision makers achieve a careful balance of the various priorities within each society, which is needed to ensure sustainable development for the benefit of both humans and the environment.

INTRODUCTION

Global human population is projected to grow hugely over the next four decades, adding 2–2.5 billion people to the current 7.1 billion (United Nations, 2008). Virtually all of the additional people will be added to the developing world, and the great majority of them will most likely be poor and hungry. The ensuing demands for water, food, and energy will intensify land-use conflicts, contribute to greenhouse gas emissions, and exacerbate threats to natural ecosystems and wildlife (Tilman *et al.*, 2001; Evans, 2009; Royal Society, 2009; Godfray *et al.*, 2010). Some scientists even warn of an impending "perfect storm" of multiple global crises (Sample, 2009). Therefore, it is imperative that we develop ways to balance our growing consumptive needs with environmental protection. This is particularly exigent in the developing tropics where population growth has been the greatest and will continue to be so for the foreseeable future, the people are poorest, and biodiversity is richest and yet most threatened globally. In this chapter, I highlight the key challenges facing land-use decision makers in the tropics, and discuss some recent research that seeks to address these challenges.

FOOD AND BIOFUEL DEMANDS

By 2050, global human population is projected to have grown from the current 7.1 billion to 9.5 billion people. These people will require more food, and better quality food containing higher proportions of meat and dairy products, which will require more land, water, and energy to produce (Tilman *et al.*, 2001; Evans, 2009; Royal Society, 2009; Godfray *et al.*, 2010). The future demand for food likely will be driven by emerging economies such as China and India, both of which have already passed the productive capacity of their own lands in the demand for natural resources, as well as in developing tropical countries in Latin America, Africa, and Southeast Asia, with the population of sub-Saharan Africa, for example, projected to double from its present 1 billion to 2 billion people during the next 40 years.

Juxtaposed alongside rising food requirements is an increase in demand for biofuels, particularly from developed nations seeking to reduce their dependency on fossil fuels and/or to meet carbon emissions reduction targets (European Parliament and Council, 2009). The production of first-generation biofuel feedstocks, such as oil palm, soy, and sugarcane, to meet this demand will require additional land to be turned over to agriculture (Koh, 2007a; Koh and Ghazoul, 2008; Koh and Wilcove, 2009; Koh and Ghazoul, 2010a).

A case in point is Indonesia: between 2000 and 2010, oil palm-cultivated land in the country expanded from 2 million to 10 million hectares (ha) (FAO, 2011). Indonesia is currently the world's largest producer (23 million tons per year) and exporter (18 million tons per year) of palm oil (USDA-FAS, 2011). Annual export of Indonesian palm oil to India, Pakistan, Bangladesh, and China, mainly for food, amounts to over 30% of Indonesia's total export volume; another ~20% is imported by the Netherlands, Germany, and Singapore, primarily for industrial uses, including biofuel production (FAO, 2011).

Global consumption and supply of palm oil have increased over the past decade; palm-oil prices have also been rising steadily (USDA-FAS, 2011). These trends possibly signify that global palm-oil demand continues to outpace supply, and/or demand remains high in spite of rising production costs. Either way, rising palm-oil prices, coupled with stable and expanding global vegetable-oil markets, reflect strong financial incentives for future oil-palm expansion. This likely will exacerbate land-use conflicts and environmental impacts, particularly in the tropics, which are increasingly being targeted for food and biofuel production (Koh, 2007a; Koh and Ghazoul, 2008; Koh and Wilcove, 2009; Koh and Ghazoul, 2010a).

FOREST CONSERVATION AND REDD+

Tropical deforestation is both a major source of greenhouse gas emissions and a leading cause of biodiversity loss (Danielsen *et al.*, 2008; Murdiyarso, Hergoualc'h and Verchot, 2010; Koh *et al.*, 2011). Financial mechanisms, such as the UN's REDD (Reducing Emissions from Deforestation and Forest Degradation) programme, have been proposed to compensate landowners and land users for the value of carbon stored in forests that would otherwise be lost through deforestation (Myers, 2007; Nepstad *et al.*, 2007; Miles and Kapos, 2008). Additionally, carbon credits generated from REDD could be a potential source of income to support biodiversity-conservation and poverty-alleviation efforts in developing countries (Laurance, 2007; Butler, Koh and Ghazoul, 2009; Venter *et al.*, 2009). A second iteration of the REDD scheme (known as REDD+), which additionally allows for reforestation and sustainable forestry, is increasingly seen as the way forward for carbon payment schemes in recognition of the economic significance of forestry to many tropical nations.

Indonesia is the world's third largest emitter of greenhouse gases and second largest emitter of greenhouse gases from land-use change and forestry, a sector that contributes to almost 30% of global emissions (WRI, 2011). The country also contains high concentrations of endemic species that are threatened with extinction (Sodhi *et al.*, 2004; Koh, 2007b; Sodhi *et al.*, 2010). Indonesia is, therefore, a global priority for actions to both reduce emissions and conserve biodiversity (Strassburg *et al.*, 2010).

In May 2010, Norway pledged US$1 billion in support of a nationwide REDD+ programme in Indonesia (Norway-Indonesia, 2010). While this agreement holds great promise by virtue of its substantial financial contribution towards reducing Indonesia's emissions and deforestation rates, the actual implementation of REDD+ still faces several social and economic challenges.

The economic impediments to REDD+ stem not only from the forfeiture of revenues derived directly from business-as-usual development options (i.e., direct opportunity costs), such as oil-palm expansion (Butler, Koh and Ghazoul, 2009; Venter *et al.*, 2009), but also from the loss of downstream economic benefits (i.e., indirect costs), which are associated with processing, marketing, and service industries (Ghazoul *et al.*, 2010a; Ghazoul, Koh and Butler, 2010b). Furthermore, there is a need to fully recognize and compensate for the foregone livelihood opportunities of forest-based communities in REDD-affected areas (i.e., less tangible costs).

The success of the Norway–Indonesia partnership, and emissions reduction efforts in general, will depend on its ability to achieve appropriate compensation that encompasses the full range of economic, social, and political implications of avoiding deforestation. These must be realistically and transparently assessed lest REDD+ mechanisms be rendered theoretically attractive but practically unrealistic. Furthermore, the success (or failure) of the Norway-Indonesia partnership could be a bellwether of outcomes in other forest-rich nations seeking REDD+ funds, such as the Democratic Republic of Congo and Colombia (www.un-redd.org). Therefore, the Indonesian case study likely will have profound implications for climate-change mitigation and conservation across the developing tropics.

WEATHERING THE STORM

To make informed land-use and development decisions in the context of multiple competing societal priorities (as described earlier), natural resource managers and policy makers in the developing tropics need access to science-based and knowledge-based support from the scientific community. Environmental and social scientists can help by developing decision-support tools that will enable decision makers to evaluate the consequences and trade-offs of pursuing alternative development options in relation to the biophysical, socioeconomic, and technical constraints and considerations within individual societies and landscapes. I provide three recent examples of such tools.

Species-area models

Species-area models have often been applied by ecologists to predict the magnitude of species extinctions resulting from forest loss (Brooks, Pimm and Collar, 1997; Brook, Sodhi and Ng, 2003). This application assumes that the original forest is replaced by a landscape matrix that is completely inhospitable to the taxa considered, despite evidence demonstrating variable effects of the matrix on populations within forest fragments (Gascon *et al.*, 1999; Ricketts, 2001; Antongiovanni and Metzger, 2005). To take account of such matrix effects, Koh and Ghazoul (2010b) recently developed a matrix-calibrated species-area model. This model accounts both for changes in original forest cover and for taxon-specific responses to each component of a heterogeneous landscape matrix. The improved model performed 13 times better (based on information-theoretic evidence ratios) than the conventional species-area model in predicting biodiversity losses across 20 tropical biodiversity hotspots. In follow-up research, Koh *et al.* (2010) further developed this model to also account for the deleterious effects of forest edges on species extinction risks.

By accounting for differences in land-use composition between landscape scenarios, these improved species-area models allow attribution of any changes in biodiversity to specific land-use transitions (Koh and Ghazoul, 2010a; Koh *et al.*, 2011). Furthermore, these theoretical models have important implications for conservation practitioners and land-use decision makers in that they allow assessments of the degree by which species extinction risks could be reduced, or biodiversity enhanced, through improvements in landscape characteristics.

Spatially explicit scenario analyses

Indonesia, the world's largest oil-palm producer, plans to double its production by 2020, with unclear environmental and land-use implications. In a recent study, Koh and Ghazoul (2010a) developed a computer model that simulated oil-palm expansion in Indonesia to evaluate the outcomes of alternative development scenarios. These scenarios separately prioritize oil-palm production, food production, forest preservation, or carbon conservation. The simulation suggests that every single-priority scenario had substantial trade-offs associated with other priorities. For example, under a pro-forest preservation scenario, oil-palm development would have little impact on Indonesia's primary forests; however, by diverting oil-palm expansion to lands suitable for rice cultivation, this strategy compromises the country's future food security. Instead, the optimal solution was a hybrid approach wherein expansion targeted degraded and agricultural lands that were most productive for oil palm, least suitable for food cultivation, and contained the lowest carbon stocks. This approach avoided any loss in forest or biodiversity, and substantially ameliorated the impacts of expansion on carbon stocks and food security. These results suggest that the environmental and land-use trade-offs associated with oil-palm expansion could be reduced through the implementation of a properly planned and spatially explicit development strategy.

Apps for conservation

By 2015 there will be 1.2 billion additional internet users in the rapidly developing "BRICI" nations of Brazil, Russia, India, China, and Indonesia. A large proportion of these people will be accessing the internet through mobile communications devices, such as Apple's iPhones and other "smart" phones, rather than personal computers (Aguiar *et al.*, 2010). Scientists can leverage on these technological and societal trends by developing web-based and/or mobile applications to maximize the accessibility of their research for the public. Such applications serve at least two important purposes: (1) to educate the public on environmental science and natural resource management; and (2) to enable land-use decision makers to evaluate the environmental and socioeconomic consequences and trade-offs of their development strategies.

An example of such an application is the Land Use Calculator (www.landusecalculator.com). Primarily targeted at land-use decision makers in the tropics, this tool allows end-users to evaluate the implications and trade-offs of alternative development scenarios by simultaneously accounting for agricultural production, economic development, carbon conservation, and biodiversity protection. The Land Use Calculator requires users to specify a few key environmental and socioeconomic parameters describing a landscape scenario, and the tool determines the implications of that scenario in terms of biodiversity, carbon stocks, greenhouse gas emissions, financial returns of the land, and even employment opportunities.

OUTLOOK

The challenges of meeting the demands of a growing human population require novel solutions. Paradoxically, some of the solutions being developed to address these challenges end up exacerbating the problem. For example, rising demands for renewable fuels and carbon credits are driven by solutions designed to mitigate global climate change. At the local level, however, agricultural expansion (to produce biofuels) and forest protection (to generate carbon credits) are in fact competing land uses. Keeping this in mind, environmental scientists and land-use decision makers should consider the widest range of societal priorities possible and, more crucially, recognize the trade-offs among these often-competing environmental and socioeconomic priorities.

It is admittedly not always easy for biologists of different sub-disciplines to work together, and even more difficult for transdisciplinary collaborations among biologists, economists, and political and social scientists. However, the challenges facing humans in the coming decades are cross-scale, transcultural, and transdisciplinary, and so too must be the strategies being developed to meet these challenges. Ultimately, scientists play a crucial role in helping decision makers achieve a careful balance of the various priorities within each society, which is needed to ensure sustainable development for the benefit of both humans and the environment.

REFERENCES

Aguiar, M., Boutenko, V., Michael, D., Rastogi, V., Subramanian, A. and Zhou, Y. (2010) *The Internet's New Billion: Digital Consumers in Brazil, Russia, India, China, and Indonesia.* http://www.bcg.com/documents/file58645.pdf (accessed March 19, 2013).

Antongiovanni, M. and Metzger, J. P. (2005) Influence of matrix habitats on the occurrence of insectivorous bird species in Amazonian forest fragments. *Biological Conservation*, **122**, 441–451.

Brook, B. W., Sodhi, N. S. and Ng, P. K. L. (2003) Catastrophic extinctions follow deforestation in Singapore. *Nature*, **424**, 420–423.

Brooks, T. M., Pimm, S. L. and Collar, N. J. (1997) Deforestation predicts the number of threatened birds in insular Southeast Asia. *Conservation Biology*, **11**, 382–394.

Butler, R. A., Koh, L. P. and Ghazoul, J. (2009) REDD in the red: palm oil could undermine carbon payment schemes. *Conservation Letters*, **2**, 67–73.

Danielsen, F., Beukema, H., Burgess, N. D., Parish, F., Brühl, C. A., Donald, P. F., Murdiyarso, D., Phalan, B., Reijnders, L., Struebig, M. and Fitzherbert, E. B. (2008) Biofuel plantations on forested lands: double jeopardy for biodiversity and climate. *Conservation Biology*, **23**, 348–358.

European Parliament and Council (2009) Directive 2009/28/EC on the Promotion of the Use of Energy from Renewable Sources. http://eur-lex.europa.eu/LexUriServ/LexUriServ.do?uri=OJ:L:2009:140:0016:0062:EN:PDF (accessed March 19, 2013).

Evans, A. (2009) *The Feeding of the Nine Billion: Global Food Security for the 21st Century*. http://www.chathamhouse.org/publications/papers/view/108957 (accessed March 19, 2013).

FAO (2011) FAOSTAT. http://faostat.fao.org (accessed March 19, 2013).

Gascon, C., Lovejoy, T. E., Bierregaard Jr, R. O., Malcolm, J. R., Stouffer, P. C., Vasconcelos, H. L., Laurance, W. F., Zimmerman, B., Tocher, M. and Borges, S. (1999) Matrix habitat and species richness in tropical forest remnants. *Biological Conservation*, **91**, 223–229.

Ghazoul, J., Koh, L. P. and Butler, R. A. (2010b) A REDD light for wildlife friendly farming. *Conservation Biology*, **24**, 644–645.

Ghazoul, J., Butler, R. A., Mateo-Vega, J. and Koh, L. P. (2010a) REDD: a reckoning of environment and development implications. *Trends in Ecology and Evolution*, **25**, 396–402.

Godfray, H. C. J., Beddington, J. R., Crute, I. R., Haddad, L., Lawrence, D., Muir, J. F., Pretty, J., Robinson, S., Thomas, S. M. and Toulmin, C. (2010) Food security: the challenge of feeding 9 billion people. *Science*, **327**, 812–818.

Koh, L. P. (2007a) Potential habitat and biodiversity losses from intensified biodiesel feedstock production. *Conservation Biology*, **21**, 1373–1375.

Koh, L. P. (2007b) Impacts of land use change on South-east Asian forest butterflies: a review. *Journal of Applied Ecology*, **44**, 703–713.

Koh, L. P. and Ghazoul, J. (2008) Biofuels, biodiversity, and people: understanding the conflicts and finding opportunities. *Biological Conservation*, **141**, 2450–2460.

Koh, L. P. and Ghazoul, J. (2010a) Spatially explicit scenario analysis for reconciling agricultural expansion, forest protection, and carbon conservation in Indonesia. *Proceedings of the National Academy of Sciences of the United States of America*, **107**, 11140–11144.

Koh, L. P. and Ghazoul, J. (2010b) A matrix-calibrated species-area model for predicting biodiversity losses due to land use change. *Conservation Biology*, **24**, 994–1001.

Koh, L. P. and Wilcove, D. S. (2009) Oil palm: disinformation enables deforestation. *Trends in Ecology and Evolution*, **24**, 67–68.

Koh, L. P., Lee, T. M., Sodhi, N. S. and Ghazoul, J. (2010) An overhaul of the species-area approach for predicting biodiversity loss: incorporating matrix and edge effects. *Journal of Applied Ecology*, **47**, 1063–1070.

Koh, L. P., Miettinen, J., Liew, S. C. and Ghazoul, J. (2011) Remotely sensed evidence of tropical peatland conversion to oil palm. *Proceedings of the National Academy of Sciences of the United States of America*, **108**, 5127–5132.

Laurance, W. F. (2007) A new initiative to use carbon trading for tropical forest conservation. *Biotropica*, **39**, 20–24.

Miles, L. and Kapos, V. (2008) Reducing greenhouse gas emissions from deforestation and forest degradation: global land-use implications. *Science*, **320**, 1454–1455.

Murdiyarso, D., Hergoualc'h, K. and Verchot, L. V. (2010) Opportunities for reducing greenhouse gas emissions in tropical peatlands. *Proceedings of the National Academy of Sciences of the United States of America*, **107**, 19655–19660.

Myers, E. C. (2007) Policies to Reduce Emissions from Deforestation and Degradation (REDD) in tropical forests: an examination of the issues facing the incorporation of REDD into market-based climate policies. *Discussion paper*. http://www.rff.org/rff/Documents/RFF-DP-07-50.pdf (accessed March 19, 2013).

Nepstad, D., Soares-Filho, B., Merry, F., Moutinho, P., Rodrigues, H. O., Bowman, M., Schwartzman, S., Almeida, O. and Rivero, S. (2007) The Costs and Benefits of Reducing Carbon Emissions from Deforestation and Forest Degradation in the Brazilian Amazon. http://www.whrc.org/policy/pdf/cop13/WHRC_Amazon_REDD.pdf#search="The Costs and Benefits of Reducing Carbon Emissions from Deforestation and Forest Degradation in the Brazilian Amazon" (accessed March 19, 2013).

Norway–Indonesia (2010) Letter of Intent between the Government of the Kingdom of Norway and the Government of the Republic of Indonesia on "Cooperation on reducing greenhouse gas emissions from deforestation and forest degradation". www.norway.or.id/PageFiles/404362/Letter_of_Intent_Norway_Indonesia_26_May_2010.pdf (accessed 19 March 2013).

Ricketts, T. H. (2001) The matrix matters: effective isolation in fragmented landscapes. *American Naturalist*, **158**, 87–99.

Royal Society (2009) Reaping the Benefits: Science and the Sustainable Intensification of Global Agriculture. http://royalsociety.org/Reapingthebenefits (accessed March 19, 2013).

Sample, I. (2009) World faces "perfect storm" of problems by 2030, chief scientist to warn. *The Guardian*, 18 March. www.guardian.co.uk/science/2009/mar/18/perfect-storm-john-beddington-energy-food-climate (accessed March 19, 2013).

Sodhi, N. S., Koh, L. P., Brook, B. W. and Ng, P. K. L. (2004) Southeast Asian biodiversity: an impending disaster. *Trends in Ecology and Evolution*, **19**, 654–660.

Sodhi, N. S., Posa, M. R. C., Lee, T. M., Bickford, D., Koh, L. P. and Brook, B. W. (2010) The state and conservation of Southeast Asian biodiversity. *Biodiversity and Conservation*, **19**, 317–328.

Strassburg, B. B. N., Kelly, A., Balmford, A., Davies, R. G., Gibbs, H. K., Lovett, A., Miles, L., Orme, C. D. L., Price, J., Turner, R. K. and Rodrigues, A. S. L. (2010) Global congruence of carbon storage and biodiversity in terrestrial ecosystems. *Conservation Letters*, **3**, 98–105.

Tilman, D., Fargione, J., Wolff, B., D'Antonio, C., Dobson, A., Howarth, R., Schindler, D., Schlesinger, W. H., Simberloff, D. and Swackhamer, D. (2001) Forecasting agriculturally driven global environmental change. *Science*, **292**, 281–284.

United Nations (2008) World Population Prospects: The 2008 Revision. http://www.who.int/pmnch/topics/2008_populationstats/en (accessed 29 March 2013).

USDA-FAS (US Department of Agriculture-Foreign Agricultural Service) (2011) *Oilseeds: World Markets and Trade*. Circular Series FOP 1–11 January 2011. https://www.fas.usda.gov/report.asp (accessed 29 March 2013).

Venter, O. E. M., Possingham, H., Dennis, R., Sheil, D., Wich, S., Hovani, L. and Wilson, K. (2009) Carbon payments as a safeguard for threatened tropical mammals. *Conservation Letters*, **2**, 123–129.

WRI (2011) *Climate Analysis Indicators Tool (CAIT) version 8.0*. http://www.wri.org/tools/cait (accessed 29 March 2013).

CHAPTER 29

BIODIVERSITY CONSERVATION PERFORMANCE OF SUSTAINABLE-USE TROPICAL FOREST RESERVES

Carlos A. Peres

Centre for Biodiversity Research, University of East Anglia, Norwich, UK

SUMMARY

Vast tracts of tropical forest have been persistently degraded by anthropogenic habitat modification and overhunting, even if they remain standing in the foreseeable future. To counteract these trends, conservation planners have typically focused on the geographic priorities of regional protected area allocation, rather than on other reserve design criteria, including reserve categories and management restrictions. Here, I argue that sustainable-use tropical forest reserves, coexisting with varying densities of human residents and their extractive activities, play a huge role in capturing representative biotas and key ecosystem services, but present decisive challenges in biodiversity conservation. I review the global emergence of sustainable-use reserves (SURs), and examine the extractive pressure exerted on natural resources by local communities as proxied by human population density, and their impacts on the most harvest-sensitive forest vertebrate fauna. Finally, I discuss a number of wildlife conservation performance issues in relation to tropical forest SURs, and how these can be managed in the best interest of landscape-scale forest biodiversity.

INTRODUCTION

Although tropical forests are home to as much as half of the world's terrestrial biota, the relative merits of different approaches to ensure their long-term survival remain highly contentious, to say the least (Robinson, 2011). Most would agree, however, that both establishing protected areas (PAs) and exercising some form of user restraint on forest resource depletion are among the most effective of all viable conservation measures. Deforestation, forest degradation, and hunting comprise some of the leading drivers of tropical forest biodiversity loss (Laurance and Peres, 2006), and de facto or de jure protection from these threats can be conferred by either enforcement of regulations or physical remoteness. Passive protection in currently inaccessible areas will, however, play an increasingly negligible role as human populations grow and expand with road infrastructure, so that the fate of tropical biodiversity will ultimately be determined by designated norms on local use restrictions and degrees of compliance with those norms.

Large-scale attempts to assess PA conservation performance have largely rested on conventional

Conservation Biology: Voices from the Tropics, First Edition. Navjot S. Sodhi, Luke Gibson, and Peter H. Raven.
© 2013 John Wiley & Sons, Ltd. Published 2013 by John Wiley & Sons, Ltd.

remote-sensing approaches quantifying spatial and/or temporal differences in rates of habitat loss and disturbance. These have concentrated on deforestation, forest fragmentation, and fire incidence, rather than on population or community-level indicators of biotic integrity (Naughton-Treves *et al.*, 2006; Gaston *et al.*, 2008). The broad-scale approaches fail to detect most forms of subcanopy anthropogenic disturbance that directly or indirectly result in biodiversity loss (Peres, Barlow and Laurance, 2006). Moreover, the effects of forest habitat loss and degradation on population extinctions and declines are highly non-linear so that assessing forest cover alone is rarely a robust proxy of the integrity of tropical forest biotas. Vast tracts of apparently intact tropical forests in both protected and unprotected areas are in fact moderately to severely degraded from overhunting (e.g., Peres and Lake, 2003), and those losses almost always lead to long-term consequences for ecosystem dynamics (Wilkie *et al.*, 2011). Important questions, with both local as well as general implications, that are usually not addressed in previous reserve performance analyses thus include whether (1) PAs that may appear to be relatively intact do indeed retain full complements of forest species diversity; and (2) how the extent of cryptic patterns of disturbance scale to human population density (HPD) in both protected and unprotected areas.

Here, I consider the global to regional emergence of SURs, with an emphasis on Amazonia, the world's largest tropical forest region. SURs are subjected to intermediate scales of structural and non-structural human disturbance, so I examine the extractive pressure exerted on natural resources by resident communities as proxied by HPD. In both protected and unprotected areas, I also quantify a response metric to hunting pressure in terms of the relative game biomass extracted from the most harvest-sensitive components of the forest vertebrate fauna. Finally, I discuss a number of conservation performance issues in relation to tropical forest SURs, and how these can be managed in the best interest of forest biodiversity.

SURS WORLDWIDE AND IN AMAZONIA

The number and extent of PAs worldwide have increased nearly 14-fold over the past 50 years, with >120 000 PAs covering ~21 million km² by 2008, corresponding to 12.2% and 6.4% of Earth's terrestrial and marine areas, respectively (WDPA, 2010). These figures are over-optimistic because levels of protection granted to PAs ("on paper" or *de facto*) diverge widely. For example, >86% of the PAs decreed worldwide admit some form of human use (Figure 29.1a), and only <0.01% of the aggregate coverage of 980 marine reserves representing ~19% of the world's coral reefs can be defined as no-take areas subjected to low risks from external threats (Mora *et al.*, 2006).

Reserve management targets are broadly classified by the IUCN, in which categories I–II (and specially Ia) can be defined as strictly protected areas (hereafter, SPAs) and categories III–VI as sustainable-use reserves (hereafter, SURs). SURs are often co-managed by local communities to support local livelihoods and preserve cultural legacies and ecosystem services. Both people- and nature-centric reserve denominations, however, can legally or illegally contain extractive communities operating under varying human population densities and management objectives "on paper." Reserves assigned to categories I–II (13,411 units) represent 38.6% of the global PA coverage, compared to 61.4% in categories III–VI (52,124 units). The dominance in terms of number and total area of SURs is much greater in the Neotropics than in the African and Asian tropics (Chape, Spalding and Jenkins, 2008), particularly considering forest PAs (Schmitt *et al.*, 2009).

SURs in Brazil are defined as "social spaces" *where natural resources can be exploited sustainably to ensure both the usufruct rights of local communities and the perpetuity of all organisms and their ecological processes* (MMA, 2000). Policy enthusiasm for SURs is a relatively recent phenomenon firmly grounded in sociopolitical demands from disenfranchised communities, rather than the interests of native biodiversity conservation per se. Conventional parks in Brazil were far more prevalent before the 1980s, when they were seen as more reliable instruments in the traditional conservation toolbox. Since 1991, however, ~63.1 millions of hectares (Mha) of new SURs have been created across Brazilian Amazonia (or 51.5% of the total conservation acreage decreed in the same period, excluding Indigenous Lands) thanks to the intensification of land struggles, the subsequent political organization of many local communities, and the emergence of the extractive reserve movement (Figure 29.1b, c). As a result, the total number and coverage of SURs and indigenous territories now outweigh those of SPAs by factors of 5.5 and 4.1, respectively.

The total non-private conservation acreage in Brazilian Amazonia including all denominations now repre-

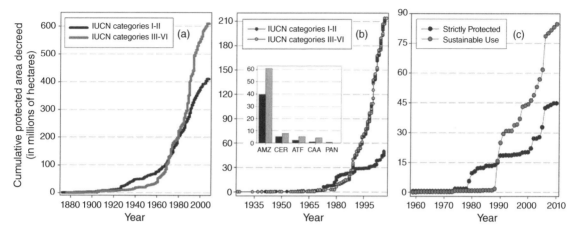

Figure 29.1 Cumulative growth in the total extent of terrestrial protected areas over time for (a) all land masses worldwide; (b) Brazil; and (c) Brazilian Amazonia. Solid lines and circles represent strictly protected reserves (IUCN categories I–II); gray lines and circles represent sustainable-use reserves (IUCN categories III–VI). Data on PA coverage worldwide (a) were derived from the WDPA (2010) database and include all reserves for which an IUCN category was assigned. Data in (b) and (c) are from the Brazilian Ministry of Environment (MMA) and the environmental agencies of all Amazonian states within Brazil. Inset (b) represents the aggregate areas (in millions of ha) of all strictly protected (black bars) and sustainable use reserves (gray bars, excluding indigenous territories) in each of the five major Brazilian biomes (AMZ: Amazonia; CER: Cerrado; ATF: Atlantic Forest; CAA: Caatinga; and PAN: Pantanal).
Source: (a) Data on PA coverage worldwide were derived from the WDPA (2010) database and include all reserves for which an IUCN category was assigned. (b and c) Protected area data obtained from the Brazilian Ministry of Environment (http:// http://www.mma.gov.br/areas-protegidas/sistema-nacional-de-ucs-snuc).

sents ~219.75 Mha encompassing 721 PAs, nearly 44% of the entire ~5 million km^2 region and over a quarter of Brazil's land area (IMAZON, 2011). This accounts for 12.6% of Earth's total terrestrial protected acreage, excluding Antarctica. Of these PAs, ~117.4 Mha are reserves managed by federal and state government agencies, whereas ~108.7 Mha are indigenous territories under the jurisdiction of FUNAI (National Indian Foundation), and 0.97 Mha are *quilombos* (designated Afro-Brazilian communal territories). Considering all types of PAs, this means that 80.4% of the existing PA coverage across Brazilian Amazonia (35% of the region) has been allocated to forest reserves where human residents are empowered to pursue their livelihoods indefinitely. Much of the long-term future of Amazonian forest biodiversity has therefore been trusted to extractive, sustainable development, and indigenous reserves. Yet little is known about the conservation performance of these reserves beyond the resolution of satellite images, and how to best co-manage them in the long-term interest of both local livelihoods and the biotic integrity of forest ecosystems.

HUMAN DENSITIES IN AMAZONIAN RESERVES

Human population density within Amazonian SPAs (0.32 ± 0.78 person/km^2, $N = 37$) is considerably lower than in either sustainable-development (16.8 ± 106.2 person/km^2, $N = 86$) and indigenous reserves (6.4 ± 31.5 person/km^2, $N = 294$). Excluding unoccupied reserves, human density is negatively related to reserve area (slope = -0.524, $R^2 = 0.596$, $N = 398$, Figure 29.2). HPD ranged over five orders of magnitude from fewer than 0.01 person/km^2 in reserves >1 Mha to over 100 persons/km^2 in reserves >00 ha. This relationship is strongly affected by broad reserve categories; indigenous territories have a steeper slope than both extractive reserves and SPAs (ANCOVA, $p < 0.001$) and exhibit a tighter population HPD:area relationship ($R^2 = 0.700$, $N = 294$), presumably because extractive reserve residents have historically received government subsidies (e.g., inflated rubber prices) and local initiatives to create a SUR in the first place often come from heavily settled areas containing more politically articulated

communities. Strictly protected reserves are often uninhabited or accommodate lower human densities (Figure
29.2), but otherwise follow the same HPD:area relationship as other human-occupied reserves (ANCOVA,
$p = 0.649$).

WILDLIFE CONSERVATION
PERFORMANCE OF SURS

Game vertebrate populations sustain the protein subsistence needs of millions of forest dwellers worldwide,
and are indisputably the most valuable non-timber
tropical forest commodity to local livelihoods within
and outside Amazonian reserves (Redford and Robinson, 1987; Peres, 2000). Indeed, managing sustainable access to sources of both terrestrial and aquatic
animal protein is a key challenge in SURs (see Figure
29.3). Fluvial and road networks also render these
reserves widely accessible to hunters (Peres and Lake,
2003), but capitalizing on this physical access depends
on local consumer demand in terms of human population sizes. I therefore indirectly assessed the degree to
which harvest-sensitive game stocks have been depleted
by overhunting across a large pool of otherwise largely
undisturbed protected and unprotected forest sites
across seven Amazonian countries. Building on a pre-

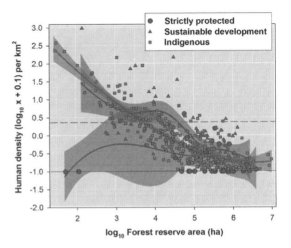

Figure 29.2 Relationships between forest reserve size and
the internal human population density for 417 protected
areas in Brazilian Amazonia. Slope of indigenous territories
($N = 294$, red squares) is significantly steeper than that of
sustainable development reserves ($N = 86$, purple triangles;
ANCOVA, $p < 0.001$), in which local communities have
historically received the highest external subsidies. Strictly
protected reserves ($N = 37$, blue circles) showing the lowest
human densities (solid black line) were effectively
unoccupied (HPD = 0). Curves and shaded areas show loess
smoothers and 95% confidence intervals. Dashed gray line
represent a threshold HPD = 1 person/km^2.

Figure 29.3 Tropical forest SURs play a critical role in regional scale biodiversity conservation, yet managing harvest-
sensitive resource populations within reserves that will become more densely settled is one of the key challenges in PA
implementation this century.

vious study (Jerozolimsky and Peres, 2003), I compiled a database – of 56,065 mammal kills from 79 hunting studies describing prey profiles from 113 settlements – obtained from a range of both published and unpublished sources, including our own studies at eight indigenous and *caboclo* settlements. The sizes and locations of these settlements and the total sampling effort (consumer-days) were known so that the site-specific mean per capita biomass harvest rate (g person^{-1} day^{-1}) could be calculated for each species. I also quantified the landscape-scale human population density associated with each hunting catchment based on the size of each village and neighboring settlements within a 20-km radius. From the 38 mammal species typically encountered in Amazonian harvest profiles, I also estimated the proportion of the overall game biomass extracted from the 10 largest bodied, lowest fecundity (slow-breeding and long-lived) game species, which are associated with population growth rates (λ) lower than 1.3. These preferred, low-λspecies comprise the first-

choice prey items pursued by hunters across the Amazon (Jerozolimsky and Peres, 2003) and the Congo basin (Fa, Ryan and Bell, 2005), but this selectivity pattern breaks down whenever their local populations become depleted, with hunters shifting to smaller bodied, higher fecundity species. Local communities in sparsely settled areas derived a much larger proportion of their game biomass from low-λspecies. On average, these species accounted for >80% of the total prey biomass harvested when human density was <0.03 person/km^2 but only <40% in areas sustaining >1 person/km^2 (Figure 29.4a). Although this does not account for the variation in habitat productivity on different game species, many heavily settled communities had shifted to smaller bodied, high fecundity species, presumably because of the local "depletion envelope" of large-bodied species. In particular, the dietary transition from low-λ to high-λspecies occurred well before human density reached 1 person/km^2, a threshold below which game hunting in tropical forests has been

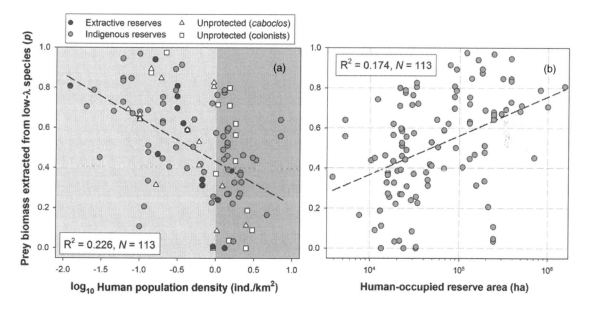

Figure 29.4 Relationships between the proportional biomass of harvest-sensitive (low-λ) game vertebrate species extracted from the aggregate number of prey items sampled at different Amazonian settlements over an entire hunting study and (a) the landscape-scale HPD centered on these settlements. Symbols (a) represent different classes of settlements in both formally protected (solid and gray circles) and unprotected areas (open triangles and squares). The forest reserves sizes that these settlements would occupy (b) were predicted on the basis of the linear relationship between HPD and reserve area for all Amazonian indigenous, extractive, and strictly protected reserves for which human population sizes are known (see Figure 29.2).

assumed to be sustainable (Robinson and Bennett, 2000). This is significant given that Amazonian subsistence hunters are often severely limited in terms of time/energy and ammunition, and regard pursuit of large-bodied species as most cost effective. Furthermore, there were no clear differences in the way these patterns of prey harvest respond to depletion history between protected and unprotected areas, or between different denominations of extractive and indigenous reserves, once differences in HPD had been taken into account.

As a heuristic exercise, I back-predicted the total size of hypothetical human-occupied reserves, on the basis of the general HPD-area relationship (where HPD > 0), for all sites where both HPD and reliable game harvest data were available. In these terms, SURs would have to be larger than 100,000 ha to maintain a sustainable harvest in terms of the landscape-scale source-sink dynamics of harvest-sensitive prey populations and sustain harvest profiles largely dominated by these species (Figure 29.4b). Only hunters from sparse settlements within reserves approaching 1 Mha would be able to focus single-mindedly on preferred, low-fecundity game species.

DISCUSSION

Protected areas of *all* sizes, denominations, and scales of protection will retain varying levels of biological integrity, but can serve critical ecosystem functions only if they can maintain a relatively intact primary habitat cover. However, only sufficiently large strictly protected reserves under largely inviolate conditions are likely to safeguard full complements of biodiversity in the long run. These reserves should be the "crown jewels" of any regional/national PA network, which inescapably will be dominated in numbers and total area by multiple-use reserves admitting varying levels of structural and non-structural human disturbance.

Human-occupied reserves worldwide range widely in their internal and external human densities, and associated pressures. Increasing habitat isolation (DeFries *et al.*, 2005) and burgeoning human densities beyond reserve boundaries (Wittemyer *et al.*, 2008) do not bode well for PA effectiveness in terms of biodiversity retention. Lowland Amazonia is atypical of many tropical forest regions in that HPDs within reserves are usually greater than those outside. Yet the degree to which natural resource extraction can be defined as sustainable is largely governed by HPD within modes

of livelihood. Patterns of faunal extraction at many Amazonian forest sites show clear signs of harvest-sensitive game depletion beyond ~0.1 person/km^2 (Figure 29.4a, b), an HPD threshold 10-fold lower than that widely considered as sustainable (Robinson and Bennett, 2000). In most Amazonian SURs, this will be further aggravated by internal population growth rates of 2.7–4.2% yr^{-1}, doubling consumer demand every 16–27 years even if net migration approximates zero.

Formal or informal community-based land management cover >370 Mha globally, and agroforestry mosaics would likely double or triple this total (Molnar, Scherr and Khare, 2004). However, the explicit mission of millions of ha in SURs of serving both the long-term interests of local communities and forest biodiversity is at best a huge challenge fraught with poorly explored contradictions. Amazonian extractive and indigenous reserves sustaining relatively dense human populations are often severely overhunted and no longer contain viable populations of large-bodied, harvest-sensitive game species (Peres and Palacios, 2007). Prey availability within village catchments largely determines patterns of game selectivity (Jerozolimski and Peres, 2003), and population densities of harvest-sensitive prey are largely a function of both habitat productivity and human densities (C. A. Peres, unpublished data). This is reflected in the markedly lower contribution of these species in prey-harvest profiles from high-HPD village catchments (Figure 29.4a). Game vertebrates represent only a single class of non-timber forest resources overexploited in many high-HPD areas, and overhunting is not the only mechanism of "cryptic" forest degradation (Peres *et al.*, 2006). Yet all reserve denominations across a wide range of human population sizes have so far been largely effective in retaining relatively intact forest cover (Nepstad *et al.*, 2006; Joppa, Loarie and Pimm, 2008; Adeney, Christensen and Pimm, 2009; Soares-Filho *et al.*, 2010). Several SURs in densely settled contexts, however, have already witnessed clear signs of deforestation and severe forest degradation (e.g., Pedlowski *et al.*, 2005). Amazonian SURs had lost 298,500 ha of their total forest area (or ~0.41%) following their creation by December 2009, compared with 108,100 ha (~0.24%) in SPAs (IMAZON, 2011). For example, >6.3% of the 970,570-ha Chico Mendes Extractive Reserve in the state of Acre has been deforested since the reserve creation in 1990, due to an 11-fold expansion in pastures to accommodate more than 10,000 head of cattle (Prado, 2010). This notorious PA was originally set aside to ensure land-tenure rights to ~3,000 families of rubber tappers and Brazil-nut collectors. However, most of these homesteads have gradually shifted from non-timber resource extraction to animal husbandry in the form of small-scale cattle ranching, ironically to partly compensate the animal protein shortfall resulting from depleted wild game populations (Rosas, 2006).

Considering PAs decreed in the Brazilian Amazon by December 2010, 45.1% and 47.3% of the aggregate area of SURs and SPAs, respectively, were under some level of anthropogenic disturbance as detectable from satellite images (IMAZON, 2011). These figures are substantial, but entirely overlook subcanopy disturbance such as hunting (Peres and Lake, 2003). Amazonian and most other tropical forest PAs are severely underfunded and understaffed, so that enforcing reserve regulations is usually prohibitive. Each of the 305 staff assigned to manage 163 state-managed PAs across Brazilian Amazonia (excluding Rondônia) on average has jurisdiction over a combined forest area of 63,520 ha (194 staff) within SPAs and 403,280 ha (111 staff) within SURs (IMAZON, 2011). This institutional capacity scenario may sound abysmal, but the ratio of staff to PA acreage is actually 3.2-fold higher than that 18 years earlier (Peres and Terborgh, 1995), despite the additional PA expansion of ~103.5 Mha over this period.

To be clear, indigenous, *caboclo* and smallholder communities across Amazonia can and do significantly aid several mainstream conservation objectives, including tribal land boundary enforcement against illegal logging and deforestation (Zimmerman *et al.*, 2001), suppression of rampant land speculation in contested landscapes (Campos and Nepstad, 2006), effective protection of communal subsistence fisheries (Almeida *et al.*, 2011), and filling the wilderness power-vacuum that would otherwise be occupied by predatory loggers and miners (Parry *et al.*, 2010). In all of these cases, local residents operating under low-governance conditions intentionally or inadvertently suppress more destructive forces of forest conversion and degradation, often driven by the enterprising sector. It does not follow, however, that complete biological communities will be maintained under the watch of traditional communities. The key question, therefore, is not whether local communities are friends or foe of conservation objectives, but what levels of biological integrity can we hope to realistically maintain under different contexts of land use and consumer population density.

Resolving the perennial parks versus people conflict is well beyond the scope of this chapter. The copious and raucous academic dispute on this topic (e.g., Terborgh and Peres, 2002; West, Igoe and Brockington, 2006) is largely counterproductive and often imparts little policy influence on government agencies managing tropical PAs.

Setting aside >70 Mha of Amazonian SURs has resulted in little spatial redistribution of human residents. In fact, the modus operandi in delineating PA boundaries and assigning management categories in Amazonia is largely based on prior human occupation, rather than biological design criteria. Yet involuntary human displacements from SURs, with or without compensation, are reprehensible, because such displacements are ethically inappropriate, monetarily prohibitive, and politically unfeasible. Rather, conditional agreements based on clear targets co-managed by local communities remain the best available option forward (Peres and Zimmerman, 2001; Ferraz, Marinelli and Lovejoy, 2008). This often requires developing local capacity through training workshops; regulating immigration where internal economic conditions are favorable; setting sustainable-harvest quotas on offtakes; and zoning no-take areas within reserve boundaries. This is both expensive and labor intensive, often deflecting competing investments away from SPAs. Yet all PA denominations are important – to different components of biodiversity with varying sensitivity to overharvesting and forest degradation – in a national PA network. However, many pro-poor conservation approaches (e.g., Kaimowitz and Sheil, 2007) should also target local communities in multifunctional landscapes outside PAs, where we are likely to lose most forest habitat faster (Nepstad *et al.*, 2006; Soares-Filho *et al.*, 2010). PAs alone will not lift millions of tropical forest dwellers out of poverty (Naughton-Treves *et al.*, 2006), nor should they be seen as key policy instruments of income generation. Yet SURs can hugely contribute to the persistence of key ecosystem services and viable populations across the entire range of species life histories (and degrees of sensitivity to disturbance and harvest) if they can be embedded within larger reserve networks that include SPAs. It is a more balanced joint approach, including both the designation of SURs and the strict protection of large forest areas remaining faunally and floristically intact, that will maximize the benefits to both local communities and the preservation of tropical forest biodiversity.

ACKNOWLEDGMENTS

This essay – written during a sabbatical at Duke University's Center for Tropical Conservation – is dedicated to the memory of Navjot Sodhi, who heartbreakingly passed away at the time of this writing. I will miss Navjot dearly as a friend, professional colleague, and fellow-warrior for the cause of tropical forest biodiversity.

REFERENCES

Adeney, J. M., Christensen, N. L. and Pimm, S. L. (2009) Reserves protect against deforestation fires in the Amazon. *PloS One*, **4**, e5014.

Almeida, O., Lorenzen, K., Mcgrath, D. G. and Rivero, S. (2011) Impacts of the comanagement of subsistence and commercial fishing on Amazon fisheries, in *The Amazon* (eds M. A. Várzea Pinedo-Vasquez, M. Ruffino, C. J. Padoch and E. S. Brondízio), Springer, Dordrecht, The Netherlands, pp. 107–117.

Campos, M. T. and Nepstad, D. C. (2006) Smallholders, the Amazon's new conservationists. *Conservation Biology*, **20**, 1553–1556.

Chape, S., Spalding, M. and Jenkins, M. D. (2008) *The World's Protected Areas: Status, Values and Prospects in the 21st Century*. University of California Press, Berkeley, CA.

DeFries, R., Hansen, A., Newton, A. C. and Hansen, M. C. (2005) Increasing isolation of protected areas in tropical forests over the past twenty years. *Ecological Applications*, **15**, 19–26.

Fa, J. E., Ryan, S. F. and Bell, D. J. (2005) Hunting vulnerability, ecological characteristics and harvest rates of bushmeat species in Afrotropical forests. *Biological Conservation*, **121**, 167–176.

Ferraz, G., Marinelli, C. E. and Lovejoy, T. E. (2008) Biological monitoring in the Amazon: recent progress and future needs. *Biotropica*, **40**, 7–10.

Gaston, K. J., Jackson, S. F., Cantú-Salazar, L. and Cruz-Piñón, G. (2008) The ecological performance of protected areas. *Annual Review of Ecology, Evolution, and Systematics*, **39**, 93–113.

IMAZON (2011) *Áreas protegidas na Amazônia brasileira: avanços e desafios*. IMAZON and Instituto Socioambiental, São Paulo, Brazil.

Jerozolimski, A. and Peres, C. A. (2003) Bringing home the biggest bacon: a cross-site analysis of the structure of hunter-kill profiles in Neotropical forests. *Biological Conservation*, **111**, 415–425.

Joppa, L. N., Loarie, S. R. and Pimm, S. L. (2008) On the protection of "protected areas". *Proceedings of the National Academy of Sciences of the USA*, **105**, 6673–6678.

Kaimowitz, D. and Sheil, D. (2007) Conserving what and for whom? Why conservation should help meet basic human needs in the tropics. *Biotropica*, **39**, 567–574.

Laurance, W. F. and Peres, C. A. (2006) *Emerging Threats to Tropical Forests*. University of Chicago Press, Chicago, IL.

MMA (Ministério do Meio Ambiente) – SNUC (Sistema Nacional de Unidades de Conservação) (2000) MMA, SNUC, Brasília, Brazil.

Molnar, A., Scherr, S. J. and Khare, A. (2004) *Who Conserves the World's Forests?* Forest Trends, Washington, D.C.

Mora, C., Andréfouët, S., Costello, M. J., Kranenburg, C., Rollo, A., Veron, J., Gaston, K. J. and Myers, R. A. (2006) Coral reefs and the global network of marine protected areas. *Science*, **312**, 1750–1751.

Naughton-Treves, L., Alvarez-Berríos, N., Brandon, K., Bruner, A., Holland, M. B., Ponce, C., Saenz, M., Suarez, L., and Treves, A. (2006) Expanding protected areas and incorporating human resource use: a study of 15 forest parks in Ecuador and Peru. *Sustainability: Science, Practice, and Policy*, **2**, 32–44.

Nepstad, D., Schwartzman, S., Bamberger, B., Santilli, M., Ray, D., Schlesinger, P., Lefebvre, P., Alencar, A., Prinz, E., Fiske, G., Rolla, A. (2006) Inhibition of Amazon deforestation and fire by parks and indigenous lands. *Conservation Biology*, **20**, 65–73.

Parry, L., Peres, C. A., Day, B. and Amaral, S. (2010) Rural-urban migration brings conservation threats and opportunities to Amazonian watersheds. *Conservation Letters*, **3**, 251–259.

Pedlowski, M. A., Matricardi, E. A. T., Skole, D., Cameron, S. R., Chomentowski, W., Fernandes, C. and Lisboa, A. (2005) Conservation units: a new deforestation frontier in the Amazonian state of Rondônia, Brazil. *Environmental Conservation*, **32**, 149–155.

Peres, C. A. (2000) Effects of subsistence hunting on vertebrate community structure in Amazonian forests. *Conservation Biology*, **14**, 240–253.

Peres, C. A. and Lake, I. R. (2003) Extent of nontimber resource extraction in tropical forests: accessibility to game vertebrates by hunters in the Amazon basin. *Conservation Biology*, **17**, 521–535.

Peres, C. A. and Palacios, E. (2007) Basin-wide effects of game harvest on vertebrate population densities in Amazonian forests: implications for animal-mediated seed dispersal. *Biotropica*, **39**, 304–315.

Peres, C. A. and Terborgh, J. W. (1995) Amazonian nature reserves: an analysis of the defensibility status of existing conservation units and design criteria for the future. *Conservation Biology*, **9**, 34–46.

Peres, C. A. and Zimmerman, B. (2001) Perils in parks or parks in peril? Reconciling conservation in Amazonian reserves with and without use. *Conservation Biology*, **15**, 793–797.

Peres, C. A., Barlow, J. and Laurance, W. F. (2006) Detecting anthropogenic disturbance in tropical forests. *Trends in Ecology and Evolution*, **21**, 227–229.

Prado, G. B. (2010) Consumo de carne bovina e pecuarização da Amazônia. MSc dissertation, Faculdade de Saúde Pública, São Paulo, Brazil.

Redford, K. H. and Robinson, J. G. (1987) The game of choice: patterns of Indian and colonist hunting in the neotropics. *American Anthropologist*, **89**, 650–666.

Robinson, J. G. (2011) Ethical pluralism, pragmatism, and sustainability in conservation practice. *Biological Conservation*, **144**, 958–965.

Robinson, J. G. and Bennett, E. L. (2000) Carrying capacity limits to sustainable hunting in tropical forests, in *Hunting for Sustainability in Tropical Forests* (eds J. G. Robinson, and E. L. Bennett), Columbia University Press, New York, NY, pp. 13–30.

Rosas, G. K. C. (2006) Pressão de caça, abundância, densidade e riqueza de mamíferos em duas áreas de coleta de Castanha-do-Brasil situadas no sudoeste do estado do Acre, Brasil. MSc thesis, Universidade federal do acre, Rio Branco, Acre, Brazil.

Schmitt, C. B., Burgess, N. D., Coad, L., Belokurov, A., Besançon, C., Boisrobert, L., Campbell, A., Fish, L., Gliddon, D., Humphries, K., Kapos, V., Loucks, C., Lysenko, I., Miles, L., Mills, C., Minnemeyer, S., Pistorius, T., Ravilious, C., Steininger, M. and Winkel G. (2009) Global analysis of the protection status of the world's forests. *Biological Conservation*, **142**, 2122–2130.

Soares-Filho, B., Moutinho, P., Nepstad, D., Anderson, A. and Rodrigues, H. (2010) Role of Brazilian Amazon protected areas in climate change mitigation. *Proceedings of the National Academy of Sciences of the USA*, **107**, 10821–10826.

Terborgh, J. and Peres, C. A. (2002) The problem of people in parks, in *Making Parks Work: Strategies for Preserving Tropical Nature* (eds J. Terborgh, C. van Schaik, M. Rao, and L. Davenport), Island Press, Washington, D.C., pp. 307–319.

WDPA (2010) The World Database on Protected Areas (WDPA). UNEP-WCMC, Cambridge, UK.

West, P., Igoe, J. and Brockington, D. (2006) Parks and peoples: the social impact of protected areas. *Annual Review of Anthropology*, **35**, 251–277.

Wilkie, D. S., Bennett, E. L., Peres, C. A. and Cunningham, A. A. (2011) The empty forest revisited. *Annals of the New York Academy of Sciences*, **1223**, 120–128.

Wittemyer, G., Elsen, P., Bean, W. T., Coleman, A., Burton, O. and Brashares, J. S. (2008) Accelerated human population growth at protected area edges. *Science*, **321**, 123–126.

Zimmerman, B., Peres, C. A., Malcolm, J. and Turner, T. (2001) Conservation and development alliances with the Kayapó of south-eastern Amazonia, a tropical forest indigenous peoples. *Environmental Conservation*, **28**, 10–22.

Chapter 30

CONCLUDING REMARKS: LESSONS FROM THE TROPICS

Luke Gibson[1] and Peter H. Raven[2]

[1]*Department of Biological Sciences, National University of Singapore, Singapore*
[2]*Missouri Botanical Garden, St. Louis, MO, USA*

What lessons can we glean from these voices from the tropics? Coming from more than 20 countries from throughout the tropics, these stories describe conservation in a variety of situations, from some of the richest countries in the world to some of the poorest, from stable democracies to war-torn dictatorships, from some of the largest countries on each continent to small island nations. In each case, the ability to deal with the richest and most complex ecosystems on earth is limited by insufficient knowledge and funds, and often by a lack of determination to deal with the issues with resolve. It is likely that more than half of the world's organisms occur in just one tropical ecosystem – tropical moist forests – and many other tropical habitats also are extraordinarily rich in species. As we have seen, the populations of most tropical countries are growing rapidly and there are large numbers of people living in poverty, so sustainability often is an elusive goal. What we are able to accomplish in conserving tropical ecosystems and the biodiversity that occurs in them will depend on our skill and determination to create a just world, one in which social justice is the norm, and one in which the people of wealthy nations share a common concern for people who live much less privileged and demanding lives than they could ever imagine.

The fate of biodiversity on small island nations may foreshadow its fate on the mainland. In his chapter, Florens describes Mauritius and Rodrigues as a "conservation laboratory for the tropics," given the extent to which humans have transformed the habitats on these islands and the unique biota that have survived on them over the past few centuries. Indeed, these two oceanic islands may present a picture of "what awaits the rest of the tropical world as the latter catches up in terms of human overpopulation, habitat destruction, and fragmentation . . ." Like Mauritius, Singapore can also be viewed as a laboratory for studies on extreme human impacts, having lost over 98% of its original forests, a trend that has caused massive extinctions (Brook, Sodhi and Ng, 2003). In Singapore, the remaining forests and natural areas are now rigorously conserved and seem likely to provide not only Singapore but probably the whole area with the last survivors of many species that were once common throughout the region. The relatively small remaining protected areas in Singapore are now safe from conversion, logging, and hunting. In contrast, the government of Mauritius has displayed a declining commitment to conserving biodiversity, and local non-governmental organizations (NGOs) that are active there have not yet adopted the kind of adaptive management that would

Conservation Biology: Voices from the Tropics, First Edition. Navjot S. Sodhi, Luke Gibson, and Peter H. Raven.
© 2013 John Wiley & Sons, Ltd. Published 2013 by John Wiley & Sons, Ltd.

be necessary to reverse the disintegration of most remaining ecosystems. Effective on-the-ground activities that involve local communities will be necessary if the unique biota of this island is to be preserved over the long run.

A third island model is that of Sri Lanka. As Pethiyagoda notes, conservation in Sri Lanka is based on the failed paradigm that the central government is the sole protector of biodiversity, whereas experiences throughout the world indicate clearly that a more participatory approach offers the only hope, with government, science, and the public all needing to be involved to make conservation work. In particular for Sri Lanka, it will be necessary to encourage trust between the central government and civil society, and to encourage gathering information of fundamental value.

In Hawaii, treated here by Atkinson and his colleagues, the spread of aliens and other forms of destruction of natural communities have been devastating over the 950 years since the arrival of people on the archipelago, and climate change promises to become an even worse problem in the future. In a state blessed with long-term funding, strong leadership, and broad public education and awareness, the ecosystems of Hawaii may be partly recovered and a proportion of the surviving endemic species saved from extinction.

The Pacific islands, extremely fragile ecologically and mostly overpopulated for their size, are viewed by Brodie *et al.* as lacking strong leaders to help the people reconnect to their land and ecosystems and to accept spiritual values that have the potential of guiding them to sustainability. Although population in the area is growing relatively slowly, the drive for increased levels of consumption is causing natural resources to be sold and ecosystems to be destroyed.

The large island of Madagascar has had a substantial history of protected areas, and new ones are being created at the present time also. As Rakotomanana *et al.* discuss, however, the people are closely dependent on the relatively limited natural ecosystems that have survived to the present, with their numbers growing very rapidly. The distribution of benefits is highly inequitable. Although the available scientific information is relatively rich, its dissemination and use by the people certainly needs improvement. Conservation appears to face a very difficult future in Madagascar.

If the islands just discussed are in different degrees models of what awaits the rest of the developing tropical world, Papua New Guinea, which constitutes over half of the area of the world's largest tropical

island, represents an exception to the trends ongoing in island states, particularly in tropical Asia. Tropical Asia is clearly the most devastated tropical region, and, according to the chapter presented here by Shearman, Papua New Guinea is about two decades behind the rest of the area in the process of liquidating its natural capital. However, its rich endowment of tropical forests is rapidly being depleted as logging operations expand and often revisit the areas whose destruction they have already begun, and mining expands everywhere in the highlands. There is little general concern with conservation, and the numerous local groups tend to believe that the resource supply is endless. Very few people are receiving benefits from the vast flow of money that is paying for the destruction of the nation's resources, which would be adequate to advance sustainability if the leaders had the will to direct some of the funds in this way. It is not clear whether the numerous corrupt government officials who are benefiting from selling off the wealth of the nation can develop the will to exercise the necessary leadership to save the situation before it is too late. International NGOs seem to have had relatively little effect on the conservation of Papua New Guinea's natural resources even though it is often featured in the literature and their campaigns. Local communities would need to be deeply involved in conservation for there to be any reasonable expectation of lasting effects.

The other half of the island, part of the archipelagic nation of Indonesia, faces similar problems with government, as Prawiradilaga and Soedjito report. After initially prohibiting mining in protected areas, Indonesia later granted permits to companies to develop openpit mines in protected forests. To resolve this hypocrisy, the government simply changed the law to allow mining in protected areas. If protected areas are to function as an effective conservation strategy, enforcement must be improved. A very serious problem in Indonesia has arisen from the granting of rights to natural resources to districts throughout the country. In those districts, everything extractable has tended to be viewed as a source of cash, and little attention has been paid to sustainability or community–biodiversity linkages.

Turning now to a greater emphasis on the principles of conservation and to mainland areas, the rapid growth of human populations in tropical countries is a source of the greatest possible concern. As we mentioned in the introduction to this collection of essays, we are currently estimated to be using about 150% of the sustainable productivity of the world (http://footprintnetwork.org) and yet 1 billion of our 7.1 billion current population is malnourished and 100 million are on the verge of starvation. This means that to attain sustainability with our current population, and without improving the lot of the poor and hungry at all, would require a planet 50% larger than ours, a truly drastic situation. With the world population projected to grow from its current 7.1 billion people to 8.1 billion by 2025 and to 9.6 billion by 2050 – almost all of the growth in the developing countries of the tropics – and essentially everyone everywhere wishing to consume more and more and to go on using current technologies, the future looks extremely bleak. It is, however, only by the application of particular practices at specific places that the problems facing our common future can be ameliorated.

Although not often mentioned in the individual essays presented here, a second major problem is that of global climate change. International bodies speak of holding the increase in global temperatures to 2°C above the background level, but very few scientists studying the situation project anything less than a 4°C increase, or even more, during the remainder of this century. Such an increase will have disastrous effects on food production, especially since it will be linked with changes in precipitation levels varying from region to region – and all of that while we are trying to overcome widespread hunger and malnutrition now, and to accommodate and feed an additional 2.5 billion people over the same period. Ecosystems will be disturbed or destroyed and many species, especially on mountains, will be driven to extinction. Sea level rise during this century will also vary from place to place, but there is general agreement that it will have risen by at least 1 meter over approximately the remainder of this century. Such an increase will have destructive effects on low-lying areas worldwide, and be especially damaging on the fragile ecosystems of small islands.

There are major differences between different tropical countries in every imaginable respect. Thus the population of sub-Saharan Africa, currently 900 million people, is projected to more than double to 2.1 billion over the next 37 years; that of South Central Asia from 1.8 billion to 2.6 billion people, India contributing the most to this growth; Southeast Asia from 600 million to 800 million people; and Latin America from 600 million to 740 million people. These projections clearly indicate major problems for all of these areas, but especially for sub-Saharan Africa, where

there is a gain of roughly 100,000 net people per day, and India, projected to grow from 1.3 billion people today to 1.7 billion, with a net gain of about 30,000 people per day. Since poverty and hunger are widespread in both of these areas, and with global climate change adversely affecting the prospects for agriculture, the future looks very difficult indeed – but we must do the very best that we can!

To cite some specifics for continental Africa, there is a pervasive theme of corruption, armed conflict, and a lack of adequate connection of conservation activities with local populations. Here as elsewhere, the appearance of REDD+ treaties and payments will call into focus the importance of local participation in whatever sales are made and whatever decisions are taken. Conservation is often set aside in the name of poverty alleviation, which may greatly deplete remaining resources. The chapter by Inogwabini and Leader-Williams on bonobo conservation in the Democratic Republic of the Congo illustrates how well our closest relatives are being protected in a settled area, better than in a park established for the purpose (but without sufficient information to set the boundaries ideally). Again, for the Cross River State of Nigeria, Abua and his colleagues emphasize the need for local involvement but good planning centrally as both necessary to achieve the desired results. Writing about Kenya, Githiru thoughtfully puts in focus the need for greater idealism, without which economic forces will always trump conservation. In Côte de Ivoire, Koné illustrates how armed conflict has shattered sporadic conservation efforts, with deforestation proceeding very rapidly in a context of little general awareness of the need for conservation, and ineffective, poorly coordinated government institutions. Ineffective laws and ineffective enforcement of those that exist in the face of widespread corruption have led to the gradual destruction of natural resources throughout Ethiopia, according to a thorough analysis by Gebresenbet and colleagues. And Tanzania, according to Tibazarwa and Gereau, replays the familiar theme of economic gain being chosen over conservation importance in almost every instance.

In Asia, these themes are repeated in various ways in different nations. Australia, an industrialized country neighboring Asia, does relatively well with the protection of its tropical communities, with Laurance arguing mainly for a better application of knowledge in the pursuit of sustainability. India, analyzed in different chapters by both Bawa and by Pandit and Kumar, has a great need to apply academic knowledge to practical problems on the ground. Population and consumption levels are growing so rapidly that effective action can only increase in difficulty as the years go by. China has many of the same problems and opportunities as Australia, and, as Liu indicates, needs to strengthen its sustainability-related initiatives. Clear threats include water scarcity, climate change, increased consumption levels, and the possible abandonment of the one-child policy.

In Latin America, more of the original ecosystems remain than in the other tropical parts of the world, and the population is generally growing more slowly as well. Mexico, for example, analyzed by Ceballos and García, is doing a good job of setting up reserves and applying a high knowledge of biodiversity to maintaining them. Ways are being explored to compensate private landowners for carrying out sustainable practices on their lands. Colombia, as Murcia *et al.* report, with its extremely high biological diversity, is suffering and will suffer more in the future from the effects of climate change, drastic changes in regional land use, and the common struggle between economic gain and the maintenance of sustainable ecosystems. The situation in Paraguay is even more drastic, as declared by Yanosky, with the natural ecosystems in the eastern half of the country largely wiped out for soybean production and the western part, the very fragile Chaco ecosystem, now being deforested rapidly by Brazilian cattle companies with the consequent loss of both indigenous rights and the unique biodiversity of the region. As Kalamandeen shows, Guyana presents a textbook example of the implementation of REDD+ agreements with insufficient attention paid so far to the rights of the indigenous people who live in the forests.

Moving now to general principles that run throughout the essays presented here, the most often repeated principle concerns the need to involve local populations in protection strategies for tropical ecosystems and biodiversity (Peres; Koh; other essays). Even in relatively wealthy countries like Singapore, Mexico, and Australia, conservation cannot succeed without such involvement, and, in fact, without tangible returns of one kind or another to the people affected. This in turn underlines the need for participatory governance of all regions, with frameworks being developed and continually improved at the national or regional level. Although a few authors have emphasized the importance of completely protected areas, we wonder how practical they may be in the face of the 2.5 billion people, mostly poor, who will almost all be added to the populations of developing countries over the next 35 years or so. Not only are pristine wildlife parks almost universally seen as relicts of colonial times, often unprotectable, but the involvement of people in harvesting goods sustainably from protected areas is increasingly seen as a given in most parts of the world.

Clever strategies and models have been advanced by many of the authors here, which, implemented locally, will have highly beneficial impacts. The adoption of hornbill families by individuals, often in urban areas, who support their conservation financially, is a fine example of what can be done to bring people together around ultimately common goals, as Poonswad and Chimchome report from Thailand. Working in Kenya, Martins has shown how learning the importance of pollinators can provide a cogent argument for conserving forest fragments in agricultural communities. This is a fine example of the way in which understanding ecosystem services can lead to local conservation efforts. In a sense, the traditional protection of charismatic species, closely linked to broad conservation efforts and to ecotourism, provides a similar example of a triggering mechanism for fostering sustainability.

Unfortunately, the activities of international conservation NGOs are not seen as particularly positive in any of the countries from which our essays were obtained. They seem often to operate at levels not accessible to local people and not meaningful on the ground. All of them are presumably engaged in a rigorous reexamination of the effectiveness of their efforts, and we hope that enhanced local involvement and participation on the ground may be one outcome of such examination.

The difficulties of operating at an international level are often exacerbated by the high levels of governmental corruption that characterize many of the countries whose problems are discussed in this volume, and which Peh summarizes in his chapter on corruption. Not only conservation, but education, health delivery, and other vital government services are often seriously and negatively affected by corruption. The rapid flow of money into many tropical countries in exchange for the one-time extraction of natural resources only tends to make the problem of corruption worse, and it often has little or no effect on the quality of life of the poor people in the countries receiving the funds.

Scientific knowledge is seen as the most appropriate basis for conservation programs generally, and it is generally poor throughout the tropical regions of the world compared with the temperate ones. Where the education system is relatively strong, as in Brazil (Pardini *et al.*) or India (Bawa), a need is expressed for the linkage between the accumulation of knowledge and its delivery in digestible form to the people who need it to inform their activities. The Brazilian program discussed here specifically addresses this problem through the implementation of practical courses at a few universities, but these are not traditionally as prized as the strictly academic ones that are traditional.

Poor organization and lack of synergy between conservation agencies within national and regional governments, and coordination between the central government and many political entities within a given country, often also pose serious obstacles to successful conservation in tropical countries. To protect its natural resources in the Ivory Coast, the Ministry of Environment resettled people from protected areas at the same time as the Ministry of Education built schools inside protected areas so that resident children have access to education. In their chapter on Ethiopia, Gebresenbet *et al.* and their local guides could not even locate the boundary while visiting a national park. The Ethiopian government faces a conflict between protecting nature and national development; the federal government grants licenses for agricultural plantations without requiring agricultural companies to consult with regional government or conservation authorities, seriously undermining local efforts to preserve natural habitats. In Tanzania, the government seeks national development by granting an Indian chemical company access to a soda ash lake, threatening an important breeding ground for flamingos in the Great Rift Valley. In Sri Lanka, a pest moth species that infects rice crops is federally protected, but pesticides that specifically target the moth are legally available and in widespread use. As we have seen, central governments having ceded forested areas to indigenous people may then turn around and sell the carbon benefits afforded them by the same forests, accepting restrictions to the use of the forests about which the indigenous people may not

be consulted at all. Such structural problems must be solved in order for conservation to work smoothly and effectively, and we hope that the experiences shared in the pages of this book may afford solid, practical examples of use to many who are laboring to conserve the magnificent diversity of this earth.

Effective conservation in the tropics will require the full and continuous participation of local people, organizations, and governments. It will require scientific results that can readily be translated into on-the-ground sustainable activities. It will require the best possible collaboration between government departments, and their ongoing interaction with the local communities that are affected by their rules. It will need the implementation of mechanisms to circumvent corruption and see that funds get to the places for which they are intended. Collectively, we must address the interlinked problems of population growth, increasing desire for ever-higher levels of consumption, and the continued use of unsustainable forms of technology. It will require people from throughout the world to respect one another and honor the potential inherent in each human being. These are tall orders, but nothing less will gain us entrance to the sustainable world that we must build for our common future.

In this collection, many voices from the tropics have spoken, sharing their perspectives on conservation issues within their own countries. We hope that their messages will be heard, so that conservation science and implementation can advance in this biologically rich, unique, and seriously threatened part of the world. To achieve these goals will require listening to the voices and cooperating respectfully in the attainment of goals that are of great importance for our common future.

REFERENCE

Brook, B. W., Sodhi, N. S. and Ng, P. K. L. (2003) Catastrophic extinctions follow deforestation in Singapore. *Nature*, **424**, 420–423.

INDEX

Page numbers in *italics* refer to figures, those in **bold** refer to tables.

Printed and bound by CPI Group (UK) Ltd, Croydon, CR0 4YY